AF541578

A Handbook on Plant Health Medicines

NIPA® GENX ELECTRONIC RESOURCES & SOLUTIONS P. LTD.

New Delhi-110 034

About the Author

Prof. S.G. Borkar, is an alumni of Indian Agricultural Research Institute, New Delhi for his Masters and Doctorate degree in Plant Pathology (1977-1983), a French government scholar for post-doctorate at INRA, Angers, France in 1984, and had his Doctorate of Science (D.Sc) in Plant Pathology in 1999 from International University, Washington, USA.

As an academician and scientist of international repute, he has 130 research publications published over a period of 40 years in 17 foreign and in 39 indian research journals besides 5 university publications and 2 book Chapters. Presented 29 scientific as well as Lead papers in national and international symposiums; published 9 books including USA publications, 6 patents, developed 4 wheat varieties, 2 documentary films, established 2 laboratories, 2 museums, mentored 30 students for their masters and doctorate degrees and received 21 national and international awards for his contribution including a plant pathogenic *klebsiella pneumoniae* strain after his name by NCBI, USA, as *klebsiella pneumoniae* strain Borkar. His Biographical note is included in Asia Pacific WHO'S WHO (vol.1, 1998) and Twentieth Century Admirable Achievers Distinguished WHO'S WHO (1999).

He served in 3 agricultural universities in 2 states of India in various academic, scientific and administrative positions including the post of Dean, Post-Graduate Institute. Lead the wheat research in peninsular India as zonal-co-ordinator at the national level for around 5 years and lead the university in plant pathology as a Head of Department for 10 years. He also rendered his services as Designated Inspection Authority of Plant Quarantine for Western Maharashtra, for government of India, for ten years while in the position of Head of Department of Plant Pathology at Mahatma Phule Agriculture University, Rahuri, Maharashtra state. After his superannuation from the university services of Mahatma Phule Agriculture University, Rahuri, in 2018, founded his own research laboratory" Endeavour Scientific Agriculture' as Startup (www.allagrisolutions.com) at Nashik in Maharashtra state, India, to serve the Indian farmers.

A Handbook on Plant Health Medicines

Suresh G. Borkar
Former Professor and Head, Department of Plant Pathology
Mahatma Phule Krishi Vidyapeeth, Rahuri, Maharashtra, India

NIPA® GENX ELECTRONIC RESOURCES & SOLUTIONS P. LTD.
New Delhi-110 034

NIPA® GENX ELECTRONIC RESOURCES & SOLUTIONS P. LTD.

101,103, Vikas Surya Plaza, CU Block
L.S.C. Market, Pitam Pura, New Delhi-110 034
Ph : +91 11 27341616, 27341717, 27341718
E-mail: newindiapublishingagency@gmail.com
www: www.nipabooks.com
For customer assistance, please contact
Phone: + 91-11-27 34 17 17
Fax: + 91-11-27 34 16 16
E-Mail: feedbacks@nipabooks.com

ISBN: 978-93-94490-39-0

Composed and Designed by NIPA®.

Dedicated to
The Farmers of the World

Acknowledgement

The author wishes to thank his doctoral degree students for their help during the preparation of the manuscript of this book. Special thanks are to my wife Dr. Sandhya Borkar, (Ph.D), and son Antriksh Borkar for their special attitude towards science and for sparing time to complete this task in the interest of world agricultural communities.

Prof. Suresh G. Borkar

Foreword

UN proclaims the Year 2020 as the "International Year of Plant Health". The year was expected to increase awareness among the public and policy makers of the importance of healthy plants and the necessity to protect them to achieve sustainable development goals.

Healthy plants are the foundation for all life, ecosystem functions and food security. Plant pests and diseases damage crops, reducing the availability of food and increasing its cost. Today up to 40 percent of global food crops produce is lost annually due to plant pests and diseases in terms of economic value, plant diseases alone cost the global economy around US dollars 220 billion annually and invasive insects around US dollars 70 billion. Therefore, the importance of plant health to enhance food security, protect the environment and biodiversity from pests; address the effect of climate change and supports the efforts to end hunger, malnutrition and poverty and boost economic development was the major thrust of the year for this celebration.

The International year of plant health was expected to increase awareness about the importance of plant health management in achieving sustainable development goals. The celebration of International Year of Plant Health 2020 is over; however not much has changed at farm levels in the management of plant health may be due to lack of adequate and appropriate information on the plant health medicines to be used to manage the particular plant health issue as is evident from the Food and Agriculture Organisation's report of 3rd April 2019 and 2nd June 2021 which indicated the same quantum of losses during these years. As per the report of Global giving (www.globalgiving.org) of 1st March 2021, on how much it would cost to end world hunger, it estimates the need of $ 7 billion to $ 265 billion each year, which is less than the amount the world agriculture loose every year due to improper management of crop pest and diseases in the absence of proper knowledge among the farming communities on plant health medicines.

The book "A Handbook on Plant Health Medicines" is a commendable work done by Professor. Suresh. G. Borkar. The book addresses the key issues of plant health and suggests measure that will help in minimising the crop losses and improving crop productivity to achieve world food security.

The book will be very useful guide and important source of reference for the policy makers, plant protection departments, plant pathologists, students and farmers in effective plant health management.

I heartily congratulate Professor Borkar and the Publisher in bringing out this valuable book.

Professor Anupam Varma
Former ICAR National Professor
Division of Plant Pathology
Indian Agricultural Research Institute
New Delhi

Preface

World agriculture and food securities are of immense significance for the survival of human race on this earth. In recent years, world agriculture is threatened due to climate change patterns, heavy floods, droughts, and pest/disease damages. In turn, this causes huge losses to agricultural produce, thereby fluctuation in the marketing prices of agricultural commodities making the citizens to suffer economically, emotionally and socially.

The repeated appearance of known pests and diseases in the geographical region may be due to inappropriate control measures adopted, development of resistance to pesticide molecule in the pathogen/pest, spurious pesticides used in the plant protection measures system, use of non-labelled pesticides, lack of awareness campaign in the farmers, marketing strategies of pesticide dealers and their lack of expertise in the recommendation of pesticide to farmers; for example, in Indian scenario, where farmers purchases pesticide directly from pesticide shop without consulting the plant protection expert, the pesticide shop owner is more interested in his commission than in the control of pest/disease and thus sells the non-labelled pesticide for the control of the pest. Secondly, the introduction of alien pests/diseases in new area causes tremendous losses in the absence of known control measures in an entirely new and different ecological environment.

This situation can be overcomed if a handy book on Plant Health Medicines is made available in the agricultural knowledge system. Though different titles of plant protection books are available, no book is as much illustrative as the present book of Plant Health Medicines. The book "A Handbook on Plant Health Medicines" is prepared with an objective to make available the scientific as well as working knowledge of useful and recommended plant protection and plant health medicines to the personnel associated with agriculture in different capacities and at different levels, and to the end-users, i.e. farmers.

The book contains three sections viz. chemical molecules for control of plant diseases, chemical molecules to control plant physiological processes and physiological disorders, and Pesticides for management of plant-insect pest. These three sections are divided into 30 chapters. The respective section has got images of all the representative diseases, pests, as well as mineral deficiencies symptoms for the identification of the cause and different molecules for their control and rectification, thus making the book more practical use oriented to all personnel associated with the agriculture sector. With its availability particularly, to pesticide shop dealers and farmers, the proper plant protection and plant health medicines will be applied in the agricultural ecosystem to manage the problem associated with plant health. Further, it will reduce the use of unnecessary pesticides/medicines in the agricultural ecosystem, thereby making the plant produce less contaminated with pesticide residues, which harms the health of the world citizens.

Having the copy of this book with the agricultural personnel particularly pesticide dealers and farmers, the task of plant health management can be taken care of to a larger extent.

Prof. Suresh. G. Borkar

Contents

Acknowledgement .... *vii*

Foreword .... *ix*

Preface .... *xi*

Section-I: Chemical Molecules for Control of Plant Diseases

1. Classification of Diseases Based on Symptomatology and Pathogen Involved (For Selection of Suitable Chemical Molecules) 3

1.1. Disease Symptoms Induced by Fungal Pathogens 3

1.2. Disease Symptoms Induced by Bacterial Pathogens 7

1.3. Disease Symptoms Induced by Nematode Pathogens 9

1.4. Disease Symptoms Induced by Viruses and Phytoplasmas 10

1.5. Disease Symptoms Induced by Algal Pathogens 12

1.6. Disease Symptoms Induced by Protozoan Pathogens 13

2. The Fungicides: Chemical Molecules for Control of Fungal Diseases 15

2.1. The Journey of Fungicide 16

2.2. The key Classes of Modern Fungicides 18

2.3. Characters of an Ideal Fungicide 21

2.4. Global Fungicide Market 22

3. Groups of Fungicides and Diseases Controlled 27

3.1. Based on Mode of Action 27

3.2. Based on General Use 27

3.3. Based on Chemical Composition 28

3.4. Systemic Fungicides 39

3.5. Organo Phosphorous Compounds 45

3.6. Piperazine 45

3.7. Phenol Derivative 45

4. New Fungicides (by Trade Name) and their Use ... 47

5. Ready-Reckoner for Disease Specific Fungicides ... 59

6. Development of Resistance in Fungal Pathogens Against the Fungicide and Fungicide Resistance Management ... 61

6.1. Mode of Action of Major Fungicides Classes and Resistance Risk Class ... 62

6.2. Risks of Fungicides Resistance ... 80

7. Fungicide Product with Their Active Ingredients and Hazard Class ... 83

8. Chemical Molecules for Control of Bacterial Diseases ... 101

8.1. Bactericides for Plant Pathogenic Bacteria ... 101

8.2. Antibiotics for Management of Plant Pathogenic Bacteria ... 112

9. Chemical Molecules for Management of Viruses and Phytoplasmas ... 117

9.1. Management of Plant Viruses ... 117

9.2. Management of Phytoplasma Diseases ... 119

10. Chemical Molecules as Nematicides for Nematode Management ... 123

10.1. Plant Parasitic Nematodes ... 123

10.2. Management of Nematode Diseases ... 124

10.3. Biological Control of Plant Parasitic Nematodes ... 132

11. Chemical Molecules for Management of Algal Infection ... 137

11.1. Algal Species Involved in Plant Infection ... 137

11.2. Symptoms of Algal Infections ... 137

11.3. Important Hosts Crops ... 138

11.4. Crop Damage ... 138

11.5. Management of *Cephaleuros* Diseases ... 139

12. Management of Protozoan Diseases ... 141

12.1. The Genus *Phytomonas* ... 141

12.2. Major Protozoan Diseases ... 143

12.3. Economic Significance ... 145

12.4. Management Approaches ... 145

13. Methods of Application of Chemical Molecules As Plant Medicines ... 147

13.1. Seed Treatment ... 147

13.2. Soil Treatment ... 150

13.3. Foliar Application........ 151
13.4. Post-harvest Treatment........ 152
13.5. Special Method of Application........ 152

14. Precautions in Selection of a Proper Plant Health Medicine 155
14.1. Adverse Effects on Crops and Useful Organisms........ 156

Section-II: Chemical Molecules to Control Plant Physiological Processes and Physiological Disorders

15. Plant Growth Regulators........ 159
15.1. Use of Plant Growth Regulators 159

16. Anti-Transpirant Chemicals to Mitigate Water-Stress........ 179
16.1. Types of Anti-transpirants........ 179

17. Chemicals to Induce Early Flowering........ 183
17.1. Potassium Nitrate 183
17.2. Calcium Nitrate 184
17.3. Fertilizer........ 184
17.4. Pesticide Including Herbicides........ 184
17.5. Plant Nutritients 184
17.6. Hormones........ 185

18. Chemicals to Control Fruit Dropping........ 187
18.1. Growth Regulators to Control Fruit Drop........ 187

19. Chemical Nutrient Deficiencies/Toxicities in Plants and Their Rectification 189
19.1. Major/Macronutrients 189
19.2. Micronutrients........ 189
19.3. Representative Symptoms of Nutrient Deficiencies/Toxicity in Plants........ 190
19.4. Chemicals Used to Rectify the Nutrient Deficiencies........ 202

Section-III: Pesticides for Management of Insect Pest

20. Classification of Insect Pest Based on Their Appearance and Feeding Habit (For Proper Insecticide Selection and Application) 207
20.1. Leaf and Foliage Eating Insect Pest........ 207
20.2. Plant Sap Sucking Insect pests........ 208
20.3. Bollworm and Fruit Borer........ 210

20.4. Shoot Borer, Stem borer and Cutworm 211
20.5. Fruit flies and Friut Moths 212
20.6. Bark eating Caterpillar 213
20.7. Leaf Tissue Scratching Insect Pest 213
20.8. Web Forming Insect Pest 214
20.9. Root Damaging Insect Pest 215

21. Insecticides for Control of Insect Pests 217
21.1. Pesticide Groups 218
21.2. Inorganic Pesticides 220
21.3. Organic Pesticides 222
21.4. Cyclodiene Insecticides 253

22. Botanical Insecticides 257
22.1. Pyrethrum and Related Compounds 257
22.2. Sabadilla (veratrine alkaloids) 262
22.3. Ryania 264
22.4. Nicotine 264
22.5. Citrus Oil Extracts 266
22.6. Other Essential Plant Oils 266
22.7. Neem 267

23. Invertebrate Animal Based Insectcide 277

24. Pesticides for Control of Slug and Snails 279

25. Ready-Reckoner for Insect Pest Specific Insecticide 281

26. Insect Attractants for Insect Pest Management 285
26.1. Herbal Attractants 285
26.2. Chemical Attractants 286
26.3. Attractant-Baited Traps to Monitor Pest Populations 287
26.4. Attractant-Baited Traps to "Trap out" Pest Populations 288
26.5. Attractants to Disrupt Insect Mating 288
26.6. Attractants in Poison Baits 289
26.7. Lights to Attract Insects 289
26.8. Colored Objects to Attract Insects 289
26.9. Other Traps 290

27. Insect Repellent for Insect Pest Management.................................. 293
27.1. Botanicals as Repellent 293

28. Chemical Molecules as Weedicides/Herbicides for Weed Management 299
28.1. Classification Based on Translocation 300
28.2. Classification Based on Time of Application.............................. 300
28.3. Classification Based on Method of Application.......................... 301
28.4. Classification Based on Specificity 301
28.5. Classification Based on Site of Action.. 302
28.6. Most Commonly used and Recommended Herbicides 304
28.7. Commonly Used and Available Herbicides in India.................... 306
28.8. Spreader and Stickers for Herbicide 309
28.9. Adjuvants, Surfactant, Wetting & Spreading Agents to Enhance Efficiency of Herbicide.. 309

29. Biological Control Agents for Disease Pest Management................ 329
29.1. What is Biological Control.. 329
29.2. Important Early Milestones in Biological Control Program......... 329
29.3. Biological Control Agents... 330
29.4. Advantage and Constraints of Myco-Biocontrol Agents 343
29.5. Strategies of Biological Pest Control 346

30. Scenario of Banned and Restricted Pesticides................................ 353
30.1. Pesticide Substances Banned and Authorised in the European Union 354
30.2. Scenario of Banned Pesticdes in USA 384
30.3. Scenario of Banned and Restricted Pesticides in China............... 387
30.4. Scenario of Pesticide Banned in Brazil....................................... 394
30.5. Status of Banned and Restricted Pesticides in India 396
30.6. Label Claim Insecticides 401

Literature Cited..411

Section-I

Chemical Molecules for Control of Plant Diseases

1

Classification of Diseases Based on Symptomatology and Pathogen Involved (For Selection of Suitable Chemical Molecules)

Plant diseases are caused by different etiological agents like Fungi, Bacteria, Viruses, Viroids, Phytoplasmas, Rickettsia like bacteria, Nematodes, Algae and Protozoan parasites. To control these specific pathogens and the diseases caused by them, a specific group of medicine/chemical molecule is required. Even these individual pathogen induces different types of symptoms based on their morphological and pathogenic variability available within them e.g. some group of fungi causes powdery mildew, other causes downy mildew while some other group causes wilt, blight, dieback and so on. The plant medicines are specific for the control of specific diseases symptoms and a group of pathogen. Therefore, the knowledge of disease symptoms and their causal agents are very important in the management of the disease caused by different plant pathogens.

1.1. Disease Symptoms Induced by Fungal Pathogens

The diseases symptoms induced by fungal plant pathogens varies and can be categories as (Photo.1).

1. Damping off of seedlings

2. Wilt

3. Root rot

4. Crown rot

5. Gummosis

6. Anthrocnose

7. **Die back**

8. **Twig blight**

9. **Leaf spot**

10. **Leaf blight**

11. **Downey mildew**

12. **Powdery mildew**

13. Rust

14. Smut

15. Bunt

16. Scab

17. Canker

18. Club root

19. Wart

20. Fruit rot

Photo.1. Types of disease symptoms induced by fungal plant pathogens (Courtesy: Dr. Borkar's Laboratory and Research Centre, Nashik, India)

1.2. Disease Symptoms Induced by Bacterial Pathogens

The disease symptoms induced by bacterial plant pathogens varies and can be categories as (Photo.2).

1. Leaf spot

2. Blight

3. Wilt

4. Canker

5. Fire blight

6. Crown gall

7. Bacterial soft rot

Photo. 2: Types of disease symptoms induced by bacterial plant pathogens (Courtesy: Dr.Borkar's Laboratory and Research Centre, Nashik, India)

1.3. Disease Symptoms Induced by Nematode Pathogens

The disease symptoms induced by different plant pathogenic nematode varies and can be categories as (Photo.3).

1. Meloidogyne Root Knot

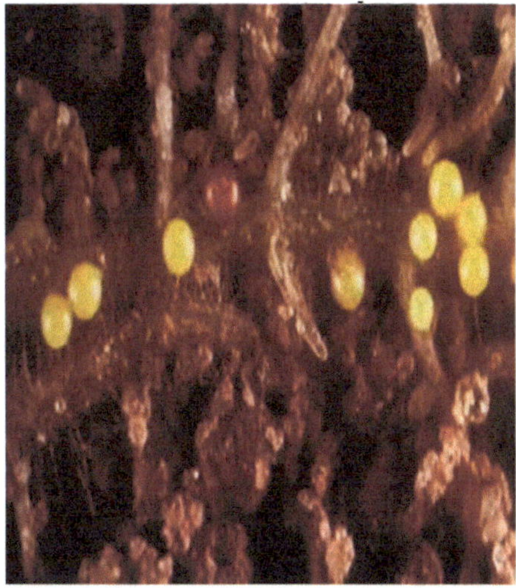

2. Heterodera cyst

3. Seed Gall nematode

4. Stem and Bulb nematode

Photo. 3: Types of disease symptoms induced by plant pathogenic nematodes (Courtesy: Dr.Borkar's Laboratory and Research Centre, Nashik, India)

1.4. Disease Symptoms Induced by Viruses and Phytoplasma

The disease symptoms induced by plant viruses and phytoplasma-like organisms (PLOs) varies and can be categories as (Photo.4).

1. Mosaic

2. Leaf roll

3. **Leaf curl**

4. **Yellow vein mosaic**

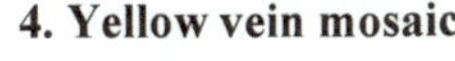

5. **Ring Spot**

6. **Little leaf**

7. Greening

Photo. 4: Types of disease symptoms induced by plant viruses and PLOs (Courtesy: Dr.Borkar's Laboratory and Research Centre, Nashik, India)

1.5. Disease Symptoms Induced by Algal pathogens

The disease symptoms induced by Algal pathogens varies with the host and affected Plant's part (Photo.5).

1.**Algal Leaf spot**

2. Infection on Tree truck

3. Algal Infection on Fruits **4. Algal infection on arecanuts**

Photo. 5: Algal pathogen on leaves, trunk, fruit and nuts. (Courtesy: Dr.Ajayasree, T .S; Kerala Agriculture University, India)

1.6 Disease Symptoms Induced by Protozoan Pathogens

The disease symptoms induced by Protozoan pathogens varies with the host and affected Plant's part (Photo.6).

1. Hartroot disease

Photo. 6: Hartroot disease induced by Protozoa in coconut palm (Courtesy: Ricardo B. Sgrillo; Centro de Pesquisas do Cacau - CEPLAC - Rod. Ilhéus-Itabuna, km 22, 45650-000, Ilhéus, BA).

2

The Fungicides Chemical Molecules for Control of Fungal Diseases

Fungi are one of the major causes of crop damage which affect crop yield, quality and revenue. Therefore to control these fungal diseases, fungicidal crop protection products are used at different stages of crop growth. Many of our current fungicides produce excellent results with respect to efficacy, crop quality, food safety, and improved cost/profit ratios of agricultural production.

The word **fungicide** is derived from two Latin words, viz., "fungus" and "credo". The word "credo" means "to kill". Thus, the **fungicide** is any chemical which has the ability to kill the fungus. Some chemical molecules do not kill the fungal pathogens, but they simply arrest the fungus growth temporarily. Such chemicals are called **fungistat** and the phenomenon of inhibiting the fungal growth temporarily is termed as **fungistatis**. Some other chemical molecules may inhibit the spore production without affecting the fungal growth and are called as **Antisporulant**. The antisporulant and fungistatic compounds though do not kill the fungi, these are important in restricting the multiplication and spread of the fungal pathogen. All these chemical molecules are also termed as **Fungitoxicants.**

Fungicides are extensively used for:

1. Protection of seed grain during storage, shipment and germination;
2. Protection of crops, berries, seedlings, flowers in the field, in storage and during shipment;
3. Suppression of mildews that attack painted surfaces;
4. Control of slime in paper pulps; and
5. Protection of carpet and fabrics in the home.

2.1. The Journey of Fungicide

Earlier development of the fungicide was the result of good observations e.g. it was observed that the seed wheat salvaged from the sea was free of bunt indicating the role of salty water in the control of bunt. This lead to the first use of brining of grain with salt water followed by liming in the middle of the 17th century to control bunt much before Tillet (1755) established that seed-borne fungi (*Tilletia tritici*, *T. laevis*) causing bunt of wheat could be controlled by seed treatments of lime, or lime and salt. Another important observation was made in France in 1885 by Millardet, who noticed that grape vines that had been sprayed with a bluish-white mixture of copper sulfate and lime to deter pilferers retained their leaves through the season, whereas the unsprayed vines lost their leaves to the fungal disease named as downy mildew. After numerous spraying experiments Millardet concluded that a mixture of copper sulfate and hydrated lime could effectively control downy mildew of grape. Until the 1940s, chemical disease control relied upon inorganic chemical preparations, frequently prepared by the user.

Up to 1940: The major products used up to 1940 are listed in Table 1. Most users prepared their own fungicides from basic recipes to use against diseases of horticultural crops particularly those of fruits and vegetables, and as seed treatments for seed borne diseases.

Table 1: Fungicides in use until 1940 [after Russell 2005]

Year	Fungicide	Primary Use
1637	Brine	Cereal seed treatment
1755	Arsenic	Cereal seed treatment
1760	Copper sulfate	Cereal seed treatment
1824	Sulfur (dust)	Powdery mildew and other pathogens
1833	Lime sulfur	Broad spectrum foliar pathogens
1885	Bordeaux mixture	Broad spectrum foliar pathogens
1891	Mercury chloride	Turf fungicide
1900	$CuOCl_2$	Especially *Phytophthora infestans*
1914	Phenylmercury chloride	Cereal seed treatment
1932	Cu_2O	Seed and broad spectrum foliar diseases
1934	Dithiocarbamates patented	Broad spectrum protectants
1940	Chloranil, Dichlone	Broad spectrum seed treatment

1940 to 1970: During the period of 1940 to 1970, there were a number of new chemical molecules introduced as fungicides (Table 2). These were dithiocarbamates and phthalimides as a major improvement over the previously used inorganic fungicides. These were more active, less phytotoxic and easier to prepare by the user.

Table 2. Key classes of fungicides introduced between 1940 and 1970.

Fungicide class	Active Ingredient and Year
Dithiocarbamate	thiram 1942, zineb, nabam 1943, maneb 1955, mancozeb 1961
Aromatic Hydrocarbon	biphenyl 1944
Phthalimide	captan, folpet 1952, captafol 1962
Fentin	fentin acetate, fentin hydroxide 1954
Antibiotic	blasticidin S 1955, kasugamycin, polyoxin 1965
Triazine	anilazine 1955
Guanidine	dodine 1957
Nitroanaline	dicloran 1960
Benzimidazoles	thiabendazole 1964, benomyl 1968, thiophanate methyl 1970
Phthalonitrile	chlorothalonil 1964
Morpholine	dodemorph 1965, tridemorph 1969
Carboxanilide	Carboxin 1966, oxycarboxin 1966

Some of these introduced chemical molecules were discovered as herbicides, e.g., triazines and nitroanalines. During this period an antibiotic was introduced to control rice blast in Japan and was classified as a systemic.The decade from 1960 to 1970 saw a rapid expansion of research and development in this sector with a rapid growth of the fungicide markets. The most widely used protectant fungicides, mancozeb and chlorothalonil, were introduced. During this period only the first broad-spectrum foliar systemic fungicide thiabendazole and the systemic seed treatment fungicide carboxin were introduced.

Beyond 1970: The more important modern fungicides introduced since 1970 are listed in Table 3 according to their group/chemical class and mode of action.

Table 3. Major fungicide groups introduced since 1970 with their most important representatives.

Group	Year	Common name of compounds	Main spectrum /uses
Inhibitors of sterol biosynthesis (triazoles if not indicated otherwise)	1973	Triadimefon, imazalil (imidazole)	broad, post-harvest & seed treatment
	1975	fenarimol (pyrimidine)	powdery mildew
	1977	Triadimenol, prochloraz (imidazole)	seed treatment cereal fungicide
	1979	propiconazole, bitertanol, fenpropimorph (morpholine)	broad broad / cereals
	1982	Triflumizole	Broad
	1983	flutriafol, diniconazole, fluzilazole, penconazole	broad
	1986	fenpropidin (morpholine) hexaconazole, cyproconazole, myclobutanil, pyrifenox (pyridine), tebuconazole	broad / cereals broad broad / leaf crops broad, foliar & seed

	1988	Difenoconazole, tetraconazole, fenbuconazole	broad, foliar & seed broad
	1990	Epoxiconazole	broad / cereals
	1992	Metconazole, fluquinconazole, triticonazole	broad broad, foliar & seed
	2002	Prothioconazole	Broad
Inhibitors of cytochrome bc1 (Qo site & strobilurin analogues if not indicated otherwise)	1992	Azoxystrobin, kresoxim-methyl	broad cereal fungicide
	1996	Famoxadone (azolone)	Oomycetes
	1998	fenamidone (azolone), trifloxystrobin	oomycetes broad
	2000	Picoxstrobin, pyraclostrobin, fluoxastrobin	cereal fungicide broad
	2001	Cyazofamid (cyanoimidazole)	Oomycetes

Other classes, various fungicides and plant activators	Common names with year of introduction	Main spectrum / uses
Dicarboximides	Iprodione (1974), vinclozolin (1975), procymidione (1976)	Botrytis, Monilinia
Phenylpyrroles	Fenpiclonil (1990), fludioxonil (1990)	broad, foliar & seed treatment
Anilinopyrimidine	Pyrimethanil (1992), cyprodinil(1994)	Broad
Melanin Biosynthesis Inhibitors(MBIs)	Tricyclazole (1975), pyroquilone (1985), carpropamide (1997)	rice/ foliage
CAA fungidices*	Dimethomorph (1988), iprovalicarb (1998), benthiavalicarb (2003), mandipropamid (2005)	Oomycetes
Defense activators	Probenazole (1979), acibenzolar-S. methyl (1996)	fungi, bacteria, viruses
	Cymoxanil (1976), fosetyl-Al (1977), propamocarb (1978),	Oomycetes
	Carbendazim (1976), fluazinam (1992)	Broad
	Quinoxyfen (1997)	powdery mildew

2.2. The key Classes of Modern Fungicides

1. Benzimidazole

Benzimidazoles were introduced in the 1960s and early 1970s as foliar fungicides, for seed treatments and use in post harvest applications. They possessed unique properties as protectants. These properties included low use rates, broad spectrum and systemicity with post-infection action that allowed for extended spray interval.

The first case of benzimidazoles resistance occurred in powdery mildew in greenhouses in 1969, one year after its introduction. By 1984, resistance had been reported in many of the pathogens against which benzimidazoles were effective. The reason for the rapid development of resistance was that these fungicides were single site inhibitors of fungal microtubule assembly during mitosis, via tubulin-benzimidazole-interactions.

The primary patent holders of this class were DuPont (Benlate), Merck, Sharp & Dohme (Mertec) and Nippon Soda (Topsin M).

2. Morpholine

Morpholine fungicides are best known for their excellent control of cereal diseases, powdery mildew on vegetables and grapes, and sigatoka of banana. During the 1980s fenpropidin and fenpropimorph from this group were key fungicides in the European cereal market, while tridemorph was used extensively for sigatoka. This class of chemical, although having seen shifts in sensitivity by some pathogens (sigatoka in Central and South America), is still in use. Morpholine fungicides are often referred to as sterol biosynthesis inhibitors (SBI).

Key patents were held by BASF (Calixin and Corbel) and Dr. R. Maag (Corbel and Tern).

3. Piperazines

The major fungicide in this group was triforine, used extensively as a home and garden product especially on roses. It was widely accepted due to its efficacy and safety to a wide range of plants. The key producer was CelaMerck (Saprol).

4. Imidazoles

The most important fungicide in this group are imazalil (Janssen) and prochloraz (Boots). The primary uses for imazalil were as a seed treatment and post harvest treatment, while prochloraz (Sportak) was used on cereals, being especially active on *Pseudocercosporella* eyespot.

5. Pyrimidines

The most important fungicides in this group are nuarimol, fenarimol and triarimol. The widely used fungicide was fenarimol (Rubigan) on pome fruit, grapes and turf.

6. Triazoles

Bayer was first to launch a triazole group of fungicide, namely triadimefon (Bayleton) in 1973. This was soon followed by triadimenol (Baytan), bitertanol (Baycor) and propiconazole (Tilt) which was launched in 1979. Numerous other triazoles have been launched since then, including prothiaconazole (Proline). This class of fungicides are highly efficient broad spectrum products, and the occurrence of resistance in the pathogen is a slow shift resulting in a decreased sensitivity to their mode of action as de-methylation inhibitors (DMI).

7. Anilides

Anilides are a diverse group of fungicides. The earliest introduction was anilazine (Dyrene), primarily as a leaf spot fungicide from Bayer and Nissan, followed by the seed treatment carboxin (Vitavax), which is highly effective on bunts, smuts and assorted Basidiomycetes such as *Rhizoctonia spp.* This was followed by the dicarboximides iprodione (Rovral) from Rhone-Poulenc, vinclozolin (Ronilan) from BASF and procymidone (Sumisclex) from Sumitomo. These fungicides had exceptional protectant activity against the fungus of the genera *Botrytis, Monilinia* and *Sclerotinia.* Combating resistance became an issue with the wide scale use of these fungicides.

The most important among this group of anilides were the phenylamide fungicides metalaxyl (Apron/ Ridomil) from Ciba–Geigy and benalaxyl (Galben) from Isagro. These, along with phosphonate fosetyl-Al (Aliette) from Rhone-Poulenc, which was also introduced in 1977, brought a completely new level of control to the Oomycetes through their systemic properties by offering protection to the plants as seed treatments, and soil or foliar applications. Oxadixyl (Sandofan) from Sandoz was a later member of the phenylamides. Syngenta (1996) with mefenoxam (Apron XL and Ridomil Gold) and Isagro (2005) with kiralaxyl have introduced the resolved isomers of metalaxyl and benalaxyl. Again, what has limited the use of the phenylamide fungicides has been the development of resistance, even though the manufacturers tried introducing combinations with protectant fungicides such as mancozeb and chlorothalonil.

The other anilide to be registered (2003) is boscalid (Emerald, Endura, Pristine) from BASF. Boscalid is registered for foliar use on a wide range of vegetables, fruits and nut crops, either alone or in a mixture with pyraclostrobin as Pristine.

8. Strobilurins

Strobilurins launched in 1996, are now the second largest chemical group of fungicides and are in widespread use on cereals and, more recently on soybeans (a market that reached $600 million in 2004). The strobilurin fungicides are broad spectrum, highly efficacious, and suitable for a wide range of crops. Some problems with disease resistance are affecting sales (e.g., *Septoria* in wheat in Europe, and the U.S. turf market). As a result, companies are adjusting the use recommendations by developing mixtures and other uses, including seed treatments.

9. Other Systemic Fungicides

Other systemic fungicides include a diverse group of products, such as: tricyclazole (Beam) launched in 1975 by Eli Lily/Dow and still widely used for control of rice blast; cymoxanil (Curzate), for downy mildew from DuPont; cyprodinil (Vanguard, Unix) on cereals and fruit crops from Syngenta; fludioxonil (Saphire, Switch, Maxim) from Syngenta; and quinoxyfen (Fortress, Quintec), for powdery mildew from Dow.

New active ingredients introduced recently are benthiavalicarb (Valbon from Certis) and mandipropamid (Revus from Syngenta) from the carboxylic acid amide (CAA) fungicide group; and fluopicolide (Infinito from Bayer), metrafenone (Flexity from BASF), proquinazid (Talius from DuPont), and zoxamide (Electis from Dow). One of the most novel new products introduced by Ciba-Geigy is acibenzolar-S-methyl (Actigard, Bion). At a use rates of 30 gr/ha (~0.4 oz/acre) or less, Actigard activates the host's systemic acquired resistance (SAR) process in many crop plants and offers broad protection against fungi, bacteria and viruses without having any direct activity on these pathogens. Actigard has performed best when incorporated into a program of chemical sprays, to increase the level of disease control as compared to when applied alone. This product has initiated a whole new field of research into utilizing peptides for controlling diseases, and other means of stimulating SAR and the Jasmonic acid pathway (JA) with chemicals and biological agents in plants. Probenazole has been used successfully against rice blast since 1979 and was later shown to activate defense mechanisms in rice.

2.3. Characters of an Ideal Fungicide

1. It should have low phytotoxicity
2. It should have long shelf life
3. Stability during dilution
4. It should be less toxic to human being, cattle, earth worms, microorganisms etc.
5. It should be a broad spectrum in its action

6. Fungicide preparation should be ready for use
7. It should have compatibility with other agrochemicals
8. It must be cheaper one
9. It should be available in different formulate
10. It should be easily transportable

2.4. Global Fungicide Market

The global fungicide market varies with the region (Fig.1 and 2).

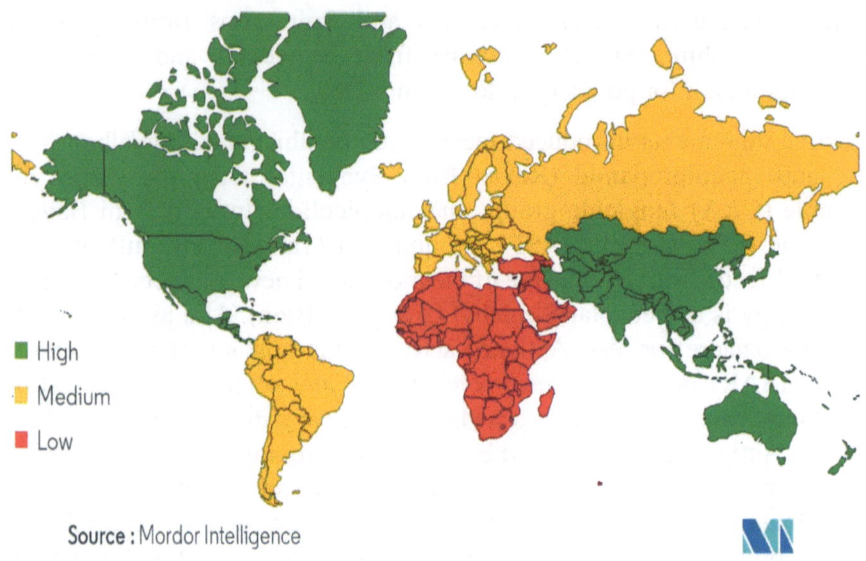

Fig. 1: Global Fungicide market in 2019.

The global fungicides market size is projected to reach USD 26.50 billion by 2027, exhibiting a CAGR of 4.78% during the forecast period (2020-2027)

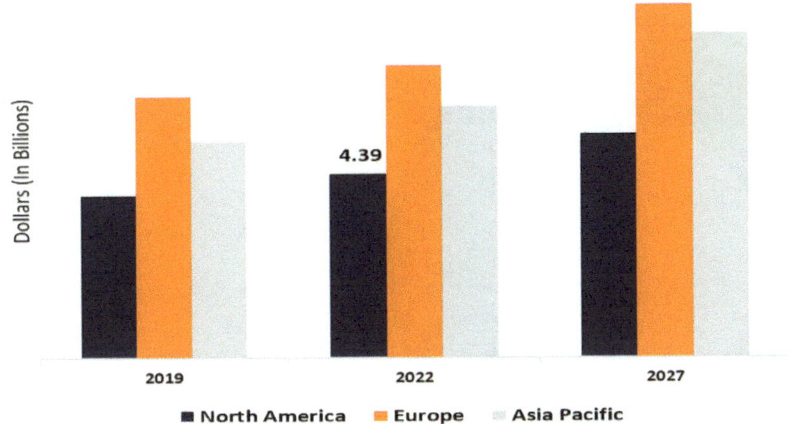

The market in the APAC region is likely to grow at the fastest rate of 5.2% in the forecast period. Countries, such as China and India, are the foremost agrarian economies in the Asia Pacific. The governments in these nations are persuading end-users for the use of fungicides through awareness programs and subsidies.

Europe dominated the market in 2019 and is likely to grow at a rate of 3.9% in the forecast period.

The market in the North American region is likely to grow at a rate of 4.1% in the forecast period. Farmers in the U.S. and Canada have increased the use of fungicides currently to fight against a host of diseases linked to fungi, in row crops.

Fig. 2: Fungicide market in different regions

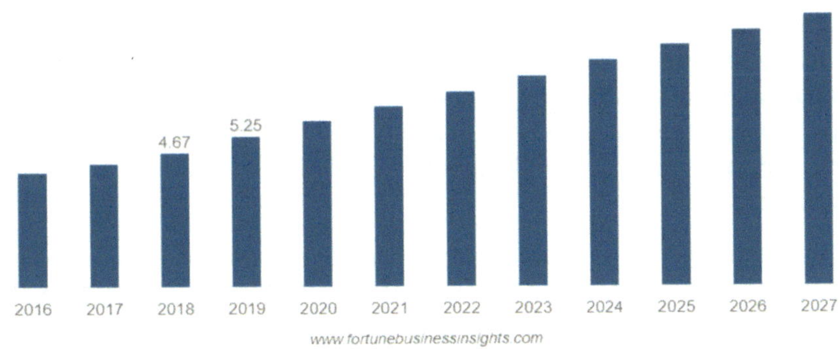

Fig. 3: Asia Pacific Fungicide Market, 2016- 2027 (USD Billion)

Asia-Pacific fungicide market size (fig.3) for 2016-17 was US$ 5.25 Billion, for 2018, it was US$ 4.67 Billion, and in 2019 it was US$ 5.25 Billion.

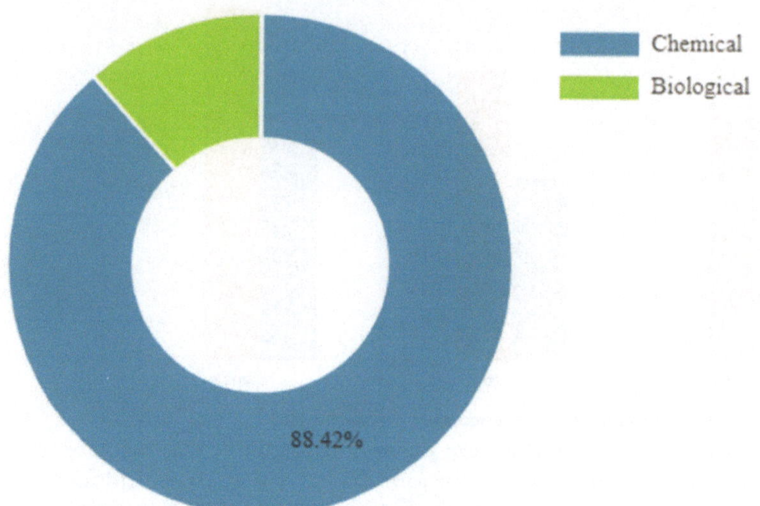

Fig. 4: Global Fungicide market share in term of Chemical and Biological Fungicide in 2019

The global chemical fungicide market in term of chemical and biological fungicide (Fig.4) contributes the share of 88.42% of chemical fungicide which constitute the Triazoles, Dithiocarbamates, Strobilurins, Inorganics Chloronitriles and others, while less than 13 % constitute the biological agents.

The expected Global fungicide market growth is predicted by CAGR at 4.78 % (for a period 2020-2027) and in term of US$, in 2019 it was 18.29 Billion and expected to US$ 26.50 billion in 2027.

Asia-Pacific currently holds the largest fungicidal market share and is predicted to retain its position throughout the forecast period. The growing agriculture industry, growing population, increasing rice cultivation, and the rising post harvest losses are the major factors driving the market in this region. China, India, Japan, and Australia are the key countries in the region that reports high fungicide consumption. Moreover, China is the leading fungicide consumer in the world and these factors would support the dominance of the region in the global market.

The European market will showcase considerable demand on account of the climate change that are leading to the onset of various diseases. In addition, the region is the key producer of fruit and vegetables and has a strong presence of global market players such as Syngenta, BASF, and Bayer. These factors are likely to boost the market in the region.

The South American market is predicted to expand at the fastest pace during the foreseeable years. The rapid increase in the consumption in the region, especially for the cultivation of cereals and oilseeds presents a resilient opportunity for the companies to increase their footprint. Brazil, Argentina, and Colombia are the key consumer in the region. However, their consumption in Brazil and Argentina is projected to grow at a stable rate while the consumption in Colombia is declining.

The Fungicide market is consolidated with more than 40,000 companies present across the globe. Among these, Syngenta AG, FMC Corporation, Bayer AG, ADAMA Limited, and BASF SE are some of the leading companies in the global market with the highest share. These companies are adopting significant strategies such as new product launches, mergers, and acquisitions to enhance their market position.

Climate change is a major concern for agricultural production across the world. The change in temperature and rise in atmospheric moisture content influence the growth of fungi and the emergence of hazardous fungal diseases. The increasing prevalence of economic losses caused during the crop growth stages owing to disease occurrence and further increase in the need for more fungicidal spray applications, is significantly contributing to the fungicidal market growth

3

Groups of Fungicides and Diseases Controlled

Fungicides can be broadly grouped based on their (i) mode of action (ii) general use and (iii) chemical composition.

3.1. Based on Mode of Action

3.1.1. Protectant

Fungicide which is effective only if applied prior to fungal infection is called a protectant, eg. **Zineb, Sulphur.** Protectant fungicides are prophylactic in their behaviour.

3.1.2. Therapeutant

Fungicide which is capable of eradicating a fungus after it has caused infection and there by curing the plant is called chemotherapeutant, eg. **Carboxin, Oxycarboxin, antibiotics like Aureofungin.** Usually chemo therapeutant are systemic in their action and affect the deep-seated infection.

3.1.3. Eradicant

Fungicide which remove pathogenic fungi from an infection court eg. **Organic mercurials, lime sulphur, dodine** etc. These chemicals eradicate the dormant or active pathogen from the host. They can remain effective on or in the host for some time.

3.2. Based on General Use

The fungicides can also be classified based on the nature of their use in managing the diseases.

3.2.1. Seed protectants: e.g. Captan, thiram, organomercurials, carbendazim, carboxin etc.

3.2.2. Soil fungicides (pre-plant): e.g. Bordeaux mixture, copper-oxy-chloride, Chloropicrin, Formaldehyde, Vapam, etc.,

3.2.3. Soil fungicides (post-plant): e.g. Bordeaux mixture, copper-oxy-chloride (for growing plants), Captan, PCNB, thiram etc.

3.2.4. Foliage and blossom fungicide: e.g. Captan, ferbam, zineb, mancozeb, chlorothalonil etc

3.2.5. Fruit protectants: e.g. Captan, maneb, carbendazim, mancozeb etc.

3.2.6. Eradicants: e.g. Organomercurials, lime sulphur, etc

3.2.7. Tree wound dressers: e.g. Bordeaux paste, Chaubatia Paste, etc.

3.2.8. Antibiotics: e.g. Actidione, Griseofulvin, Streptomycin, Streptocycline, etc.,

3.2.9. General purpose spray and dust formulations.

3.3. Based on Chemical Composition

The chemical available for plant disease control runs into hundreds, however, all are not equally safe, effective and popular. Major group of fungicides used include salts of toxic metals and organic acids, organic compounds of sulphur and mercury, quinines and heterocyclic nitrogenous compounds. Copper, mercury, zinc, tin and nickel are some of the metals used as base for inorganic and organic fungicides. The non-metal substances include, sulphur, chlorine, phosphorous etc. The fungicides can be broadly grouped as follows and discussed in detail.

1. Copper Fungicides,
2. Sulphur Fungicides
3. Mercury Fungicides

3.3.1. Copper Fungicide

The fungicidal action of copper was mentioned as early as 1807 by Prevost against wheat bunt disease *(Tilletia caries),* but its large scale use as a fungicide started in 1885 after the discovery of Bordeaux mixture by Millardet in France. The mixture of copper sulphate and lime was effective in controlling downy mildew of grapevine caused by *Plasmopara viticola* and later, late blight of potato *(Phytophthora infestans).*

Some other copper sulphate preparations later developed were Bordeaux paste, Burgundy mixture and Cheshnut compound which are all very effectively used in the control of several plant diseases. In addition some preparations of copper-oxy-chloride are also used. These are all insoluble copper compounds very successfully used in managing several leaf diseases and seedling diseases in nursery. Different copper based preparations and fungicides are mentioned below:

3.3.1.1. Copper sulphate preparations

a. Bordeaux Mixture

In 1882, Millardet in France (Bordeaux University) accidently observed the efficacy of the copper sulphate against the downy mildew of grapes caused by *Plasmopara viticola.* When copper sulphate was mixed with lime suspension, it effectively checked the disease incidence. The mixture of copper sulphate and lime was named as "Bouillie Bordelaise" (Bordeaux Mixture). The original formula developed by Millardet contains 5 lbs of CuSO4 + 5lbs of lime +50 gallons of water. The chemistry of Bordeaux mixture is complex and the suggested reaction is:

$$CuSO_4 + Ca\,(OH)_2 = Cu(OH)_2 + CaSO_4$$

The ultimate mixture contains a gelatinous precipitate of copper hydroxide and calcium sulphate, which is usually sky blue in colour. Cupric hydroxide is the active principle and is toxic to fungal spores.

In metric system, to prepare 1% Bordeaux mixture, 1 kg of copper sulphate is powdered and dissolved in 50 litres of water. Similarly, 1 kg of lime is powdered and dissolved in another 50 litres of water. Then copper sulphate solution is slowly added to lime solution with constant stirring or both the solutions may be poured simultaneously to a third container and mixed well. The ratio of copper sulphate to lime solution determines the pH of the mixture. The mixture prepared in the above said ratio gives neutral or alkaline mixture.

If the quality of the lime is inferior, the mixture may become acidic. **If the mixture is acidic, it contains free copper which is highly phytotoxic resulting in scorching of the plants**. Therefore, it is highly essential to test the presence of free copper in the mixture before applied.

There are several methods to test the neutrality of the mixture, which are indicated below:

(i) **Field Test**: Dip a well polished knife or a sickle in the mixture for few minutes. If reddish deposit appears on the knife/sickle, it indicates the acidic nature of the mixture.

(ii) **Litmus paper test**: The colour of blue litmus paper must not change when dipped in the mixture.

(iii) **pH paper test**: If the paper is dipped in the mixture, it should show neutral pH.

(iv) **Chemical test**: Add a few drops of the mixture into a test tube containing 5 ml of 10% potassium ferrocyanide. If red precipitate appears, it indicates the acidic nature of the mixture. If the prepared mixture is in the acidic range, it can be brought to neutral or near alkaline condition by adding some more lime solution into the mixture.

Bordeaux mixture preparation is cumbersome and the following precautions are needed during preparation and application.

(i) The solution should be prepared in earthen or wooden or plastic vessels. Avoid using metal containers for the preparation, as it is corrosive to metallic vessels.

(ii) Always copper sulphate solution should be added to the lime solution, reverse the addition leads to precipitation of copper and resulted suspension is least toxic.

(iii) Bordeaux mixture should be prepared fresh every time before spraying. In case, the mixture has to be stored for a short time or a day, jaggery can be added at the rate of 1 kg/100 litres of the mixture.

(iv) Bordeaux mixture is sometimes phytotoxic to apples, peaches, rice varieties like IR8 and maize varieties like Ganga Hybrid 3.

b. Bordeaux paste

Bordeaux Paste consists of same constituents as that of Bordeaux mixture, but it is in the form of a paste as the quantity of water used is too little. It is nothing but 10 percent Bordeaux mixture and is prepared by mixing 1 kg of copper sulphate and 1 kg of lime in 10 litres of water. The method of mixing solution is similar to that of Bordeaux mixture. It is a wound dresser and used to protect the wounded portions, cut ends of trees etc., against the infection by fungal pathogens.

c. Burgundy mixture

It is prepared in the same way as Bordeaux mixture, except the lime is substituted by sodium carbonate. So it is called as "Soda Bordeaux". It was developed in Burgundy (France) in 1887 by Mason. The usual formula contains 1 kg of copper sulphate and 1 kg of sodium carbonate in 100 litres of water. It is a good substitute for Bordeaux mixture and used in copper-sensitive crops.

d. Cheshnut compound

Cheshnut compound is usually prepared by mixing 2 parts of copper sulphate and 11 parts of ammonium carbonate. This formula was suggested by Bewley in the year 1921. The two salts are well powdered, mixed thoroughly and stored in an air tight container for 24 hours before being used. The ripened mixture is used by dissolving it in water at the rate of 3 g/litre. The mixture is dissolved initially in a little hot water and volume is made up with cold water and used for spraying.

3.3.1.2. Copper Carbonate Preparation

A. Chaubatia Paste

Chaubatia paste is another wound dressing fungicide developed by Singh in 1942 at Government Fruit Research Station, Chaubatia in the Almora district of Uttar Pradesh. It is usually prepared in glass containers or chinaware pot, by mixing 800g of copper carbonate and 800g of red lead in a litre of raw linseed oil or lanolin. This paste is usually applied to pruned parts of apple, pear and peaches to control several diseases. The paste has the added advantage that it is not easily washed away by rain water.

B. Copper Carbonate Containing Fungicides

Trade name- Fungimar, Perenox, Copper Sandoz, Copper 4% dust, Perecot, Cuproxid, Kirti copper.

Diseases controlled- Cuprous oxide is a protective fungicide, used mainly for seed treatment and for foliage application against blight, downy mildew and rusts.

3.3.1.3. Copper Oxychloride Containing Fungicide

Trade name- Blitox, Cupramar 50%WP, Fytolan, Bilmix 4%, Micop D-06, Micop w-50, Blue copper 50, Cupravit, Cobox, Cuprax, Mycop.

Diseases Controlled - It is a protective fungicide, controls *Phytophthora infestans* on potatoes and several leaf spot and leaf blight pathogens in field.

3.3.2. Sulphur Containing Fungicides

Use of sulphur in plant disease control is probably the oldest one and can be classified as inorganic sulphur and organic sulphur.

3.3.2.1. Inorganic Sulphur

Inorganic sulphur is used in the form of elemental sulphur or as lime sulphur. Elemental sulphur can be either used as dust or wettable sulphur, later being more widely used in plant disease control. Sulphur is best known for its effectiveness against powdery mildew of many plants, but also effective against certain rusts, leaf blights and fruit diseases.

Sulphur fungicides emit sufficient vapour to prevent the growth of the fungal spores at a distance from the area of deposition. This is an added advantage in sulphur fungicides as compared to other fungitoxicants.

3.3.2.1.1. Elemental Sulphur

i. Sulphur dust

Trade Name - Sulphur dust, Cosan, Wetsulf, Microsul .

Diseases Controlled – Sulphur dust is a contact and protective fungicide, normally applied as sprays or as dust. It is generally used to control powdery mildews of fruits, vegetables, flowers and tobacco.This is also effective against apple scab (*Venturia inaequallis*) and rusts of field crops.

ii. Lime Sulphur (Calcium polysulphide)

It can be prepared by boiling 9 Kg of rock lime and 6.75Kg of sulphur in 225 litres of water.

Disease controlled- Lime Sulphur is effective against powdery mildews as a protective fungicide.

3.3.2.1.2. Organic Sulphur

(Dithiocarbamates)

Organic compounds of sulphur are now widely used in these days. All these compounds, called as "carbamate fungicides", are derivatives ofDithiocarbamic acid. Dithiocarbamates are broadly grouped into two, based on the mechanism of action. These groups are viz.

1. Monoalkyl Dithiocarbamates (eg. Zineb, Maneb) and
2. Dialkyl Dithiocarbamates (eg. Thiram, Ziram, Mancozeb, Nabam, Vapam and Ferbam)

i. Monoalkyl dithiocarbamate

Trade Name- Hexathane 75% WP, Dithane Z-78, Funjeb, Lonocol, Parzate C

Disease controlled -It is used to protect foliage and fruits of a wide range of crops.

1. Zineb

Chemical name- (Zinc ethylene bisdithiocarbamate)

Trade name- Du Pont Fungicide A, Polyram.

Disease controlled - It is effective against diseases such as early and late blight of potato and tomato, downy mildews and rusts of cereals, blast of rice, fruit rot of chilly etc.

2. Maneb

Chemical name- (Manganese ethylene bisdithiocarbamate)

Trade name- Dithane M22, Manzate WP, MEB

Disease controlled - These are protective fungicide used to control many fungal diseases of field crops, fruits, nuts, ornamentals and vegetables, especially blights of potatoes and tomatoes, downy mildews of vines, anthracnose of vegetables and rusts of pulses.

3. Mancozeb

Chemical name- (Maneb +Zinc ion)

Trade name- Dithane M45, Indofil M45, Manzeb.

4. Nabam (DSE)

Chemical name- (Di Sodium ethylene bisdithiocarbamate)

Trade name- Chembam, Dithane A-40, Dithane D-14, Parzate Liquid.

Disease controlled- Nabam is primarily used for foilar application against leaf spot pathogens of fruits and vegetables. Soil applications were also reported to have a systemic action on *Pythium, Fusarium* and *Phytophthora.* It is also used to control algae in paddy fields.

5. Vapam (SMDC)

Chemical name- (Sodium methyl dithiocarbamate)

Trade name- Vapam, VPM, Chemvape, 4-S Karbation, Vita Fume.

Disease controlled -It is a soil fungicide and nematicide with fumigant action. It is also reported to have insecticidal and herbicidal properties. It is effective against damping off disease of papaya and vegetables and wilt of cotton. It is also effective against nematode infestation in citrus, potato and root knot nematodes in vegetables.

ii. **Dialkyl Dithiocarbamate**

1. Ziram

Chemical name- (Zinc dimethyl dithiocarbamate)

Trade name- Cuman L. Ziram, Ziride 80 WDP, Hexazir 80% WP, Corozate,

Fukiazsin, Karbam white, Milbam, Vancide 51Z, Zerlate, Ziberk, Zitox 80% WDP.

Disease controlled - Ziram is a protective fungicide for use on fruit and vegetables crops against fungal pathogens including apple scab. It is non phytotoxic except to zinc sensitive plants. It is highly effective against anthracnose of beans, pulses, tobacco & tomato, and also against rusts of beans etc.

2. Ferbam

Chemical name- (Ferric dimethyl dithiocarbamate)

Trade name- Coromat, Febam, Ferberk, Femate, Fermate D, Fermicide, Hexaferb 75% WP, Karbam Black, Ferradow.

Disease controlled - Ferbam is mainly used for the protection of foliage against fungal pathogens of fruits and vegetables including *Taphrina deformans* of peaches, anthracnose of citrus, downy mildew of tobacco and apple scab.

3. Thiram

Chemical name- (Tetra methyl thiram disulphide)

Trade name- Thiride 75 WDP, Thiride750, Thiram 75% WDP, Hexathir,

Normerson, Panoram 75, Thiram, TMTD, Arasan, Tersan 75, Thylate, Pomarsol, Thiuram.

Disease controlled - It is used for seed treatment both, as dry powder or as a slurry. It is a protective fungicide, also suitable for application to foliage to control *Botrytis spp.* on lettuces, ornamental, soft fruits and vegetables, rust on ornamentals and *Venturia pirina* on pears. It is also effective against soil borne pathogens like *Pythium, Rhizoctonia* and *Fusarium.*

3.3.3. Mercury Fungicides

Mercury fungicides can be grouped as inorganic and organic mercury compounds. Both the groups are highly fungitoxic and were extensively used as seed treatment chemicals against seed borne diseases.

Inorganic compounds show bactericidal property also. However, due to their residual toxicity in soil and plants and their extreme toxicity nature to animal and human beings, the use of mercury fungicides is beings discouraged. In most of the countries, the use of mercury fungicides is banned and in countries like India, the use of mercury fungicides is restricted only in seed treatment for certain crops. The list of diseases against which mercury fungicides used are listed below.

3.3.3.1. Inorganic Mercury

1. Mercuric chloride

Trade name- Merfusan, Mersil

Disease controlled -It is used for treating potato tubers and propagative materials of other root crops

2. Mercurous chloride

Trade name- M-C Turf fungicide

Disease controlled - Mercurous chloride is limited to soil application in crop protection use because of its phytotoxicity.

3.3.3.2. Organomercurials

a. **Methoxy ethyl mercury Chloride.**

   Trade name- Agallol, Aretan, Emisan, Ceresan wet (India)

b. **Phenyl mercury chloride,**

   Trade name- Ceresan Dry (India), Ceresol, Leytosan.

c. **Ethyl mercury chloride-**

   Trade name- Ceresan (USA)

d. **Tolyl mercury acetate –**

   Trade name- Agrosan GN.

Disease controlled -These are used mainly for treatment of seeds and planting materials. These fungicides are used for seed treatment by dry, wet or slurry method. For seed treatment 1% metallic mercury is applied at 0.25% concentration.

3.3.4. Heterocyclic Nitrogen Compounds, Quinones and Miscellaneous Fungicides

3.3.4.1. Heterocyclic Nitrogen Compounds

Heterocyclic nitrogen compounds are mostly used as foliage and fruits protectants. Some compounds are very effectively used as seed dressers. Some of the commonly used fungicides are listed below.

1. Captan (Kittlesons's Killer)

Chemical name- (N-trichloromethyl thio-4-cyclohexene-1, 2- dicarboximide).

Trade name- Captan 50W, Captan 75 W, Esso Fungicide 406, Orthocide 406, Vancide 89, Deltan, Merpan, Hexacap.

Disease controlled - It is a seed dressing fungicide used to control diseases of many fruits, ornamental and vegetable crops against rots and damping off.

2. Captafol

Chemical name- (Cis-N-1, 1, 2, 2-tetra chloro hexane1,2- dicarboximide)

Trade name- Foltaf, Difolatan, Difosan, Captaspor, Foleid, Sanspor.

Disease controlled - It is a protective fungicide, widely used to control foliage and fruit diseases of tomato, coffee, potato.

3. Glyodin

Trade name- Glyoxaliadine, Glyoxide, Glyodin, Glyoxide Dry, Glyodex 30% liquid and 70% WP.

Disease controlled - It has a narrow spectrum of activity. As a spray, it controls apple scab and cherry leaf spot.

4. Folpet (Folpet)

Chemical name- [N- (trichloromethyl-thi)]

Trade name- phthalimide Phartan, Acryptan, Phaltan, Folpan, Orthophaltan.

Disease controlled - It is also a protective fungicide used mainly for foliage application against leaf spots, downy and powdery mildews of many crops.

3.3.5. Benzene Compounds

Many aromatic compounds have important anti-microbial properties and have been developed as fungicides. Some important benzene compounds commonly used in plant disease control are listed below.

1. Quintozene (PCNB)

Trade name- Brassicol, Terraclor, Tritisan 10%, 20%, 40% D and 75% WP, PCNB 75% WP.

Disease controlled - It is used for seed and soil treatment. It is effective against *Botrytis, Sclerotium, Rhizoctonia and Sclerotiniam spp*

2. Dichloran

Trade name- Botran 50% WP and 75% WP, Allisan.

Disease controlled - It is a protective fungicide and very effective against *Botrytis, Rhizopus and Sclerotinia* spp.

3. Fenaminosulph

Chemical name- (Sodiumdimethylamino Benzenediazosulphonate)

Trade name- Dexon 5% G and 70% WP

Disease controlled- It is very specific in protecting germinating seeds and growing plants from seeds borne as well as soil borne infection of *Phythium, Aphanomyces* and *Phytophthora* spp.

4. Dinocap

Chemical name- (2, 4-dinitro-6- octyl phenylcrotonate)

Trade name- Karathane, Arathane, DNOPC, Mildex, Crotothane, Crotothane 25% WP, Crotothane 48% Liq.

Disease controlled -It is a non-systemic acaricide and fungicide recommended to control powdery mildews on various fruits and ornamentals. It is also used for seed treatment.

3.3.6. Quinone Fungicides

Quinone are present naturally in plants and they exhibit anti-microbial activity. Some compounds are successfully developed and used in the plant disease control. Quinones are very effectively used for seed treatment and two commonly used fungicides are listed below:

1. Chloronil

Chemical name- (2, 3, 5, 6- tetrachloro- 1, 4-benzoquinone)

Trade name- Spergon

Disease controlled -Chloronil is mainly used as a seed protectant against smuts of barely and sorghum and bunt of wheat.

2. Dichlone

Chemical name- (2, 3-dichloro- 1, 4- napthoquinone)

Trade name- Phygon, Phygon XL WP.

Disease controlled - Dichlone has been used widely as seed protectant.

This is also used as a foliage fungicide, particularly against apple scab and peach leaf curl.

3.3.7. Organo – Phosphorous fungicide

1. Ediphenphos (Edifenphos)

Chemical name- (O-ethyl-S-diphenyldithiophosphate)

Trade name- Hinosan 50% EC and 2% D.

Disease controlled - It has a specific action against *Pyricularia oryzae* (Rice blast). It is also effective against *Corticium sesakii* and *Cochliobolus miyabeanus* in rice.

3.3.8. Organo Tin Compounds

Several other organic compounds containing tin, lead, etc. have been developed and successfully used in plant disease control. Among them, organo- tin compounds are more popular and effective against many fungal diseases. These compounds also show anti- bactericidal properties. Some of the organo-tin compounds commonly used are listed below.

1. Fentin hydroxide

Chemical name- (TPTH- Triphenyl tin hydroxide)

Trade name- Du-Ter WP 20% or 50% WP. Du-Ter Extra-WP, Farmatin 50 WP, Du- Terforte WP, Tubotin.

Disease controlled - It is a non-systemic fungicide recommended for the control of early and late blight of potato, leaf spot of sugar beet, blast of rice and tikka leaf spot of ground nut.

2. Fentin acetate

Chemical name- (TPTA-Triphenyl tinacetate)

Trade name- Brestan WP 40% and 60% WP.

Disease controlled - It is a non-systemic fungicide recommended to control *Ramularia* spp.on celery and sugar beet anthracnose and downy mildew.

3. Fentin Chloride

Chemical name- (TPTC- Triphenyl tinchloride)

Trade name- Brestanol 45% WP, Tinmate.

Disease controlled- It is effective against *Cercospora* leaf spot of sugar beet and paddy blast.

3.4. Systemic Fungicides

There has been substantial development in systemic fungicides since late 1960s. Any compound capable of being freely translocated in the plant after its penetration/absorption is called **systemic fungicide** and controls a fungal infection which has become systemic in the host system. Thus, a systemic fungicide could eradicate established infection and protect the new parts of the plant to succumb to the infection. Several systemic fungicides have been used as seed dressing to eliminate seed infection. These chemicals, however, have not been very successful in the cases of trees and shrubs.

On the basis of chemical structures these are classified as Benzimidazoles, Thiophanates, Oxathiins and related compounds, pyrimidines, morpholines, organo-phosphorus compounds and miscellaneous group.

3.4.1. Oxathiins and Related Compounds

Oxathiins were the earliest developed compounds. This group of systemic fungicide is also called as carboxamides, carboxylic acid anillides, carboxa-anillides or simply as anillides which are effective only against the fungi belong to *Basidiomycotina* and *Rhizoctonia solani.* Some of the chemicals developed are:

1. Carboxin

Chemical name- (5, 6-dihydro- 2- methyl-1-4-oxathiin-3-carboxanilide)

Trade name- Vitavax 10% D, Vitavax 75% WP, Vitavax 34% liq. Vitaflow.

Disease controlled - It is systemic fungicide used for seed treatment of cereals against bunts and smuts, especially loose smut of wheat.

2. Oxycarboxin

Chemical name- (5, 6- dihydro-2-methyl- 1, 4- oxathiin-3- carboxanilide-4,4-dioxide)

Trade name- Sicarol.

Disease controlled - It is a systemic fungicide used for the treatment of rust diseases of cereals, pulses, ornamentals, vegetables and coffee

3. Pyracarbolid

Chemical name- (2-methyl- 5, 6-dihydro- 4H-Pyran-3- carboxylic acid anilide).

Trade name- Plantvax 5G, Plantvax 5% liq. Plantvax 1.5 EC, 10% dust, 75 WP.

Disease controlled - It controls rusts, smuts of many crops and *Rhizoctonia solani,* but is slightly more effective than Carboxin.

3.4.2. Benzimidazoles

The chemicals of this group show a very broad spectrum activity against a variety of fungi. However, they are not effective against bacteria as well as fungi belongs to *Mastigomycotina.* Two types of fungicidal derivates of benzimidazoles are known. The first type of derivates includes fungicides such as thiabendazole and fuberidazole. The fungicidal moiety of the second type is methyl-2-benzimidazole carbamate (MBC). The fungicides of this group may be simple MBC such as carbendazim or a complex from such as benomyl, which transforms into MBC in plant system.

1. Benomyl

Chemical name- (Methyl – 10 (butyl carbomyl)-2benzimidazole carbamate).

Trade name - Benlate 50 WP, Benomyl, Bavistin 50 WP, MBC, Dersol, B.Sten 50, Zoom, Tagstin, Agrozim,

Disease controlled - It is a protective and eradicative fungicide with systemic activity, effective against a wide range of fungi

2. Carbendazim (MBC)

Chemical name- (Methyl -2- benzimidazole carbamate).

Trade name- Jkenstin.

Disease controlled - It is very effective against rice blast, apple scab, powdery mildew of cereals, rose, cucurbits and apple and Diseases caused by *Verticillium and Rhizoctonia.* It is also used as pre-and post-harvest sprays or dips for the control of storage rots of fruits and vegetables. Carbendazim is a systemic fungicide controlling a wide range of fungal pathogens of field crops, fruits, ornamentals and vegetables. It is used as spray, seedling dip, seed treatment, soil drench and as post-harvest treatment of fruits. It is very effective against wilt diseases especially, banana wilt. It controls effectively the sigatoka leaf spot of banana, turmeric leaf spot and rust diseases in many crops.

3. Thiabendazole (TBZ)

Chemical name- (2, 4-thiazolyl benzimidazole)

Trade name- Thiabendazole, Mertect, Tecto, Storite.

Disease controlled - It is a broad spectrum systemic fungicide effective against many major fungal diseases. It also controls pathogenic fungal species of *Botrytis, Ceratocystis, Cercospora, Colletotrichum, Fusarium, Rhizoctonia, Sclerotinia, Septoria and Verticillium.* It is also effective for the post-harvest treatment of fruits and vegetables to control storage diseases.

4. Fuberidazole

Chemical name- (2, (2-Furyl) - benzimidazole).

Trade name- Voronit.

Disease controlled - It is used for the treatment of seeds against diseases caused by *Fusarium,* particularly *F.nivale* on rye and *F.culmorum* of peas

3.4.3. Thiophanates

These compounds represent a new group of systemic fungicides based on thiourea. They are the derivatives of thioallophanic acid. These fungicides contain aromatic nucleus which is converted into benzimidazole ring for their activity. Hence, thiophanates are often classified under benzimidazole group and the biological activity of thiophanates resembles of benomyl. Two compounds are developed under this group

1. Thiophanate

Chemical name- (1, 2 – bis (ethyl carbonyl-2- thioureido) benzene).

Trade name- Topsin 50 WP, Cercobin 50 WP, Enovit.

Disease controlled - It is effective against *Venturia* spp. on apple and pear crops, powdery mildews, *Botrytis* and *Sclerotinia* spp. on various crops.

2. Thiophanate - methyl

Chemical name- (1, 2 bis (3 methoxycarbonyl- 2-thioureido) benzene).

Trade name- Topsin-M70 WP, Cercobin-M 70 WP, Envovit-methyl, Mildothane.

Disease controlled - It is effective against a wide range of fungal pathogens, including *Venturia spp.* on apples and pears, *Mycosphaerella musicola* on bananas, powdery mildews on apples, cucurbits, pears and vines, *Pyricularia oryzae* on rice, *Botrytis and Cerospora* on various crops.

3.4.4. Morpholines

1. Tridemorph

Chemical name- (2-6 - dimethyl- 4-cyclo – tridecyl morpholine).

Trade name- Calixin 75 EC, Bardew, Beacon

Disease controlled - It is an eradicant fungicide with systemic action, being absorbed through foliage and roots to give some protective action. It controls powdery mildew diseases of cereals, vegetables and ornamentals. It is highly effective against *Mycosphaerella, Exobasidium*

3.4.5. Pyrimidines, Pyridines, Piperidines and Imidazole

1. Triadimefon

Chemical name- (1-(4-chlorophenoxy)-3, 3-dimethyl-1-(1-2-triazol- 1yl) butan-2-one).

Trade name- Bayleton, Amiral

Disease controlled- It is very effective against powdery mildews and rusts of several crops.

2. Triadimenol

Chemical name- (1-(4-Chlorophenoxyl-3, 3-dimethyl-1(1, 2, 4- triazol-1- yl) butan-2-ol)

Trade name- Baytan

Disease controlled - It is very effective against powdery mildews and rusts of several crops.

3. Bitertanol

Chemical name- (B-(1-1-biphenyl-4-yloxy-a- (1-1-dimethyl-ethyl-1-H-1, 2-4- triazole-1-ethanol)

Trade name- Baycor

Disease controlled - It is highly effective against rusts and powdery mildew of a variety of crops. It is also used against *Venturia* and *Monilinia* on fruits and *Cereospora* leafspots of groundnut and banana.

4. Etridiazole

Chemical name- (5-ethaoxy-3- trichloromethyl, 1, 2-4-thiadiazole).

Trade name- Terrazole 25% EC, Koban, Pansol EG, Pansol 4% DP, Turban WP, Terracoat Aaterra.

Disease controlled - It is very effective against *Phytophthora* and *Pythium* spp. and seeding diseases of cotton, groundnut, vegetables, fruits and ornamentals.

3.4.6. Hydroxy Pyrinidines

1. Ethirimol

Chemical name- (5-butyl 2- ethyl amino-4-hydrop-6- methyl pyrimidine).

Trade name: Milliatem 80 WDP, Milcurb Super, Milgo

Disease controlled - It is effective against powdery mildew of cereals, cucumber and other field crops and ornamentals.

2. Dimethirimol

Chemical name- (5-butyl 2-dimethylamino-4- hydroxy-6-methypyrimidine).

Trade name- Milcurb

Disease controlled - It is very effective against powdery mildews of chrysanthemum and cucurbits

3.4.7. Furan derivatives

1. Furcarbanil

Chemical name- (2-5-dimethyl-3-furanilide)

Disease controlled - It is used as seed or soil application. It systemically control bean rust and is being used as a seed dressing fungicide against loose smut of wheat and barley.

2. Cyclafuramid

Chemical name- (N-cyclohexyl-2, 5- dimethyl furamide).

Disease controlled - It is effective against bunts, smuts and rusts of cereals, as well as coffee rust, blister blight of tea, smut, red rot of sugarcane, *Fusarium* wilt of tomato, *Rhizoctonia* on tomato, potato, groundnut, rice as well as *Armillaria* sp. on rubber.

3.4.8. Benzanilide derivative

1. Mebenil

Chemical name- (2-methyl benzanilide)

Disease controlled –It is effective against yellow rust on wheat and barley *(P. striiformis)* and brown rust on barley *(P.hordei).* It also have direct fungitoxic activity against *Sclerotium rolfsli* and *Rhizoctonia.*

3.4.9. Other Systemic Fungicides

1. Metalaxyl

Chemical name- (methyl-DLN-(2, 6- dimethylphenyl-N-) 2-methoxyacetyl).

Trade Name- Apron 35 SD, Ridomil Beam,

Diseases Controlled - It is highly effective as a seed dressing fungicide against fungal pathogens of the order Peronosporales.

2. Metalaxyl + Mancozeb

Trade name- Ridomil MZ 72 WP (8% Metalaxyl + 64%Mancozeb)

Disease controlled – It is a fungicide with systemic and contact action and effective against damping-off, root rots, stem rots, and downy mildew of grapes and millets.

3. Tricyclazole

Chemical name- (5-methyl-1, 2, 4 triazole (3,4b)- benzothiazole).

Disease controlled - It is specifically used against paddy blast fungus, *P. oryzae*

4. Fosetyl Al

Chemical name- (Aluminium – Trisaluminium)

Trade name- Bim Alliette 80 WP

Disease controlled - It is a very specific fungicide for Oomycetous fungi, especially against *Pythium* and *Phytophthora*

3.5. Organo Phosphorous Compounds

1. Pyrazophos

Chemical name- (2-0-0- Diethylthionophosphoryl) -5- methyl-

6-carbethoxypyrazolo-(1- 5a) pyrimidine).

Trade Name- Afugan, Curamil, WP Missile EC,

Diseases Controlled- It is used to control powdery mildews of cereals, vegetables, fruits and ornamentals.

2. Iprobenphos (IBP)

Chemical name- (S-benyzl-0-0-bis-isopropyl-phosphorothiate).

Trade name: Kitazin 48% EC, Kitazin 17G, Kitazin 2% D.

Disease controlled - It is used to control *Pyricularia oryzae* and sheath blight of rice.

3.6. Piperazine

1. Triforine

Chemical name- (N, N-bis-(1- foramido-2, 2, 2- trichloroethyl- piperazine).

Trade name- Saprol-EG, Fungitex

Disease controlled - It is effective against powdery mildew, scab and other diseases of fruits and rust on ornamentals and cereals.

3.7. Phenol Derivative

1. Choloroneb

Chemical name - (1-4-dichloro- 2, 5-dimethoxybenzene).

Trade name - Saprol-EG, Fungitex. Demonsan 65 WP, Tersan SP, Turf fungicide

Disease controlled - It is active against storage diseases of fruits. It is highly fungistatic to *Rhizoctonia* spp., moderately to *Pythium* spp. and poorly to *Fusarium* spp. It is used as a supplemental seed treatment for beans and soybean to control seedling disease.

4

New Fungicides (by Trade Name) and their Use

Actigard 50 WG (acibenzolar-S-methyl)

Plant defense activator used for bacterial diseases and Downy Mildews.

ActinoGrow *(Streptomyces lydicus* WYEC)

Biological soil and seed treatment for *Fusarium, Rhizoctonia, Pythium,* and *Phytophthora.*

Actino-Iron[OG] *(Streptomyces lydicus* WYEC)

Biological soil and seed treatment for *Fusarium, Rhizoctonia, Pythium,* and *Phytophthora* with added iron.

Actinovate (*Streptomyces lydicus*)

Biological for greenhouse use, only for vegetable crops.

Agclor 310

It is a commercial bleach solution registered for use to control post-harvest rots of vegetables.

Agri-Fos (potassium salts of phosphorus acid)

A fungicide active against *Pythium, Phytophthora*, and Downy Mildew.

Agrimycin-17 (Streptomycin sulfate)

A bactericide.

AgriPhage (bacteriophage)

A bactericide effective for bacterial speck on peppers and tomatoes.

Aliette WDG (fosetyl Al)

A fungicide active against *Pythium*, *Phytophthora*, and the Downy Mildews.

Alude (phosphorous acid)

A fungicide active against *Pythium*, *Phytophthora*, and the Downy Mildews, labelled for greenhouse transplant production.

Apron XL (mefenoxam)

A seed treatment against *Pythium* and *Phytophthora* seed rot and damping-off and systemic Downy Mildews of certain crops.

Armicarb 100 (potassium bicarbonate)

Powdery mildew.

BadgeX2^{OG} (copper oxychloride & copper hydroxide)

A bactericide and fungicide.

Ballad Plus Biofungicide^{OG} (*Bacillus pumilus* QST 2808)

Sweet corn fungal foliar diseases.

Basicop (copper sulfate)

A bactericide containing basic copper sulfate as the active ingredient.

Bio-Save 10 LP^{OG} (Pseudomonas syringae ESC-10)

Post-harvest decay of potato.

BiotenWPOG (*Trichoderma asperellum, T. gamsii*)

Biological soil treatment for most crops.

Blocker 4F (PCNB)

Soil-borne diseases of brassicas, beans and peas, garlic, tomatoes, and pepper.

Botran (dichloran)

A fungicide for *Botrytis* diseases, other fruit rots, *Sclerotinia* and *Sclerotium* diseases.

Bumper (propiconazole)

Corn diseases.

Bravo Weather Stik (ultrez & ZN) (chlorothalonil)

A broad spectrum fungicide.

Cabrio (pyraclostrobin)

A broad spectrum fungicide for bulb, cucurbit, fruiting, and root vegetables.

Camelot (copper octonate)

Copper product labelled for greenhouse use on vegetable transplants.

Camelot O^{OG} (copper octonate)

Copper product labelled for greenhouse use on vegetable transplants.

CeaseOG (*Bacillus subtilis* QST 713)

Biological protectant fungicide.

Champ WG^{OG} (copper hydroxide)

Copper product.

Chloronil 720 (chlorothalonil)

A broad spectrum fungicide.

Companion (*Bacillus subtilis* GB03)

Plant growth promoting rhizobacteria.

Contans^{OG} (*Coniothyrium minitans*)

Sclerotinia sclerotiorum and *Sclerotinia minor.*

Curzate 60 DF (cymoxanil)

For late blight of potato and tomato and Downy Mildew of cucurbits and lettuce.

Cuprofix Ultra (basic copper sulfate)

A broad spectrum fungicide.

Cuproxat Ultra (basic copper sulfate)

A broad spectrum fungicide.

DiTera DF^{OG} (*Myrothecium verrucaria* AARC-0255)

Plant parasitic nematodes.

Dithane (various formulas) (mancozeb)

A broad spectrum, protectant fungicide.

Decree (fenhexamid)

Botrytis control in greenhouse transplants.

Double Nickel^{OG} (*Bacillus amyloliquefaciens*):

Microbial fungicide.

Endura (boscalid)

A protectant fungicide for legumes, brassicas, bulb vegetables, fruiting vegetables, lettuce, and root and tuber vegetables.

Evolve (thiophanate-methyl & mancozeb)

Potato seed piece treatment.

Flint (trifloxystrobin)

A strobilurin fungicide with broad-spectrum activity.

Fontelis (penthiopyrad)

A fungicide with broad host clearance for leaf spots, blights, anthracnose, and *Sclerotinia* diseases.

Forum (dimethomorph)

A fungicide for use against *Phytophthora* and Downy Mildews of bulb, cucurbit and fruiting vegetables, lettuce, potatoes, and tomatoes.

Fosphite (phosphorus acid)

A phosphorous acid fungicide active against *Pythium*, *Phytophthora*, and the Downy Mildews.

Funga Stop L & G (citric acid, mint oil)

Fungal diseases.

Galltrol-A (*Agrobacterium radiobacter* K84)

Crown gall.

Gavel (zoxamide and mancozeb)

A broad-spectrum protectant fungicide for disease control in potatoes, cucurbits, and tomatoes.

GC-3^{OG} (cottonseed oil, corn oil, garlic extract)

Powdery mildew.

GEM 500 SC (trifloxystrobin)

Broad spectrum fungicide.

Headline (pyraclostrobin)

A broad spectrum (strobilurin) fungicide for use in legumes, corn, and tuberous and corm vegetables.

Headline AMP (pyraclostrobin & metconazole)

For plant health in potato.

Inspire Super (difenoconazole plus cyprodinil)

Powdery mildew.

JMS Stylet-Oil^{OG} (paraffinic oil)

Fungal diseases and aphid transmitted viruses.

Kocide 3000 (cupric hydroxide)

Broad spectrum bactericide and fungicide.

Kodiak (*Bacillus subtilis* GB03)

Biological seed treatment.

Kumulus DF (sulfur)

Broad spectrum fungicide, particularly for powdery mildew.

K-Phite

A phosphorous acid fungicide active against *Pythium*, *Phytophthora*, and the Downy Mildews, also labelled for greenhouse transplant production.

Luna Privilege (fluopyram)

Leaf spots and Powdery mildew of dried beans, potato, and watermelon.

Luna Experience (fluopyram & tebuconazole)

Powdery mildew on watermelon only.

Luna Tranquility (fluopyram & pyrimethanil)

Early blight and Powdery mildew of potato.

ManKocide (copper plus mancozeb)

A broad spectrum fungicide and bactericide.

Manzate (mancozeb)

A broad spectrum fungicide.

Mastercop (copper sulfate pentahydrate)

A broad spectrum fungicide and bactericide.

Maxim (fludioxonil)

A seed treatment fungicide for certain diseases of potato.

Merivon Xemium (pyraclostrobin & fluaxapoxad)

Powdery mildew and other diseases.

Mertect 340-F (thiabendazole)

A seed treatment.

Micora (mandipropamid)

Oomycete fungicide labelled for greenhouse use.

Micro Sulf^{OG} (sulfur)

Powdery mildew.

Microthiol D^{OG} (sulfur)

A protectant fungicide particularly useful for Powdery Mildews.

Mildew Cure^{OG} (cottonseed, corn, and garlic oil)

Powdery mildew.

MilStop^{OG} (potassium bicarbonate)

Powdery mildew and other foliar diseases.

M-Pede^{OG} (insecticidal soap)

Insecticide/fungicide labelled for greenhouse use.

MycoStop^{OG} (*Streptomyces griseoviridis* K61)

Biological seed or soil treatment.

Natria^{OG} (*Bacillus subtilis* QST 713)

Biological protectant fungicide.

Nutrol (monopotassium phosphate)

Protectant fungicide for powdery mildew.

Omega (fluazinam)

Late Blight, Downy Mildews, and *Sclerotinia* control.

Organocide^{OG} (sesame oil)

Powdery mildew.

Oso 5% SC (polyoxin D)

Broad spectrum fungicide for foliar and soil-borne diseases.

OxiDate 2.0^{OG} (hydrogen dioxide)

Preventive bioicide.

Penncozeb (various formulas) (mancozeb)

A dry flowable formulation of mancozeb for control of fungi.

Phyton 35 (copper sulfate)

Copper sulfate product labelled for greenhouse use on vegetable transplants.

Prophyt (potassium phosphite)

A fungicide for *Pythium, Phytophthora*, and Downy Mildew.

Phostrol (potassium phosphite)

A fungicide for *Pythium, Phytophthora*, and Downy Mildew.

Phyta-Guard EC (rosemary and clove oil)

Powdery mildew and bacterial diseases.

PlantShield HC^{OG} (*Trichoderma harzianum* Rifai strain KRL-AG2)

Biological soil treatment.

Polyram (Polyram)

For early and late blight in potatoes.

Presidio (fluopicolide)

A locally systemic fungicide effective against *Phytophthora* and Downy Mildews of bulb, cucurbit, fruiting, and leafy vegetables.

PreStop^{OG} (*Gliocladium catenulatum* J1446)

Biological soil treatment.

Previcur Flex (propamocarb)

A fungicide for Oomycetes. Previcur should be mixed with Bravo, Maneb or Mancozeb to prevent development of resistance.

Priaxor (fluxapyroxad & pyraclostrobin)

For disease control in beans and corn.

Pristine (pyraclostrobin and boscalid)

For use in bulb vegetables and carrots.

Procure (triflumizole)

Powdery Mildew on Brassica, cucurbits, and leafy vegetables.

Promax^{OG} (thyme oil)

Soil-borne diseases and plant parasitic nematodes.

Quadris (azoxystrobin)

A strobilurin fungicide with broad spectrum activity.

Quadris Opti (azoxystrobin & chlorothalonil)

Broad spectrum fungicide for dry beans, cucurbit vegetables, potatoes, tomatoes, and onions.

Quadris Top (azoxystrobin & difenaconazole)

Broad spectrum fungicide.

Quilt (azoxystrobin & propiconazole)

Broad spectrum fungicide for use in carrots, celery, corn, and onions.

Quilt Xcel (azoxystrobin & propiconazole)

Broad spectrum fungicide for use in carrots, celery, corn, and onions.

Quintec (quinoxyfen)

Fungicide for control of powdery mildew in cucurbits.

Rally (myclobutanil)

A fungicide for powdery mildews and rusts of vegetable crops.

Ranman (cyazofamid)

Effective against *Phytophthora* and Downy Mildew in cucurbits, tomatoes, and potatoes.

Reason 500 SC (femadione)

A fungicide for use against *Phytophthora*, Downy Mildew, and white rust on tuberous and corm vegetables, tomatoes, bulb vegetables, lettuce, and cucurbit vegetables.

Regalia^{OG} (extract of *Reynoutria sachalinensis*)

Plant defense activator for fungal and bacterial diseases.

Revus (mandipropamid)

For use against Downy Mildew on peppers, brassica, bulb crops, cucurbits, and leafy vegetables.

Revus Top (mandipropamid plus difenoconazole)

Broad spectrum fungicide for potatoes and tomatoes.

Rhapsody^{OG} *(Bacillus subtilis* QST 713)

Biological protectant fungicide.

Ridomil Gold (mefenoxam)

A fungicide active against *Pythium*, *Phytophthora*, and the Downy Mildews.

Ridomil/Gold Bravo SC (mefenoxam plus chlorothalonil)

Broad spectrum fungicide containing 4.4% metalaxyl and 72% chlorothalonil effective against both lower and true fungi.

Ridomil MZ 72 (mefenoxam plus mancozeb)

Broad spectrum fungicide containing 8% metalaxyl and 64% mancozeb effective against both lower and true fungi.

Ridomil Gold EC/Copper (mefenoxam plus copper)

Broad spectrum fungicide containing 4.8% metalaxyl and 60% copper hydroxide effective against both lower and true fungi.

RootShield WP^{OG} (*Trichoderma harzianum* Rifai strain KRL-AG2)

Biological soil treatment.

RootShield WPPlus (*Trichoderma harzianum* T-22, *T. virens* G-41)

Biological soil treatment.

Rovral 4 F (iprodione)

For control of *Rhizoctonia* and *Sclerotinia* on lettuce.

Scala SC (pyrimethanil)

Protective fungicide for bulb, tuberous, and corm vegetables.

SerenadeMax^{OG} (*Bacillus subtilis* QST 713)

Biological protectant fungicide.

Sil-Matrix (potassium silicate)

Broad spectrum preventive fungicide.

SoilGard 12G^{OG} (*Gliocladium (Trichoderm) virens* GL-21)

Biological soil treatment.

Sonata^{OG} (*Bacillus pumilus* QST 2808)

Biological protectant fungicide.

Sporan EC (rosemary, clove, and thyme oils)

Contact fungicide.

Suffoil-X^{OG} (petroleum oils)

Fungicide, insecticide, and miticide labelled for greenhouse transplant production.

Sulfur DF^{OG} (sulfur)

Powdery mildew.

Switch 62.5 WG (cyprodinil & fludioxonil)

A protective fungicide for use in beans, brassica, carrot, herbs, leafy vegetables, and onions.

T-22 HC^{OG} (*Trichoderma harzianum* Rifai strain KRL-AG2)

Biological soil treatment.

Taegro ECOOG (*B. subtilis* var. *amyloliquefaciens* FZB24)

Biological for soil-borne and bacterial diseases in cucurbits, leafy vegetables, and fruiting vegetables.

Tanos (famoxadone and cymoxanil)

A penetrant fungicide with locally systemic and curative activities against Downy Mildew and late blight diseases.

Terraclor 400 (PCNB)

A fungicide active against soil borne true fungi labeled for greenhouse transplant production.

Thiram 42-S (thiram)

A seed treatment fungicide with wide host range.

Tilt (propiconazole)

A protective fungicide for beets, carrots, celery, onions, and corn.

Topsin-M 70 WSP (thiophanate-methyl)

A systemic fungicide with broad spectrum control.

Tops-MZ-Gaucho (thiophanate-methyl & mancozeb & cymoxanil)

Fungicide and insecticide for potato seed piece treatment.

Torino (cyflufenamid)

Powdery mildew of cucurbits and strawberries.

Trilogy^{OG} (neem oil)

Various fungal diseases.

Uniform (azoxystrobin & mefenoxam)

For soil borne diseases.

Veranda O (polyoxin D)

For use in greenhouse transplant production.

Vivando (metrofenone)

For powdery mildew.

XeroTol 2.0^{OG} (hydrogen dioxide)

Preventive biocide labelled for greenhouse use, diseases of potatoes and bulb, brassica, cucurbit, fruiting, and leafy vegetables.

Zampro (ametoctradin & dimethomorph)

A fungicide for Downy mildew and *Phytophthora*

5

Ready-Reckoner for Disease Specific Fungicides

There are several fungicidal molecules available in the market, some of which are specific for certain diseases/pathogens while others are broad spectrum and can be used against a range of diseases. However, for proper control measures a knowledge of disease specific fungicide are very important. Ready-Reckoner for disease specific fungicide (Table 4) makes available the option for selection of alternative fungicide in the management of specific disease.

Table 4: Ready-Reckoner for disease specific fungicides.

Note: Please read Label claim before use of fungicide for the disease host plant.

Powdery mildews	Hexaconazole, Azoxystrobin, metrofenone, cyflufenamid, triflumizole, myclobutanil, quinoxyfen, sulphur, Dinocap, Carbendazim, , Flusilazole, Triadimefon, Tebuconazole, Thiophanate Methyl.
Downey mildews	Ametoctradin, Dimethomorph, famoxadone + cymoxanil, metalaxyl, potassium phosphate, mandipropamid, cyazofamid, propamocarb, fluopicolide, fluazinam, copper carbonate, Cymoxanil+Mancozeb, Zineb.
Anthracnose	Chlorothalonil, thiophanate-methyl, Ziram, Fentin acetate, Copper sulfate, Copper hydroxide, Captan+ Hexaconazole, Zineb.
Blight and leaf spot	Tubenoconazole, thiophanate-methyl, mancozeb, cymoxanil, propiconazole, Copper sulphate, Iprodione, Kresoxin-Methyl, Kitazin, Metarim, Ziram, Famoxadone, copper hydroxide, azoxystrobin, fluxapyroxad, pyraclostrobin, fluazinam, copper carbonate, copperoxychloride, Carbendazim, Zineb, Myclobutanil.
Wilt	azoxystrobin , mefenoxam, potassium phosphate, Bordeaux mixture, copper-oxy-chloride, Captan, thiram, Nabam,
Scab	Sulphur dust, Ziram, Glydon, Carbendazim, thiophanate-methyl.
Damping off	Etridiazole, choloroneb, Fosetyl Al, Copper sulphate, Captan, Mancozeb,
Fruit rot	Carbendazim, Thiabendazole, Endomycin, Azoxystrobin, Copper sulphate, Chlorothalonil, Difenoconazole, Hexaconazole, Kitazin, Mancozeb, Tebuconazole, Zineb, Captan+Hexaconazole.
Root rot	iprodione, Pyracarbid, mancozeb,

Smut	Ferbam, Carboxin, Pyracarbolid, furcarbanil.
Rust	Femadione, copper carbonate, Zineb, maneb, Oxycarboxin, Pyracarbolid, Sulphur, Lime Sulphur, Triadimefon.
Die-back	Difenoconazole, Myclobutanil, propineb.
Collar rot	Mancozeb
Grey leaf mold	Zineb
Buck eye rot	Mancozeb, Propineb.
Post-harvest diseases	Benomyl, Carbendazim, Thiabendazole

6

Development of Resistance in Fungal Pathogens Against the Fungicide and Fungicide Resistance Management

Development of fungicide resistance in a fungal disease pathogen is a major concern for some of the important diseases in their management. This can be overcome by designing of fungicide use strategies to manage the threat of fungicide resistance. There have been considerable efforts by industry to conduct research in the areas of mode of action, resistance risk, field monitoring for baseline sensitivity and sensitivity variations in treated fields. Based on the results, the fungicide use strategies are designed that reduce the risk of fungicide resistance build-up or in other words, the loss of efficacy of whole fungicide classes.This threat of fungicide resistance and furthermore the development of cross-resistance to other fungicides in such fungicide resistance strains/ pathogens often exists to related products of different manufacturers. This has lead to a close collaboration among them in FRAC (Fungicide Resistance Action Committee) to work on such issues of fungicide resistance. Results from research in mode of action, resistance risk and field monitoring are pooled and strategies are developed to minimize the risk of resistance build-up. FRAC has produced several monographs on various aspects of fungicide resistance and has grouped the available fungicides according to various criteria that facilitate the understanding of the resistance risk of the different fungicide groups (Table 5).

6.1. Mode of Action of Major Fungicides Classes and Resistance Risk Class

Table 5: Mode of action of major fungicides classes, their FRAC code and resistance risk class.

FRAC Code	Chemical Class	Mode of action / inhibition	Resistance risk
1	Benzimidazoles	Beta-tubulin biosynthesis	High
2	Dicarboximides	NADH cytochrome c reductase in lipids	High
3	Azoles, Pyrimidines	C-14 demethylation in sterol biosynthesis	Medium
4	Phenylamides	RNA polymerase	High
5	Morpholines	>8 and >7 isomerase and >14 reductase in sterol biosynthesis	low-medium
7	Carboxamides	Succinic acid oxidation	Medium
9	Anilinopyrimidine	Methionine biosynthesis	Medium
11	Strobilurins	Mitochondrial synthesis in cytochrome bc1	High
16	Various chemistry	Melanin biosynthesis inhibitors (two sites)	Medium
40	Carboxylic acid amides	Cell wall formation in Oomycetes	low-medium
M1	Inorganics	Multisite contact	Low
M3	Dithiocarbamates	Multisite contact	Low
M5	Phthalimides	Multisite contact	Low

The FRAC code

The Fungicide Resistance Action Committee (**FRAC**) developed a **code** of numbers and letters that can be used to distinguish the different fungicide groups based on their mode of action. This **code** is known as the **FRAC Code** and is now included on fungicide labels.

FRAC codes can help delay plant pathogens to develop resistance to fungicides by Rotating fungicides with different FRAC codes. Code numbers on fungicide labels, called "FRAC" groups, can help you develop chemical rotations that delay fungicide-resistance in fruits and vegetable diseases.

DMI fungicides (Group 3) are a natural rotation partner with strobilurin fungicides (Group 11) for many foliar diseases, especially powdery mildews and rusts. DMI fungicides offer unique application opportunities with chemigation (Eagle, Hoist and Terraguard), total release (Fungaflor TR) and drenching (Terraguard).

The fungicides belong to FRAC group 7 inhibit complex II of the fungal mitochondrial respiration by binding and blocking SDH (succinate dehydrogenase) -mediated electron transfer from succinate to ubiquinone.

For grouping of fungicide and resistance risk various parameters are considered which are as follows:

MOA Code: Different letters (A to P, with added numbers) are used to distinguish fungicide groups according to their biochemical mode of action (MOA) in the biosynthetic pathways of plant pathogens. The grouping was made according to processes in the metabolism starting from nucleic acids synthesis (A) to secondary metabolism, e.g. melanin synthesis (I), followed by host plant defence inducers (P), recent molecules with an unknown mode of action and unknown resistance risk (U, transient status, until information about mode of action and mechanism of resistance becomes available), and chemical multi-site inhibitors (M). Fungicidal compositions of biological origin are grouped according to the main mode of action within the respective pathway categories. A more recently introduced category "Biologicals with multiple modes of action" (BM) is used for agents from biological origin showing multiple mechanisms of action.

Target Site and Code: If available, the biochemical mode of action is given. In several cases the precise target site may not be known, however, a grouping within a given pathway / functional cluster is still possible. Grouping can also be made due to cross resistance profiles within a group or in relation to other groups.

Group Name: The Group Names listed are based on chemical relatedness of structures which are accepted in literature (e.g. The Pesticide Manual). They are based on different sources (chemical structure, site of action, first important representative in group).

Chemical or Biological Group: Grouping is based on chemical considerations. Nomenclature is according to IUPAC and Chemical Abstract name. Taxonomic information may be used for agents of biological origin.

Common name: BSI/ISO accepted (or proposed) common name for an individual active ingredient expected to appear on the product label as definition of the product.

Comments on Resistance: Details are given for the (molecular) mechanism of resistance and the resistance risk. If field resistance is known to one member of the Group, it is most likely but not exclusively valid that cross resistance to other group members will be present. There is increasing evidence that the degree of cross resistance can differ between group members and pathogen species or even within species. For the latest information on resistance and cross resistance status of a pathogen / fungicide combination, it is advised to contact local FRAC representatives, product manufacturer's representatives or crop protection advisors. The intrinsic risk for resistance evolution to a given fungicide group is estimated to be low, medium or high according to the principles described in FRAC Monographs 1, 2 and 3. Resistance management is driven by intrinsic risk of fungicide, pathogen risk and agronomic risk (see FRAC pathogen risk list).

FRAC Code: List© 2021 Page 3 of 17. Similar classification lists of fungicides have been published by T. Locke on behalf of FRAG – UK (Fungicide Resistance, August 2001), and by P. Leroux (Classification des fongicides agricoles et résistance, Phytoma, La Défense des Végétaux, No. 554, 43-51, November 2002). FRAC Code Numbers and letters are used to distinguish the fungicide groups according to their cross-resistance behaviour. This code should be used to define the "FUNGICIDE GROUP" code, e.g. on product labels. The numbers were assigned primarily according to the time of product introduction to the market. The letters refer to P = host plant defence inducers, M = chemical multi-site inhibitors, U = unknown mode of action and unknown resistance risk, and BM = biologicals with multiple modes of action. Reclassification of compounds based on new research may result in codes to expire. This is most likely in the U – section when the mode of action gets clarified. These codes are not re-used for new groups; a note is added to indicate reclassification into a new code. Table 6 elaborates different fungicide group, their mode of action, resistance and cross-resistance with FRAC code

Table 6: Fungicide groups, mode of Action, resistance and cross resistance with FRAC code

MoA	Target Site and Code	Group Name	Chemical or Biological Group	Common Name	Comments	FRAC Code
A: nucleic acids metabolism	A1 RNA polymerase I	PA – fungicides (PhenylAmides)	acylalanines	benalaxyl, benalaxyl-M (=kiralaxyl), furalaxyl, metalaxyl, metalaxyl-M (=mefenoxam).	Resistance and cross resistance well known in various Oomycetes but mechanism unknown. High risk. See FRAC Phenylamide Guidelines for resistance management	4
			oxazolidinones	oxadixyl		
			butyrolactones	ofurace		
	A2 adenosin- deaminase	hydroxy- (2-amino-) pyrimidines	hydroxy- (2-amino-) pyrimidines	bupirimate, dimethirimol, ethirimol	Medium risk. Resistance and cross resistance known in powdery mildews. Resistance management required.	8
	A3 DNA/RNA synthesis (proposed)	heteroaromatics	isoxazoles	hymexazole	Resistance not known.	32
			isothiazolones	octhilinone		
	A4 DNA topoisomerase type II (gyrase)	carboxylic acids	carboxylic acids	oxolinic acid	Bactericide. Resistance known. Risk in fungi unknown. Resistance management required.	31

MoA	Target Site And Code	Group Name	Chemical Or Biological Group	Common Name	Comments	Frac Code
B: Cytoskeleton and motor protein	B1 ß-tubulin assembly in mitosis	MBC - fungicides (Methyl Benzimidazole Carbamates)	benzimidazoles	benomyl, carbendazim, fuberidazole, thiabendazole	Resistance common in many fungal species. Several target site mutations, mostly E198A/G/K, F200Y in β-tubulin gene. Positive cross resistance between the group members. Negative cross resistance to N-phenyl carbamates. High risk. See FRAC Benzimidazole Guidelines for resistance management.	1
			thiophanates	thiophanate, thiophanate-methyl		
	B2 ß-tubulin assembly in mitosis	N-phenyl carbamates	N-phenyl carbamates	diethofencarb	Resistance known. Target site mutation E198K. Negative cross resistance to benzimidazoles. High risk. Resistance management required.	10
	B3 ß-tubulin assembly in mitosis	benzamides	toluamides	zoxamide	Low to medium risk. Resistance management required.	22
		thiazole carboxamide	ethylamino-thiazole-carboxamide	ethaboxam		
	B4 cell division (unknown site)	phenylureas	phenylureas	pencycuron	Resistance not known.	20
	B5 delocalisation of spectrin-like proteins	benzamides	pyridinylmethyl-benzamides	fluopicolide, fluopimomide	Resistant isolates detected in grapevine downy mildew. Medium risk. Resistance management required	43
	B6 actin/myosin/fimbrin function	cyanoacrylates	aminocyanoacrylates	phenamacril	Resistance known in Fusarium graminearum. Target site mutations in the gene coding for myosin-5 found in lab studies. Medium to high risk. Resistance management required.	47
		aryl-phenyl-ketones	benzophenone	metrafenone	Less sensitive isolates detected in powdery mildews (Blumeria and Sphaerotheca) Medium risk. Resistance management required. Reclassified from U8 in 2018	50
			benzoylpyridine	pyriofenone		

MoA	Target Site and Code	Group Name	Chemical or Biological Group	Common Name	Comments	Frac Code
C. respiration	C1 complex I NADH oxido-reductase	pyrimidinamines	pyrimidinamines	diflumetorim	Resistance not known.	39
		pyrazole-MET1	pyrazole-5- carboxamides	tolfenpyrad		
		Quinazoline	quinazoline	fenazaquin		
	C2 complex II: succinate-dehydro-genase	SDHI (Succinate-dehydrogenase inhibitors)	phenyl-benzamides	benodanil, flutolanil, mepronil	Resistance known for several fungal species in field populations and lab mutants. Target site mutations in sdh gene, e.g. H/Y (or H/L) at 257, 267, 272 or P225L, dependent on fungal species. Resistance management required. Medium to high risk. See FRAC SDHI Guidelines for resistance management.	7
			phenyl-oxo-ethyl thiophene amide	isofetamid		
			pyridinyl-ethyl-benzamides	fluopyram		
			phenyl-cyclobutyl-pyridineamide	cyclobutrifluram		
			furan- carboxamides	fenfuram		
			oxathiin- carboxamides	carboxin, oxycarboxin		
			thiazole- carboxamides	thifluzamide		
			pyrazole-4- carboxamides	benzovindiflupyr, bixafen, fluindapyr, fluxapyroxad, furametpyr, inpyrfluxam, isopyrazam, penflufen, penthiopyrad, sedaxane		
			N-cyclopropyl-N- benzyl-pyrazole-carboxamides	isoflucypram		
			N-methoxy-(phenyl-ethyl)-pyrazole-carboxamides	pydiflumetofen		
			pyridine- carboxamides	boscalid		
			pyrazine- carboxamides	pyraziflumid		

MoA	Target Site and Code	Group Name	Chemical or Biological Group	Common Name	Comments	Frac Code
C. respiration	C3 complex III: cytochrome bc1 (ubiquinol oxidase) at Qo site (cyt b gene)	QoI-fungicides (Quinone outside Inhibitors)	methoxy-acrylates	azoxystrobin, coumoxystrobin, enoxastrobin, flufenoxystrobin, picoxystrobin, pyraoxystrobin	Resistance known in various fungal species. Target site mutations in cyt b gene (G143A, F129L) and additional mechanisms. Cross resistance shown between all members of the Code 11 fungicides. High risk. See FRAC QoI Guidelines for resistance management.	11
			methoxy-acetamide	mandestrobin		
			methoxy-carbamates	pyraclostrobin, pyrametostrobin, triclopyricarb		
			oximino-acetates	kresoxim-methyl, trifloxystrobin		
			oximino-acetamides	dimoxystrobin, fenaminstrobin metominostrobin, orysastrobin		
			oxazolidine-diones	famoxadone		
			dihydro-dioxazines	fluoxastrobin		
			imidazolinones	fenamidone		
			benzyl-carbamates	pyribencarb		
		QoI-fungicides (Quinone outside Inhibitors; Subgroup A)	tetrazolinones	metyltetraprole	Resistance not known. Not cross resistant with Code 11 fungicides on G143A mutants. High risk. See FRAC QoI Guidelines for resistance management.	11A

MoA	Target Site and Code	Group Name	Chemical or Biological Group	Common Name	Comments	Frac Code
C: respiration (continued)	C4 complex III: cytochrome bc1 (ubiquinone reductase) at Qi site	QiI - fungicides (Quinone inside Inhibitors)	cyano-imidazole	cyazofamid	Resistance risk unknown but assumed to be medium to high (mutations at target site known in model organisms). Resistance management required. No spectrum overlap with the Oomycete-fungicides cyazofamid and amisulbrom	21
			sulfamoyl-triazole	amisulbrom		
			picolinamides	fenpicoxamid, florylpicoxamid		
	C5 uncouplers of oxidative phos-phorylation		dinitrophenyl-crotonates	binapacryl, meptyldinocap, dinocap	Resistance not known. Also acaricidal activity.	29
			2,6-dinitro-anilines	fluazinam	Low risk. However, resistance claimed in Botrytis in Japan.	
			(pyr.-hydrazones)	ferimzone	Reclassified to U 14 in 2012.	
	C6 inhibitors of oxidative phos-phorylation, ATP synthase	organo tin compounds	tri-phenyl tin compounds	fentin acetate, fentin chloride, fentin hydroxide	Some resistance cases known. Low to medium risk.	30
	C7 ATP transport (proposed)	thiophene-carboxamides	thiophene-carboxamides	silthiofam	Resistance reported. Risk low.	38
	C8 complex III: cytochrome bc1 (ubiquinone reductase) at Qo site, stigmatellin binding sub-site	QoSI fungicides (Quinone outside Inhibitor, stigmatellin binding type)	triazolo-pyrimidylamine	ametoctradin	Not cross resistant to QoI fungicides. Resistance risk assumed to be medium to high (single site inhibitor). Resistance management required.	45

MoA	Target Site and Code	Group Name	Chemical or Biological Group	Common Name	Comments	Frac Code
D: amino acids and protein synthesis	D1 methionine biosynthesis (proposed) (cgs gene)	AP - fungicides (Anilino- Pyrimidines)	anilino-pyrimidines	cyprodinil, mepanipyrim, pyrimethanil	Resistance known in Botrytis and Venturia, sporadically in Oculimacula. Medium risk. See FRAC Anilinopyrimidine Guidelines for resistance management.	9
	D2 protein synthesis (ribosome, termination step)	enopyranuronic acid antibiotic	enopyranuronic acid antibiotic	blasticidin-S	Low to medium risk. Resistance management required.	23
	D3 protein synthesis (ribosome, initiation step)	hexopyranosyl antibiotic	hexopyranosyl antibiotic	kasugamycin	Resistance known in fungal and bacterial (P. glumae) pathogens. Medium risk. Resistance management required.	24
	D4 protein synthesis (ribosome, initiation step)	glucopyranosyl antibiotic	glucopyranosyl antibiotic	streptomycin	Bactericide. Resistance known. High risk. Resistance management required.	25
	D5 protein synthesis (ribosome, elongation step)	tetracycline antibiotic	tetracycline antibiotic	oxytetracycline	Bactericide. Resistance known. High risk. Resistance management required.	41

MoA	Target Site and Code	Group Name	Chemical or Biological Group	Common Name	Comments	Frac Code
E: signal transduction	E1 signal transduction (mechanism unknown)	aza-naphthalenes	aryloxyquinoline	quinoxyfen	Resistance to quinoxyfen known. Medium risk. Resistance management required. Cross resistance found in Erysiphe (Uncinula) necator but not in Blumeria graminis.	13
			quinazolinone	proquinazid		
	E2 MAP/Histidine- Kinase in osmotic signal transduction (os-2, HOG1)	PP-fungicides (Phenyl Pyrroles)	phenylpyrroles	fenpiclonil, fludioxonil	Resistance found sporadically, mechanism speculative. Low to medium risk. Resistance management required.	12
	E3 MAP/Histidine- Kinase in osmotic signal transduction (os-1, Daf1)	dicarboximides	dicarboximides	chlozolinate, dimethachlone ,iprodione, procymidione, vinclozolin	Resistance common in Botrytis and some other pathogens. Several mutations in OS-1, mostly I365S. Cross resistance common between the group members. Medium to high risk. See FRAC Dicarboximide Guidelines for resistance management.	2

MoA	Target Site and Code	Group Name	Chemical or Biological Group	Common Name	Comments	Frac Code
F: lipid synthesis or transport / membrane integrity or function	F1	formerly dicarboximides				
	F2 phospholipid biosynthesis, methyltransferase	phosphoro- thiolates Dithiolanes	phosphoro-thiolates dithiolanes	edifenphos, iprobenfos (IBP), pyrazophos, isoprothiolane	Resistance known in specific fungi. Low to medium risk. Resistance management required if used for risky pathogens.	6
	F3 cell peroxidation (proposed)	AH-fungicides (Aromatic Hydrocarbons) (chlorophenyls, nitroanilines) heteroaromatics	aromatic hydrocarbons 1,2,4-thiadiazoles	biphenyl, chloroneb, dicloran, quintozene, PCNB, tecnazene, TCNB, tolclofos-methyl, etridiazole	Resistance known in some fungi. Low to medium risk. Cross resistance patterns complex due to different activity spectra.	14
	F4 cell membrane permeability, fatty acids (proposed)	Carbamates	carbamates	iodocarb propamocarb prothiocarb	Low to medium risk. Resistance management required.	28
	F5	formerly CAA-fungicides				
	F6 microbial disrupters of pathogen cell membranes	formerly Bacillus amyloliquefaciens strains (FRAC Code 44); reclassified to BM02 in 2020				
	F7 cell membrane disruption	formerly extract from Melaleuca alternifolia (tea tree oil) and plant oils (eugenol, geraniol, thymol) FRAC Code 46, reclassified to BM01 in 2021				
	F8 ergosterol binding	Polyene	amphoteric macrolide antifungal antibiotic from Streptomyces natalensis or S. chattanoogensis	natamycin (pimaricin)	Resistance not known. Agricultural, food and topical medical uses.	48
	F9 lipid homeostasis and transfer/storage	OSBPI oxysterol binding protein homologue inhibition	piperidinyl-thiazole-isoxazolines	oxathiapiprolin fluoxapiprolin	Resistance risk assumed to be medium to high (single site inhibitor). Resistance management required. (Previously U15).	49

MoA	Target Site and Code	Group Name	Chemical or Biological Group	Common Name	Comments	Frac Code
G: sterol biosynthesis in membranes	G1 C14- demethylase in sterol biosynthesis (erg11/cyp51)	DMI-fungicides (DeMethylation Inhibitors) (SBI: Class I)	piperazines	triforine	There are big differences in the activity spectra of DMI fungicides. Resistance is known in various fungal species. Several resistance mechanisms are known incl. target site mutations in cyp51 (erg 11) gene, e.g. V136A, Y137F, A379G, I381V; cyp51 promotor; ABC transporters and others. Generally wise to accept that cross resistance is present between DMI fungicides active against the same fungus. DMI fungicides are Sterol Biosynthesis Inhibitors (SBIs), but show no cross resistance to other SBI classes. Medium risk. See FRAC SBI Guidelines for resistance management.	3
			pyridines	pyrifenox, pyrisoxazole		
			pyrimidines	fenarimol, nuarimol		
			imidazoles	imazalil, oxpoconazole, pefurazoate, prochloraz, triflumizole		
			triazoles triazolinthiones	azaconazole, bitertanol, bromuconazole, cyproconazole, difenoconazole, diniconazole, epoxiconazole, etaconazole, fenbuconazole fluquinconazole, flusilazole, flutriafol hexaconazole, imibenconazole, ipconazole, mefentrifluconazole, metconazole, myclobutanil, penconazole, propiconazole, simeconazole, tebuconazole, tetraconazole, triadimefon, triadimenol, triticonazole, prothioconazole		
	G2 D^{14}-reductase and $D^{8}\rightarrow^{7}$-isomerase in sterol biosynthesis (erg24, erg2)	amines ("morpholines") (SBI: Class II)	morpholines	aldimorph, dodemorph, fenpropimorph, tridemorph	Decreased sensitivity for powdery mildews. Cross resistance within the group generally found but not to other SBI classes. Low to medium risk. See FRAC SBI Guidelines for resistance management.	5
			piperidines	fenpropidin, piperalin		
			spiroketal-amines	spiroxamine		
	G3 3-keto reductase, C4- de-methylation (erg27)	KRI fungicides (Keto Reductase Inhibitors) (SBI: Class III)	hydroxyanilides	fenhexamid	Low to medium risk. Resistance management required.	17
			amino-pyrazolinone	fenpyrazamine		
	G4 squalene-epoxidase in sterol biosynthesis (erg1)	(SBI class IV)	thiocarbamates	pyributicarb	Resistance not known, fungicidal and herbicidal activity.	18
			allylamines	naftifine, terbinafine	Medical fungicides only.	

MoA	Target Site and Code	Group Name	Chemical or Biological Group	Common Name	Comments	Frac Code
H: cell wall biosynthesis	H3	Formerly glucopyranosyl antibiotic (validamycin)			reclassified to U18	26
	H4 chitin synthase	polyoxins	peptidyl pyrimidine nucleoside	polyoxin	Resistance known. Medium risk. Resistance management required.	19
	H5 cellulose synthase	CAA-fungicides (Carboxylic Acid Amides)	cinnamic acid amides	dimethomorph, flumorph, pyrimorph	Resistance known in Plasmopara viticola but not in Phytophthora infestans. Cross resistance between all members of the CAA group. Low to medium risk. See FRAC CAA Guidelines for resistance management.	40
			valinamide carbamates	benthiavalicarb, iprovalicarb, valifenalate		
			mandelic acid amides	mandipropamid		
I: melanin synthesis in cell wall	I1 reductase in melanin biosynthesis	MBI-R (Melanin Biosynthesis Inhibitors – Reductase)	isobenzo-furanone	fthalide	Resistance not known.	16.1
			pyrrolo-quinolinone	pyroquilon		
			triazolobenzo- thiazole	tricyclazole		
	I2 dehydratase in melanin biosynthesis	MBI-D (Melanin Biosynthesis Inhibitors – Dehydratase)	cyclopropane-carboxamide	carpropamid	Resistance known. Medium risk. Resistance management	16.2
			carboxamide	diclocymet		
			propionamide	fenoxanil		
	I3 polyketide synthase in melanin biosynthesis	MBI-P (Melanin Biosynthesis Inhibitors – Polyketide synthase)	trifluoroethyl-carbamate	tolprocarb	Resistance not known. Additional activity against bacteria and fungi through induction of host plant defence	16.3

MoA	Target Site and Code	Group Name	Chemical or Biological Group	Common Name	Comments	Frac Code
P: host plant defence induction	P 01 salicylate-related	benzo- thiadiazole (BTH)	benzo-thiadiazole (BTH)	acibenzolar-S-methyl	Resistance not known.	P 01
	P 02 salicylate-related	benzisothiazole	benzisothiazole	probenazole (also antibacterial and antifungal activity)	Resistance not known.	P 02
	P 03 salicylate-related	thiadiazole-carboxamide	thiadiazole- carboxamide	tiadinil, isotianil	Resistance not known.	P 03
	P 04 polysaccharide elicitors	natural compound	polysaccharides	laminarin	Resistance not known.	P 04
	P 05 anthraquinone elicitors	plant extract	complex mixture, ethanol extract (anthraquinones, resveratrol)	extract from Reynoutria sachalinensis (giant knotweed)	Resistance not known.	P 05
	P 06 microbial elicitors	microbial	bacterial Bacillus spp.	Bacillus mycoides isolate J	Resistance not known.	P 06
			fungal Saccharomyces spp.	cell walls of Saccharomyces cerevisiae strain LAS117		
	P 07 phosphonates	phosphonates	ethyl phosphonates	fosetyl-Al	Few resistance cases reported in few pathogens. Low risk. Reclassified from U33 in 2018	P07
				phosphorous acid and salts		
	P 08 salicylate-related	isothiazole	isothiazolylmethyl ether	dichlobentiazox	activates SAR both up- and downstream of SA. Resistance not known.	P 08

MoA	Target Site and Code	Group Name	Chemical or Biological Group	Common Name	Comments	Frac Code
U: Unknown mode of action (U numbers not appearing in the list derive from reclassified fungicides)	unknown	cyanoacetamide-oxime	cyanoacetamide-oxime	cymoxanil	Resistance claims described. Low to medium risk. Resistance management required.	27
	formerly phosphonates (FRAC code 33), reclassified to P 07 in 2018					
	unknown	phthalamic acids	phthalamic acids	tecloftalam (Bactericide)	Resistance not known.	34
	unknown	benzotriazines	benzotriazines	triazoxide	Resistance not known.	35
	unknown	benzene-sulfonamides	benzene-sulphonamides	flusulfamide	Resistance not known.	36
	unknown	pyridazinones	pyridazinones	diclomezine	Resistance not known.	37
	formerly methasulfocarb (FRAC code 42), reclassified to M 12 in 2018					
	unknown	phenyl- acetamide	phenyl-acetamide	cyflufenamid	Resistance in Sphaerotheca. Resistance management required	U 06
	cell membrane disruption (proposed)	guanidines	guanidines	dodine	Resistance known in Venturia inaequalis. Low to medium risk. Resistance management recommended.	U 12
	unknown	thiazolidine	cyano-methylene-thiazolidines	flutianil	Resistance in Sphaerotheca and Podosphaera xanthii. Resistance management required	U 13
	unknown	pyrimidinone-hydrazones	pyrimidinone-hydrazones	ferimzone	Resistance not known (previously C5).	U 14
	complex III: cytochrome bc1, unknown binding site (proposed)	4-quinolyl- acetate	4-quinolyl-acetates	tebufloquin	Not cross resistant to QoI. Resistance risk unknown but assumed to be medium. Resistance management required.	U 16
	Unknown	tetrazolyloxime	tetrazolyloximes	picarbutrazox	Resistance not known. Not cross resistant to PA, QoI, CAA.	U 17
	Unknown (Inhibition of trehalase)	glucopyranosyl antibiotic	glucopyranosyl antibiotics	validamycin	Resistance not known. Induction of host plant defense by trehalose proposed (previously H3).	U 18

MoA	Target Site and Code	Group Name	Chemical or Biological Group	Common Name	Comments	Frac Code
Not specified	Unknown	diverse	diverse	mineral oils, organic oils, inorganic salts, material of biological origin	Resistance not known.	NC
M: Chemicals with multi-site activity	multi-site contact activity	inorganic (electrophiles)	inorganic	copper (different salts)	Also applies to organic copper complexes	M 01
		inorganic (electrophiles)	inorganic	sulphur		M 02
		dithiocarbamates and relatives (electrophiles)	dithio-carbamates and relatives	amobam ,ferbam, mancozeb, maneb, metiram, propineb, thiram, zinc, thiazole, zineb, ziram		M 03
		phthalimides (electrophiles)	phthalimides	captan, captafol, folpet		M 04
		chloronitriles (phthalonitriles) (unspecified mechanism)	chloronitriles (phthalonitriles)	chlorothalonil		M 05
		sulfamides (electrophiles)	sulfamides	dichlofluanid, tolylfluanid		M 06
		bis-guanidines (membrane disruptors, detergents)	bis-guanidines	guazatine, iminoctadine		M 07
		triazines (unspecified mechanism)	triazines	anilazine	generally considered as a low risk group without any signs of resistance developing to the fungicides.	M 08
		quinones (anthraquinones) (electrophiles)	quinones (anthraquinones)	dithianon		M 09
		quinoxalines (electrophiles)	quinoxalines	chinomethionat / quinomethionate	reclassified from U42 in 2018	M 10
		maleimide (electrophiles)	maleimide	fluoroimide		M 11
		thiocarbamate (electrophiles)	thiocarbamate	methasulfocarb		M 12

Moa	Target Site	Group Name	Chemical Or Biological Group	Common Name	Comments	Frac Code
BM: Biologicals with multiple modes of action: Plant extracts	multiple effects on ion membrane transporters; chelating effects	plant extract	polypeptide (lectin)	extract from the cotyledons of lupine plantlets ("BLAD")	Resistance not known. (previously M12).	BM 01
	affects fungal spores and germ tubes, induced plant defense	plant extract	phenols, sesquiterpenes, triterpenoids, coumarins	extract from Swinglea glutinosa	Resistance not known.	
	cell membrane disruption, cell wall, induced plant defense mechanisms	plant extract	terpene hydrocarbons, terpene alcohols and terpene phenols	extract from Melaleuca alternifolia (tea tree oil)	Resistance not known. (previously F7)	
				plant oils (mixtures): eugenol, geraniol, thymol		

Moa	Target Site	Group Name	Chemical or Biological Group	Common Name	Comments	Frac Code
BM: Biologicals with multiple modes of action: Microbial	multiple effects described (examples, not all apply to all biological groups): competition, mycoparasitism, antibiosis, membrane disruption by fungicidal lipopeptides, lytic enzymes, induced plant defence	microbial (strains of living microbes or extract, metabolites)	fungal *Trichoderma* spp.	Trichoderma atroviride strain I-1237, strain LU132, strain SC1 strain SKT-1, strain 77B	Resistance not known Bacillus amyloliquefaciens reclassified from F6, Code 44 in 2020 synonyms for Bacillus amyloliquefaciens are Bacillus subtilis and B. subtilis var. amyloliquefaciens (previous taxonomic classification).	BM 02
				Trichoderma asperellum strain T34, strain kd		
				Trichoderma harzianum strain T-22		
				Trichoderma virens strain G-41		
			fungal *Clonostachys* spp.	Gliocladium catenulatum strain J1446		
				Clonostachys rosea strain CR-7		
			fungal *Coniothyrium* spp.	Coniothyrium minitans strain CON/M/91-08		
			fungal *Talaromyces* spp.	Talaromyces flavus strain SAY-Y-94-01		
			fungal *Saccharomyces* spp.	Saccharomyces cerevisae strain LAS02		
			bacterial *Bacillus* spp.	Bacillus amyloliquefaciens strain QST713, strain FZB24, strain MBI600, strain D747, strain F727, strain AT-332		
				Bacillus subtilis strain AFS032321, strain Y1336, strain HAI-0404		
			bacterial *Pseudomonas* spp.	Pseudomonas chlororaphis strain AFS009		
			bacterial *Streptomyces* spp.	Streptomyces griseovirides strain K61		
				Streptomyces lydicus strain WYEC108		

6.2. Risks of Fungicides Resistance

The use of any fungicide to a crop or food raises the question of risks and benefits. This may be concerning to the risk of fungicidal toxicity to the user in the field, or its residues on the crop products to the consumers, besides its impact on the whole environment and the ecosystem in which the crops are growing. As a consequence more and more studies are required before a fungicide can be used on different crop diseases, leading to enormous development costs. This leads industry to concentrate on the major crops for big markets, while minor crops with smaller markets are increasingly left out and is urgent need of effective fungicides. In the U.S. the IR-4 program has been established to provide safe and effective pest management solutions for specialty crop growers. In December 2007 the UN-FAO held a global minor use summit along with IR-4 and EPA to establish global residue zones and standard data requirements.

Overall, most analyses concluded that the benefits of fungicides are far outweigh than the risks, if they are used carefully and according to the label recommendations. Currently more than 80% of the fruit and vegetable crops grown in the U.S. receive a fungicidal application every season. The benefit of fungicide use in the U.S. agriculture is to boost farm income by nearly $13 billion annually. The alternatives proposed by organic farmers, who are opposed to intensive farming altogether, exclude the use of synthetic fungicides, but allow the use of copper and sulfur based inorganic fungicides. There is still an ongoing debate as to whether traditionally or organically grown products are safer for the consumer, when the fungicide is not applied and the fungal contaminated produce are likely to contain the mycotoxins. Field resistance of a fungus to a fungicide occurs when the fungicidal use concentration is less than the recommended one or the reduced efficacy is due to the presence of resistant strains of the target fungus that was initially absent but develop subsequently in the struggle to life response in the presence of fungicide. This phenomenon affects most fungicide groups including some multisite compounds (e.g. mercurials, dodine). However, the most prominent cases are encountered with systemic fungicides, which often possess a single biochemical target site within fungal cells.

Field resistance against the fungicide in the fungal pathogen can occur more or less rapidly after a fungicide is introduced in the host pathogen interaction system, depending on genetic of the pathogen, epidemiological factors prevalent during the disease progression, the type of fungicide compound, the dose applied and mode of its application. For example, metalaxyl resistance in potato blight developed in only one or two years, whereas carboxin resistance in barley loose smut took about 15 years to develop. In the simplest cases that correspond to qualitative or major gene resistance, the number of phenotypes remains limited. Generally, sensitive isolates are clearly distinguishable from resistant isolates. Major gene resistance is observed with benzimidazoles, dicarboximides, and phenylamides. Usually a single mutation is responsible for resistance, which is often determined by reduced binding of fungicides to their respective target. This mechanism is not involved for organophosphorus compounds used against *Pyricularia oryzae*. Here resistance is due to reduced activation within fungal cells. Both mathematical models designed to predict the evolution of qualitative resistance in the field and experimental data show that resistance build-up is fastest when frequent, highly resistant and fit mutants of a rapidly reproducing pathogen are exposed to a highly effective, persistent fungicide that is frequently applied. This last point is important because one of the main recommendations for delaying field resistance is to limit the number of applications of at-risk fungicides. This advice implies rotation of compounds that do not show any positive cross-resistance.

Another strategy to counter the development of fungicide resistance is the use of mixtures of fungicides belonging to different groups. For example, two-way mixtures of phenylamides and multisite fungicides were developed to counter the development of resistance against downy mildew of vine and potato blight. Application of such mixtures normally controlled Plasmopara viticola, even when phenylamide-resistant strains are present at high frequencies. This two-way strategy is not effective against *P. infestans*, but some three-way mixtures (e.g. metalaxyl + fentin acetate + maneb; oxadixyl + cymoxanil + maneb) seem to be satisfactory. Another anti-resistance strategy mixes compounds that are negatively correlated with respect to cross resistance was a combination of commercially available molecule which contained the benzimidazole, carbendazim and the phenylcarbamate, diethofencarb. In 1987 this mixture was introduced against *B. cinerea* in French vineyards, but selection of a new phenotype resistant to both compounds (especially in Champagne) has restricted its use.

In field studies on resistance to fungicides such as DMIs and the 2-aminopyrimidine ethirimol in powdery mildews and dodine in apple scab, it is not possible to clearly distinguish sensitive from resistant populations. Dose-response curves from field situations are generally continuous, reflecting the presence of many phenotypic levels of resistance. This resistance is described as a quantitative or polygenic trait. Reduced efficacies of the fungicides are recorded when the frequencies of the most resistant strains are high enough. Genetic analyses suggest that control of resistance is probably polygenic, with the genetic factors involved having an additive effect; the mechanisms of resistance are not yet known. Positive cross-resistance to DMIs generally occurs in powdery mildew of cereals and of cucumber, but not with the mildew of grape vine. However, rotations of various DMIs could select strains with the largest spectra of cross-resistance. In addition, fungi are able to develop both types of resistance to DMIs (qualitative and quantitative), depending upon the fungal species.

Despite voluminous data on fungicide resistance collected over the past 20 years, reliable predictions about the risk of resistance to new fungicides are difficult to make. For example, the selection of resistant mutants in the laboratory or the inability to produce such mutants in in vitro, are not sufficient to determine what will happen in the field. Faced with this uncertainty, most companies try to develop new compounds for use in mixture with others, especially multisite inhibitors. Industrial strategies to combat the development of resistance against the fungicide are coordinated by the Fungicide Resistance Action Committee (FRAC), sponsored by the International Group of National Associations of Agrochemical Manufacturers. The undesirable side-effects of fungicides have reduced the number of active ingredients that are available, particularly in Europe where chemical control of some diseases is becoming very difficult, e.g. grey mold in Champagne vineyards or eyespot in winter wheat (since benzimidazoles were replaced by prochloraz). The situation is less critical when several types of fungicides are available against a given disease- such as sheath blight and blast of rice. In general, to prolong the useful life of any fungicide, it should only be used when necessary, and in combination with other methods of disease control (integrated pest control). The development of successful disease-forecasting systems may also help to optimize the use of fungicides and thereby the development of fungicidal resistance.

7

Fungicide Product with Their Active Ingredients and Hazard Class

A same active ingredient are available in the market with different trade name and are the product of different companies. These may also differ in their formulation and hazard class (table 7), the knowledge of which is necessary in the plant disease management in view of their effect on the pathogen, ecosystem and environment.

Table 7: Fungicide product, active ingredient, formulation and manufacturing company.

Trade name	Active ingredient	Formulation	Hazard Class	Company name
Telone C-17	1,3 dichloropropene/ chloropicrin	AL	4	Dow AgroSciences
Graftseal	2,5-dichlorobenzoic acid methyl ester/ 8 hydroxyquinoline	wax	4	Dalvern Products
Agriwax	2,5-dichlorobenzoic methyl ester/ 8-hydroxyquinolene	SP	4	Agrinema
Groenwaks	8-hydroxyquinolene	wax	4	Wynland Lab
Crown Gall Inoculant	*Agrobacterium radiobacter*	WP	4	Stimuplant
Avogreen	*Bacillus subtilis* isolate B246	SC	3	RE at UP
Alrose Complete Crop	QAC/Bardac/TBTO	SL	2	Alrose Chemicals
Crop Dusting Sulphur	Sulfur	CB	3	Sunwood Chem
Trichoplus	*Trichoderma harzianum*	WP	4	Biological Control Products
Eco-T	*Trichoderma harzianum*	WP	3	Plant Health Products
Trichodex WP	*Trichoderma harzianum strain* T39	WP	2	Makhteshim-Agan SA (Pty) Ltd

Trade name	Active ingredient	Formulation	Hazard Class	Company name
Bion 50 WG	acibenzolar-S-methyl	WG	3	Syngenta South Africa (Pty) Ltd
Brilliant SL	ammonium phosphate	SL	3	Hyper Agrochemicals (Pty) Ltd
Brilliant SL	ammonium phosphate	SL	3	Nippon Africa
Bio-Protector	ascorbic acid	SL	4	Agro Organics
Amistar	Azoxystrobin	SC	3	Syngenta South Africa (Pty) Ltd
Ortiva	Azoxystrobin	SC	3	Syngenta South Africa (Pty) Ltd
Heritage	Azoxystrobin	WG	3	Syngenta South Africa (Pty) Ltd
Quadris 50 WG	Azoxystrobin	WG	3	Syngenta South Africa (Pty) Ltd
Shelter	*bacillus subtilis* DB 101	SC	3	Dagutat Biolab
Artemis	*bacillus subtilis* DB 102	SC	3	Dagutat Biolab
Galben M	benalaxyl/mancozeb	WP	3	Sipcam South Africa (Pty) Ltd
Tairel	benalaxyl/mancozeb	WP	3	Sipcam South Africa (Pty) Ltd
Demeter	Benomyl	WP	3	Volcano Agrosciences (Pty) Ltd
Benomyl 500 WP	Benomyl	WP	3	Universal Crop Protection (Pty) Ltd
Benomyl WP	Benomyl	WP	3	Dow AgroSciences
Spotless	Benomyl	WP	4	Unisun
Rosecare 3	bifenthin/myclobutanil	EC	3	Efekto
Baycor 300 DC	Bitertanol	DC	3	Bayer SA
Prev-Gard	Borax	XX	3	Efekto
Prev-Grape	Borax	XX	4	Citrus Oil Products (SA) (Pty) Ltd
Prev-Veg	Borax	XX	3	Citrus Oil Products (SA) (Pty) Ltd
Eco Rod	boric acid equivalent	XX	4	Groundline
Vika Rod	boric acid equivalent	XX	4	DataForce Trading T/A Vikela Supplies
Cantus WG	Boscalid	WG	4	BASF South Africa (Pty) Ltd
Collis	boscalid/kresoxim-methyl	SC	4	BASF South Africa (Pty) Ltd

Trade name	Active ingredient	Formulation	Hazard Class	Company name
Granit EC	Bromuconazole	EC	3	Bayer (Pty) Ltd
Granit Plus	bromuconazole/ carbendazim	SC	3	Bayer
Granit CM	bromuconazole/ carbendazim	SC	3	Bayer
Nimrod	Bupirimate	EC	2	Makhteshim-Agan SA (Pty) Ltd
Oscar	bupirimate/ hexaconazole	EC	2	Makhteshim-Agan SA (Pty) Ltd
Captan Flo	Captab		4	Tomen Corporation
Captan FS	Captab	FS	4	Sub-Sahara
Captan FS	Captab	FS	4	Tomen Corporation
Merpan SC	Captab	FS	o	Makhteshim-Agan SA (Pty) Ltd
Royalcap FS (Formerly Captab SC/Captab FS)	Captab	FS	3	Crompton Chemical (Pty) Ltd
Prune Wound Seal (WPK)	Captab	PA	4	WPK Landbou
Captab 500 SC	Captab	SC	3	Dow Agrosciences
Merpan 50 SC	Captab	SC	4	Makhteshim-Agan SA (Pty) Ltd
Pruntect	Captab	SL	3	Dalvern Products
Merpan 80 WDG	Captab	WDG	4	Makhteshim-Agan SA (Pty) Ltd
Captab WP	Captab	WP	3	Universal Crop Protection (Pty) Ltd
Kaptan WP	Captab	WP	3	Applied Chemical Products
Royalcap 800 WDG	Captab	WS	3	Crompton Chemical (Pty) Ltd
Merpan	Captab	wdg	4	Makhteshim-Agan SA (Pty) Ltd
Bavistin SC	Carbendazim	SC	4	BASF South Africa (Pty) Ltd
Bendazid 500 SC	carbendazim	SC	4	Plaaskem (Gouws & Scheepers)
Knowin 500 SC	Carbendazim	SC	4	Plaaskem (Gouws & Scheepers)
Duett	carbendazim/ epoxiconazole	SC	3	BASF South Africa (Pty) Ltd
Alto Combi	carbendazim/ cyproconazole	SC	3	Syngenta South Africa (Pty) Ltd

Trade name	Active ingredient	Formulation	Hazard Class	Company name
Punch C	carbendazim/ flusilazole	SC	2	DuPont de Nemours
Punch-Xtra (DuPont)	carbendazim/ flusilazole	SC	2	DuPont de Nemours
Propazole Plus 250 SC	carbendazim/ propiconazole	SC	2	Universal Crop Protection (Pty) Ltd
Propazim 250 SC	carbendazim/ propiconazole	SC	2	Villa Crop Protection (Pty) Ltd
Folicur C 300 SC	carbendazim/ tebuconazole	SC	3	Bayer SA
Rambo	carbendazim/ triadimefon	SC	2	Rotam RSA (Pty) Ltd
Rambo	carbendazim/ triadimefon	SC	2	Bitrad Consulting
Anchor Red	carboxim/thiram	FS	4	Crompton Chemical (Pty) Ltd
Vitavax Neutral	carboxin/thiram	FS	3	Crompton Chemical (Pty) Ltd
Vitavax Plus	carboxin/thiram	FS	3	Crompton Chemical (Pty) Ltd
Harvest Wash WT	chlorine dioxide	SL	3	BTC Products and Services
Bravo Plus	Chlorothalonil	SC	3	Syngenta South Africa (Pty) Ltd
Chloroflo 500 SC	Chlorothalonil	SC	3	Plaaskem (Gouws & Scheepers)
Chloronil 500 SC	Chlorothalonil	SC	3	Dow Agrosciences
Chloronil 500 SC	Chlorothalonil	SC	3	Universal Crop Protection (Pty) Ltd
Chloronil Plus 500 SC	Chlorothalonil	SC	3	Villa Crop Protection (Pty) Ltd
Chlorothalonil	Chlorothalonil	SC	3	Nialcor
Clortosip L SC	Chlorothalonil	SC	3	Sipcam South Africa (Pty) Ltd
Fungistop SC	Chlorothalonil	SC	3	Sipcam South Africa (Pty) Ltd
Matrix	Chlorothalonil	SC	3	Bitrad Consulting
Mycoguard	Chlorothalonil	SC	3	Natural Plant Protection
Bravo 500	Chlorothalonil	SC	3	Efekto
Bravo 720	Chlorothalonil	SC	3	Syngenta South Africa (Pty) Ltd
Chlortosip 720 SC	Chlorothalonil	SC	3	Sipcam South Africa (Pty) Ltd

Trade name	Active ingredient	Formulation	Hazard Class	Company name
Daconil Weather Sticks	Chlorothalonil	SC	3	Syngenta South Africa (Pty) Ltd
Phytomax	Chlorothalonil	SC	3	Sipcam South Africa (Pty) Ltd
Odeon 720 SC	Chlorothalonil	SC	3	Makhteshim-Agan SA (Pty) Ltd
Fungistop DF	Chlorothalonil	WG	3	Sipcam South Africa (Pty) Ltd
Odeon 825 WG	Chlorothalonil	WG	3	Makhteshim-Agan SA (Pty) Ltd
Clortosip WP	Chlorothalonil	WP	3	Sipcam South Africa (Pty) Ltd
Rothalonil 750 WP	Chlorothalonil	WP	3	Rotam
Ruby	chlorothalonil/ cymoxanil	SC	3	Sipcam South Africa (Pty) Ltd
Garden Fungicide	chlorpyrifos/ deltamethrin/ thiophanate methyl	AL	3	Bayer (Pty) Ltd
Copper Count-N	copper ammonium acetate	SL	3	Hygrotech Seed
Copper-Flow-Plus	copper ammonium acetate	SL	3	Natural Crop Protection (Pty) Ltd
Villa Fungi Flow	copper ammonium acetate	SL	3	Villa Crop Protection (Pty) Ltd
Fungicop	copper ammonium acetate	SL	3	Plaaskem (Gouws & Scheepers)
Liquicop Liquid Fungicide	copper ammonium carbonate	SL	3	Hygrotech Seed
Copper Clear	copper amonium acetate	SL	3	Ag-Chem Africa
CopStar 120 SC	copper hydroxide	SC	2	Ag-Chem
CungFu 538 SC	copper hydroxide	SC	2	Plaaskem
Kocide	copper hydroxide	WG	3	Plaaskem(Gouws & Scheepers)
Proxan	copper hydroxide/ dichlorophen	SC	2	Plaaskem (Gouws & Scheepers)
Copper oxychloride DF	copper oxychloride	WG	3	Protea Industrial Chemicals
Copperoxychloride WG	copper oxychloride	WG	3	Cropchem
Oxi-Cop WG	copper oxychloride	WG	3	Kenchem Trading
Cuprozin 35 WP	copper oxychloride	WP	3	Agri-Registration CC
Cop Tech	copper oxychloride	WP	3	Hygrotech

Trade name	Active ingredient	Formulation	Hazard Class	Company name
Copper Oxychloride	copper oxychloride	WP	3	Villa Crop Protection (Pty) Ltd
Copper Oxychloride Ekstra	copper oxychloride	WP	3	Chempack (Pty) Ltd
Copper Oxychloride WP	copper oxychloride	WP	o	Unisun
Copper Oxychloride WP	copper oxychloride	WP	3	Universal Crop Protection (Pty) Ltd
Copper Oxychloride WP (WPK)	copper oxychloride	WP	3	WPK Agriculture
Copper oxychloride WP	copper oxychloride	WP	3	Protea Industrial Chemicals
Cuprox Super	copper oxychloride	WP	3	Natural Plant Protection
Cuprox WP	copper oxychloride	WP	3	Natural Plant Protection
Demildex	copper oxychloride	WP	3	Delta Chemicals
Kopperchlor WP	copper oxychloride	WP	3	Wenchem
Rust	copper oxychloride	WP	4	Kombat
Suncop	copper oxychloride	WP	3	Sunwood Chemicals
Sunwood Copperoxychloride	copper oxychloride	WP	3	Sunwood Chemicals
Supercup	copper oxychloride	WP	3	Hyper Agrochemicals (Pty) Ltd
Supracop 85 WP	copper oxychloride	WP	3	Densia Trading cc
Virikop	copper oxychloride	WP	3	Efekto
Copper oxychloride	copper oxychloride	WP	3	Cropchem
Oxi-Cop WP	copper oxychloride	WP	3	Kenchem Trading
Rose and Garden Dust	copper oxychloride/ mercaptothion/sulphur	DP	3	Efekto
Copsul Dust	copper oxychloride/ sulfur	DP	3	Applied Chemical Products
Cu-Sul Dust	copper oxychloride/ sulfur	DP	3	JCJ Agri Chem
Sunwood Coppersulfa	copper oxychloride/ sulfur	DP	4	Sunwood Chemicals
Copper Sulphur	copper oxychloride/ sulphur	DP	0	Orchard Suppliers
Copflo Super	copper sulphate (basic)	SC	2	Ag-Chem
Bordeaux Mixture	copper sulphate / lime	WP	3	Applied Chemical Products

Trade name	Active ingredient	Formulation	Hazard Class	Company name
Cuprofix Disperss	copper sulphate/ lime	WG	3	Chempack (Pty) Ltd
Funguran-OH	cupric hydroxide/ copper equivalent	WP	3	Agri Registrations
Hydrox	cupric hydroxide/ copper equivalent	WP	3	Agropharm
Zetanil	cymoxanil / mancozeb	WP	3	Sipcam South Africa (Pty) Ltd
Equation Pro	cymoxanil/ famoxadone	WG	3	DuPont
Copcanil	cymoxanil/copper oxychloride	WP	3	Sipcam South Africa (Pty) Ltd
Tanos	cymoxanil/famoxadone	WG	3	DuPont De Nemours
Blighter	cymoxanil/mancozeb	WP	3	Agropharm
Controller	cymoxanil/mancozeb	WP	3	RT Chemicals
Curzate Pro	cymoxanil/mancozeb	WP	3	Du Pont de Nemours
Fungarrest	cymoxanil/mancozeb	WP	3	Universal Crop Protection (Pty) Ltd
Optimo	cymoxanil/mancozeb	WP	3	Dow AgroSciences
Milraz 76 WP	cymoxanil/Propineb	WP	4	Bayer SA
Alto Amble 460 SC	cyproconazole/ carbendazim	SC	3	Syngenta South Africa (Pty) Ltd
Artea	cyproconazole/ propiconazole	EC	3	Syngenta South Africa (Pty) Ltd
Chorus 50 WG	Cyprodinil	WG	4	Syngenta South Africa (Pty) Ltd
Switch 62,5 WG	cyprodinil/fludioxonil	WG	4	Syngenta South Africa (Pty) Ltd
Basamid Granular	Dazomet	GR	3	BASF
Xanbac D	Dichlorophen	EC	2	Plaaskem (Gouws & Scheepers)
Allisan	Dicloran	WP	4	Viking Distributors
Terminator	didecyl dimethyl ammonium chloride	SL	2	Agricultural Production Systems
Sporekill	didesyl dimethyl ammonium chloride	SL	3	ICA International Chemicals (Pty) Ltd
Score 250 EC	Difenaconazole	EC	3	Syngenta South Africa (Pty) Ltd
Pronto 250 Ec	Difenoconazole	EC	3	Meridian Agrochemical Co. (Pty) Ltd
Dividend 030 FS	Difenoconazole	FS	4	Syngenta South Africa (Pty) Ltd

Trade name	Active ingredient	Formulation	Hazard Class	Company name
Eria 187,5 SC	difenoconazole/ carbendazim	SC	3	Syngenta South Africa (Pty) Ltd
Sovrin Flo	Dimethomorph	SC	4	Meridian Agrochemical Co. (Pty) Ltd
Sovrin WP	Dimethomorph	WP	4	Meridian Agrochemical Co. (Pty) Ltd
Acrobat WG	dimethomorph/ mancozeb	WG	3	BASF South Africa (Pty) Ltd
Acrobat MZ	dimethomorph/ mancozeb	WP	3	BASF South Africa (Pty) Ltd
Karathane 350 EC	Dinocap	EC	3	Dow AgroSciences
Karathane 18,25 WP	Dinocap	WP	3	Dow AgroSciences
Sabithane 400 EC	dinocap/myclobutanil	EC	3	Dow AgroSciences
No Scald DPA	Diphenylamine	EC	3	UAP Crop Care
Shield Liquid DPA	Diphenylamine	EC	3	Chempack (Pty) Ltd
Repulse 5,75 GR	disulfoton/triadimenol	GR	2	Bayer SA
Delan 500 SC	Dithianon	SC	2	BASF South Africa (Pty) Ltd
Meltatox	Dodemorph	EC	4	BASF South Africa (Pty) Ltd
Syllit 400 SC	Dodine	SC	3	Hyper Agrochemicals (Pty) Ltd
Opus	Epoxiconazole	SC	3	BASF South Africa (Pty) Ltd
Westfalia Biocoat	fatty acids, alkohols, anti-oxidants	SC	2	Hans Merenski
Verita Fenomen	fenamidone/fosetyl-AL	WG	3	Bayer
No-Blite	fenamidone/mancozeb	WG	3	Bayer
Rubigan	Fenarimol	EC	2	Klub M5
Indar 50 EW	Fenbuconazole	EW	3	Dow AgroSciences
Teldor 500 SC	Fenhexamid	SC	4	Bayer
Celest 100 FS	Fludioxonil	FS	4	Syngenta South Africa (Pty) Ltd
Celest XL White	Fludioxonil	FS	3	Syngenta South Africa (Pty) Ltd
Celest XL 035 FS	fludioxonil/ metalaxyl-M (mefenoxam)	FS	4	Syngenta South Africa (Pty) Ltd
Nustar 100 EW	Flusilazole	EW	2	DuPont

Trade name	Active ingredient	Formulation	Hazard Class	Company name
Olymp 100 EW	Flusilazole	EW	2	Du Pont de Nemours Int SA
Capitan 250 EW	Flusilazole	EW	2	DuPont de Nemours Int SA
Impact	Flutriafol	SC	3	Cheminova
Early Impact	flutriafol/carbendazim	SC	2	Cheminova
Vincit FS	flutriafol/thiabendazole	FS	3	Exel Medical and Chemical
Folpan 500 SC	Folpet	SC	3	Makhteshim-Agan SA (Pty) Ltd
Folpan 80 WDG	Folpet	WG	3	Makhteshim-Agan SA (Pty) Ltd
Folpan WP	Folpet	WP	4	Makhteshim-Agan SA (Pty) Ltd
Folpan 80 WP	Folpet	WP	4	Makhteshim-Agan SA (Pty) Ltd
Alliete WG	fosetyl-Al	WG	3	Bayer
Mikal-M	fosetyl-Al/mancozeb	WP	3	BAYER
Fongarid TB	Furalaxyl	TB	4	Syngenta South Africa (Pty) Ltd
Fongarid 25 WP	Furalaxyl	WP	4	Syngenta South Africa (Pty) Ltd
Protect	Furfural	EC	1b	Illovo
Guazatine 200 SL	Guazatine	SL	2	PW Landboudienste
Kenopel 200 SL	Guazatine	SL	2	Makhteshim-Agan SA (Pty) Ltd
Citricure	Guazatine	SL	2	ICA Laboratories
Ultracure	Guazatine	SL	2	Natural Crop Protection (Pty) Ltd
CitriWax	Guazatine	SL	2	ICA Laboratories
Guazatine 400 SL	Guazatine	SL	2	PW Landboudienste
Kenopel 400 SL	Guazatine	SL	2	Makhteshim-Agan SA (Pty) Ltd
Anvil	hexaconazole	SC	3	Syngenta South Africa (Pty) Ltd
Richter	Hexaconazole	SC	3	Volcano Agrosciences (Pty) Ltd
Neocil-Plus	hydroxyquinoline sulphate/octhilinone	PA	2	Algro-Chem CC
Fungazil 500 EC	Imazalil	EC	2	Dow AgroSciences
Fungazil 800 EC	Imazalil	EC	2	Dow AgroSciences

Trade name	Active ingredient	Formulation	Hazard Class	Company name
Magnate 800 EC	Imazalil	EC	2	Makhteshim-Agan SA (Pty) Ltd
Sanazil 800 EC	Imazalil	EC	2	Dow Agrosciences
Fungazil 750 SP	Imazalil	SP	2	Dow AgroSciences
Magnate Sulphate 750 WSP	Imazalil	SP	3	Makhteshim-Agan SA (Pty) Ltd
Penless	Imazalil	SP	2	Unisun
Magnate Sulphate	imazalil sulphate	SG	3	Makhteshim-Agan SA (Pty) Ltd
Imazalil Sulphate	imazalil sulphate	SP	3	Universal Crop Protection (Pty) Ltd
Imazalil Sulphate	imazalil sulphate	SP	2	Meridian Agrochemical Co. (Pty) Ltd
Imazapen 750 SP	imazalil sulphate	SP	2	Ag-Chem Africa
Magnate Sulphate 750 SG	imazalil sulphate	WG	3	Dow AgroSciences
Rhapsodie	imazalil/iprodione	FS	3	Bayer
Rovral Talc	Iprodione	DP	3	Bayer (Pty) Ltd
Rovral DS	Iprodione	DS	4	Bayer
Iprodione 255	Iprodione	SC	4	Erintrade cc t/a RT Chemicals
Rovral Flo	Iprodione	SC	4	Bayer
Prodione	Iprodione	SC	3	Plaaskem (Gouws & Scheepers)
Rovral Aquaflo	Iprodione	SC	4	Bayer (Pty) Ltd
Rovral Sulphur Dust	iprodione/sulphur	DP	3	Bayer (Pty) Ltd
Melody Care	iprovalicarb/ propineb	WG	4	Bayer
Melody Duo 69 WG	iprovalicarb/ propineb	WG	4	Bayer
Ardent 50 SC	kresoxim-methyl	SC	3	Makhteshim-Agan SA (Pty) Ltd
Stroby WG	kresoxim-methyl	WG	4	BASF South Africa (Pty) Ltd
Sancozeb (Formerly Dithane M-45 Dust)	Mancozeb	DP	3	Dow AgroSciences
Ifax	Mancozeb	DS	3	Dow AgroSciences
Crater 455 SC	Mancozeb	SC	3	Volcano Agrosciences (Pty) Ltd
Dithane 750 WG Neotec	Mancozeb	WG	3	Dow AgroSciences

Trade name	Active ingredient	Formulation	Hazard Class	Company name
Dithane WG	Mancozeb	WG	3	Efekto
Penncozeb WG	Mancozeb	WG	3	Total South Africa
Penncozeb WG	Mancozeb	WG	3	BASF South Africa (Pty) Ltd
Chempack Mancozeb	Mancozeb	WP	3	Sipcam South Africa (Pty) Ltd
Crater	mancozeb	WP	3	Volcano Agrosciences (Pty) Ltd
Dithane M-45	Mancozeb	WP	3	Dow AgroSciences
Dithane M-45 (Kudu)	Mancozeb	WP	o	Starke-Ayres
Dithane M-45 800 WP NT	Mancozeb	WP	3	Dow Agroscience SA (Pty)
Mancozeb	Mancozeb	WP	3	Natural Plant Protection
Mancozeb (WPK)	Mancozeb	WP	3	WPK Landbou
Mancozeb 800 WP	Mancozeb	WP	3	Volcano Agroscience (Pty) Ltd
Mancozeb 800 WP	Mancozeb	WP	3	Ag-Chem Africa (Pty) Ltd
Mancozeb WP	Mancozeb	WP	3	Cropchem
Mancozeb WP	Mancozeb	WP	3	Cropchem
Mancozeb WP	Mancozeb	WP	4	Kombat
Miceb Super WP	Mancozeb	WP	3	Sipcam South Africa (Pty) Ltd
Sancozeb WP (Formerly Santhane)	Mancozeb	WP	3	Dow AgroSciences
Tridex	Mancozeb	WP	3	Total South Africa
Unisun Mancozeb	Mancozeb	WP	3	Unisun
Unizeb 800 WP	Mancozeb	WP	3	Universal Crop Protection (Pty) Ltd
Vondozeb	Mancozeb	WP	3	Natural Crop Protection (Pty) Ltd
Rolim 700 WP	mancozeb/ metalaxyl	WP	3	Hyper Agrochemicals (Pty) Ltd
Micexanil	mancozeb/cymoxanil	WP	3	Sipcam South Africa (Pty) Ltd
Ridomil Gold Peptite	mancozeb/mefenoxam	WG	3	Syngenta South Africa (Pty) Ltd
Mancolax 700 WP	mancozeb/metalaxyl	WP	3	Plaaskem(Gouws & Scheepers)
Metasan 750 WP	mancozeb/metalaxyl	WP	3	Dow Agrosciences

Trade name	Active ingredient	Formulation	Hazard Class	Company name
Metazeb 700 WP	mancozeb/metalaxyl	WP	3	Villa Crop Protection (Pty) Ltd
Metazeb 700 WP	mancozeb/metalaxyl	WP	3	Universal Crop Protection (Pty) Ltd
Electis WG	mancozeb/zoxamide	WG	3	Dow AgroSciences
Barrier 450 SC	maneb/ procymidone/ zinc oxide	SC	3	Plaaskem (Gouws & Scheepers)
Manager SC	maneb/zinc oxide	SC	3	Applied Agricultural Products
Trimangol SC	maneb/zinc oxide	SC	3	Natural Plant Protection
Apron XL	metalaxy-M	ES	2	Syngenta South Africa (Pty) Ltd
Milor 5GR	metalaxyl	GR	4	Rotam
Rolim 5 GR	metalaxyl	GR	3	Hyper Agrochemicals (Pty) Ltd
Eclipse ST	metalaxyl 350 g/kg	WP/WS	3	Meridian Agrochemical Co. (Pty) Ltd
Kickback 5GR	metalaxyl 50 g/kg	GR	4	Bitrad Consulting
Ridomil Gold	metalaxyl M (mefenoxam)	EC	2	Syngenta South Africa (Pty) Ltd
Ridomil Gold Flo 537,5 SC	metalaxyl-M (mefenoxam), chlorothalonil	SC	3	Syngenta South Africa (Pty) Ltd
Ridomil Gold MZ	metalaxyl-M(mefenoxam)/ mancozeb	WP	4	Syngenta South Africa (Pty) Ltd
Ridomil Gold 2,5 G	metalaxyl-m	GR	4	Syngenta South Africa (Pty) Ltd
Kickback 700 WP	metalaxyl/mancozeb	WP	3	Bitrad Consulting
Milor MZ 700 WP	metalaxyl/mancozeb	WP	3	Rotam
Sanlaxyl 700 WP	metalaxyl/mancozeb	WP	3	Dow Agrosciences
Ag-Fume	metam-sodium	SL	2	Ag-Chem Africa (Pty) Ltd
Matafume	metham sodium		2	Chemfit
Nemasol	metham-sodium	SL	2	UCB
Methyl Bromide	methy bromide/ chloropicrin	GA		Novel Idea Trading
Methyl Bromide	methyl bromide	GA	1a	Landkem
Metabrom	methyl bromide/ chloropicrin	GA	1a	Landkem

Trade name	Active ingredient	Formulation	Hazard Class	Company name
Methyl bromide	methyl bromide/ chloropicrin	GA	1a	Mebrom Chemicals (Pty) Ltd
Methyl Bromide	methyl bromide/ chloropicrin	VP	1a	Universal Crop Protection
Polyram DF-WG	metiram	WG	4	BASF South Africa (Pty) Ltd
BP Crop Oil	mineral oil	EC	4	BP SA (Pty) Ltd
Cropchem 86% Winter Oil	mineral oil (heavy)	OIL	3	Cropchem
BAC-Oil SO	mineral oil (heavy-narrow)	EC	3	Lowveld/Laeveld Agrochem
Citrex	mineral oil (medium)		3	BASF South Africa (Pty) Ltd
BP Medium Spray Oil	mineral oil (medium)	EC	3	BP South Africa
Medium Spray Oil	mineral oil (medium)	EC	3	Sunwood Chem
BP Light Spray Oil	mineral oil (medium)	EC	3	BP Southern Africa
Nutroguard	mono potassium phosphate	SP	3	Ocean Agriculture (Pty) Ltd
Systhane 125 EC	myclobutanil	EC	2	Dow AgroSciences
Systhane EC	myclobutanil	EC	2	Efekto
Rally 200 EW	myclobutanil	EW	3	Dow AgroSciences
Systhane 20 EW	myclobutanil	EW	3	Dow AgroSciences
Rally 400 WP	myclobutanil	WP	3	Dow AgroSciences
Systhane 400 WP	myclobutanil	WP	3	Dow AgroSciences
Kanguard 940	organic plant acids	SL	3	Kannar CC
Kirchhoffs Margaret Roberts Organic Fungicide	organic plant acids	SL	4	Ball Straathof (Pty) Ltd
Lonlife	organic plant acids	SL	3	Kannar CC
Plantvax EC	oxycarboxin	EC	3	Crompton Chemical (Pty) Ltd
Plantvax WP	oxycarboxin	WP	3	Crompton Chemical (Pty) Ltd
Citrole 60	paraffinic complex (light mineral oil)	EC	4	Total SA
Wenfinex	parafinic complex (mineral oil)	EC	4	Wenkem SA
Excalibur 200 EW	penconazole	EW	3	Villa Crop Protection (Pty) Ltd
Penconazole 200 EW	penconazole	EW	3	Universal Crop Protection (Pty) Ltd

Trade name	Active ingredient	Formulation	Hazard Class	Company name
Monceren SC	pencycuron	SC	4	Bayer SA
Rootmaster	phosphorous acid	SP	2	Ocean Agriculture
Polysun 320	polysulphide sulphur	AL	4	Unisun
Lime Sulphur	polysulphide sulphur	SC	3	Starke-Ayres
Lime Sulphur	polysulphide sulphur	SL	3	Applied Chemical Products
Lime Sulphur	polysulphide sulphur	SOLN	3	Efekto
Rootmaster 400 SL	potassium phophite (= phosphorous acid 400 g/l)	SL	3	Ocean Agriculture
Fighter	potassium phosphate	SL	3	Ag - Chem Africa (Pty) Ltd
Phosphite 400 SL	potassium phosphate	SL	3	Universal Crop Protection (Pty) Ltd
Phosphite 400 SL	potassium phosphate	SL	3	Villa Crop Protection (Pty) Ltd
Avoguard 500 SL	potassium phosphate	SL	3	Ocean Agriculture
Phosguard 400 SL	potassium phosphite (= phosphorous acid 400 g/l)	SL	3	Ocean Agriculture
Rootmaster 200 SL Foliar	potassium phosphite (phosphorous acid equivalent)	SL	3	Ocean Agriculture
Hyperphos	potassium phosphite (phosphorous acid)	SL	3	Hyper Agrochemicals (Pty) Ltd
Phytex	potassium phosphonate (phosphorous acid equivalent)	SL	3	Horticura CC
Chronos 450 SC	prochlorax zinc complex	SC	3	Makhteshim-Agan SA (Pty) Ltd
Chronos 45 EC	prochloraz	EC	2	Makhteshim-Agan SA (Pty) Ltd
Mirage 45 EC	prochloraz	EC	2	Makhteshim-Agan SA (Pty) Ltd
Omega	prochloraz	EC	2	Bayer
Octave DP	prochloraz manganese chloride complex	DP	3	Bayer
Octave	prochloraz manganese chloride complex	WP	3	Bayer
Chronos 50 WP	prochloraz zinc complex	WP	4	Makhteshim-Agan SA (Pty) Ltd
Hit 250 SC	procymidone	SC	3	Dow AgroSciences
Rot 'n Spot	procymidone	SC	4	Ball Straathof

Trade name	Active ingredient	Formulation	Hazard Class	Company name
Sumisclex SC	procymidone	SC	4	Philagro South Africa (Pty) Ltd
Sumivit	procymidone	SC	3	Philagro South Africa (Pty) Ltd
Hit 500 SC	procymidone	SC	3	Dow Agrosciences
Sumisclex Sulphur Dust	procymidone/sulphur	DP	3	Philagro South Africa (Pty) Ltd
Proton 500	profenofos	EC	2	Agropharm
Previcur-N	propamocarb HCl	SL	2	Bayer
Proplant	propamocarb HCl	SL	3	Natural Plant Protection
Tattoo	propamocarb HCl/ mancozeb	SC	3	Bayer (Pty) Ltd
Banner	propiconazole	EC	3	Syngenta South Africa (Pty) Ltd
Bumper 250 EC	propiconazole	EC	3	Makhteshim-Agan SA (Pty) Ltd
Fungi Rid	propiconazole	EC	3	Kombat
Propicon 250 EC	propiconazole	EC	2	Volcano Agrosciences (Pty) Ltd
Propizole 250 EC	propiconazole	EC	2	Universal Crop Protection (Pty) Ltd
Propizole 250 EC	propiconazole	EC	2	Villa Crop Protection (Pty) Ltd
Sanazole 250 EC	propiconazole	EC	2	Dow AgroSciences
Tilt	propiconazole	EC	3	Syngenta South Africa (Pty) Ltd
Propicon 500 EC	propiconazole	EC	2	Volcano Agrosciences (Pty) Ltd
Antracol 70 WG	Propineb	WG	4	Bayer
Antracol 70 WP	Propineb	WP	3	Bayer SA
Bayleton -A-Powder	propineb/triadimefon	WP	3	Bayer
Solanacure	*pseudomonas resinovorans*	WP	3	Agricultural Research Council
Cabrio	pyraclostrobin	EC	2	BASF South Africa (Pty) Ltd
Cabrio Top	pyraclostrobin/metiram	WG	3	BASF South Africa (Pty) Ltd
Scala	pyrimethanil	SC	3	Bayer
Megacide	quazatine	SL	2	ICA International
Legend 250 SC	quinoxyfen	SC	3	Dow AgroSciences

Trade name	Active ingredient	Formulation	Hazard Class	Company name
PCNB 750	quintozene	WP	3	Plaaskem (Gouws & Scheepers)
Terraclor 75 WP	quintozene	WP	3	Crompton Chemical (Pty) Ltd
Bio-Build	salicylic acid	SC	3	Agro Organics
Latitude	silthiopham	FS	3	Monsanto
SOPP Solution	sodium- o-phenol phenate(Na salt)	SL	3	Acti -Chem SA
Prosper 500 EC	spiroxamine	EC	2	Bayer
Folicur Plus 383 EW	spiroxamine/ tebuconazole	EW	2	Bayer
Dusting Sulphur	sulfur	DP	3	Ag - Chem Africa (Pty) Ltd
Vine and Dusting Sulphur	sulfur	DP	3	Applied Chemical Products
Fumigation Sulphur	sulfur	GE	3	JCJ Agri Chem
Flowable Sulphur	sulfur	SC	3	Applied Chemical Products
Microthiol Special	sulfur	WG	3	Total South Africa
Striker	sulfur	WG	3	Chempack (Pty) Ltd
				WPK Agriculture
				Sunwood Chemicals
				Starke Ayres
				Universal Crop Protection (Pty) Ltd
				SA Dried Fruit Co-op
				Sunwood Chemicals

Abbreviation for formulations: EC= Emulsifiable concentrate, **WP**= Wettable powder, **SL**= Soluble liquid concentrate, **SP**= Soluble powder, **SC**= Suspension concentrate **WS**= Water soluble, **WG/WDG**= Water dispersible granules, **AL**= Other Liquids to applied undiluted, **XX**= Others, **DG**= Dispersible concentration, **Wax**= Wax containing fungicide (for grafts), **TB**= Tablet, **FS**= Flowable concentrate for seed treatment, **PA**= Paste, **DP**= Dustable powder, **DS**= Powder for dry seed treatment, **GR**= Granules, **SOLN**= Fungicide Solutions, **GA**= Gas, **VP**= Vapour releasing product, **EW**= Emulsion; oil in water.

Hazard class or Toxicity class

This refers to a classification system for pesticides that has been created by a national or international government-related or -sponsored organization. It addresses the acute toxicity of agents such as soil fumigants, fungicides, insecticides, miticides, nematicides or rodenticides.

Assignment to a toxicity class is based typically on results of acute toxicity studies such as the determination of LD_{50} values in animal experiments, notably rodents, via oral, inhaled, or external application. The experimental design measures the acute death rate of an agent. The toxicity class generally does not address issues of other potential harm of the agent, such as bioaccumulation, issues of carcinogenicity, teratogenicity, mutagenic effects, or the impact on reproduction.

Regulating agencies may require that packaging of the agent be labeled with a **signal word**, a specific warning label to indicate the level of toxicity. The World Health Organization (WHO) names four toxicity classes:

Class I – a: extremely hazardous

Class I – b: highly hazardous

Class II: moderately hazardous

Class III: slightly hazardous

The system is based on LD50 determination in rats, thus an oral solid agent with an LD50 at 5 mg or less/kg bodyweight is Class Ia, at 5–50 mg/kg is Class Ib, LD50 at 50–2000 mg/kg is Class II, and at LD50 at the concentration more than 2000 mg/kg is classified as Class III. Values may differ for liquid oral agents and dermal agents.

In United States, the United States Environmental Protection Agency **(EPA) uses four toxicity classes in its** toxicity category rating. Classes I to III are required to carry a signal word on the label. Pesticides are regulated in the United States **primarily by the** Federal Insecticide, Fungicide, and Rodenticide Act **(FIFRA).**

The Toxicity Class I

It refers the pesticide as most toxic; requires signal word: "Danger-Poison", with skull and crossbones symbol, possibly followed by:"Fatal if swallowed", "Poisonous if inhaled", "Extremely hazardous by skin contact--rapidly absorbed through skin", or "Corrosive--causes eye damage and severe skin burns"

Class I materials are estimated to be fatal to an adult human at a dose of less than 5 grams (less than a teaspoon).

The Toxicity Class II

It refers the pesticide as moderately toxic; requires Signal word: "Warning", possibly followed by:"Harmful or fatal if swallowed", "Harmful or fatal if absorbed through the skin", "Harmful or fatal if inhaled", or "Causes skin and eye irritation"

Class II materials are estimated to be fatal to an adult human at a dose of 5 to 30 grams.

The Toxicity Class III

It refers the pesticide as slightly toxic; Signal word: Caution, possibly followed by: "Harmful if swallowed", "May be harmful if absorbed through the skin", "May be harmful if inhaled", or "May irritate eyes, nose, throat, and skin"

Class III materials are estimated to be fatal to an adult human at some dose in excess of 30 grams.

The Toxicity Class IV

It refers the pesticide as practically nontoxic; no Signal Word required since 2002

In India, the Indian standardized system of toxicity labels (1 to 4) **for pesticides uses a 4-color system (red, yellow, blue, green) to plainly label containers with the toxicity**/hazard class of the contents. Toxicity labels viz. Red label, Yellow label, blue label and green label are mandatory labels employed on a pesticide containers in India identifying the level of toxicity of the contained pesticide as per the insecticide Act 1968 and the insecticide rule of 1971.

8

Chemical Molecules for Control of Bacterial Diseases

8.1. Bactericides for Plant Pathogenic Bacteria

8.1.1. Copper Based Preparations

A. Bordeaux mixture

It is used in the control of bacterial leaf spots, blights, and cankers mostly on fruits crops and ornamentals. The Bordeaux mixture, however, can cause burning of leaves or russeting of fruit such as apples when applied in cool, wet weather. The phytotoxicity of Bordeaux is reduced by increasing the ratio of hydrated lime to copper sulfate. Copper is the only ingredient in the Bordeaux mixture that is toxic to pathogens and, sometimes, to plants, whereas the role of lime is primarily that of a "safener."

Bordeaux mixture achieves its effect by means of the copper ions (Cu^{2+}) of the mixture. Bordeaux mixture must be used preventively, before the onset of the disease infection. Bordeaux mixture can be prepared by using differing proportions of the components. In preparing it, the $CuSO_4$ and the lime are dissolved separately in water and then mixed. Calcium oxide (burnt lime) and calcium hydroxide (slaked lime) give the same end result, since an excess of water is used in the preparation. Generally 1% Bordeaux mixture is used for spraying on the bacterial pathogen infected crop plants.

For dormant sprays, concentrated Bordeaux is made by the formula 10:10:100. The most commonly used formula for Bordeaux is 8:8:100. For spraying young, actively growing plants, the amounts of copper sulfate and hydrated lime are reduced, and the formulas used may be 2:2:100, 2:6:100, and so on. For plants known to be sensitive to Bordeaux, a much greater concentration of hydrated lime may be used, as in the formula 8:24:100.

Thorough coverage of the spray on the plants is necessary. The Bordeaux spray continues to adhere well to the plant during rain, though in the long term it is washed off by rain.

Bordeaux has many positive and negative features. It is a highly effective bactericide (table.8) and fungicide that is used to manage several plant diseases. The material sticks to and remains active on plant surfaces even during typical wet PNW winters. Generally it is used as a dormant spray because it may burn young juvenile tissues. The ingredients must be mixed in the right order and with mechanical agitation of the tank to avoid the formation of a sprayer clogging precipitate. Bordeaux cannot be mixed ahead of use because it deteriorates on standing.

Table 8: Bactericidal potential of different concentration of Bordeaux mixture under in vitro against Xanthomonas.

Bordeaux mixture	Concentration (%)	Bactericidal potential	MIC
	0.2	–	
	0.1	–	0.1 %
	0.05	+	
	0.025	+	
	0.01	+	

+ = Bacterial growth; – = inhibition of bacterial growth

Source: Ajayasree et.al, 2018. J. Advanced Research in Biotechnology. 3(2):1-5.

Some does and don’ts

- Use the Bordeaux mixture soon after preparation. It should not be stored for further use.
- Do not use metallic containers for preparing copper sulphate solution and lime suspension. Use a wooden stick for stirring to get a homogenous mixture. Never use metallic stick.
- Do not use the Bordeaux mixture in combination with any other chemical or pesticide e.g. In citrus, zinc sulphate spray should be given, keeping a gap of at least one week from that of the Bordeaux mixture.
- To avoid choking of the nozzle, it is advisable to strain the Bordeaux mixture through a cloth or a sieve before putting it into the spray tank.
- The Bordeaux mixture tends to sediment easily. Therefore, it’s stirring while using is desirable.
- It is not advisable to spray the Bordeaux mixture on fruit-laden trees as the spray may inflict russetting, especially on apple and pear.
- In exceptionally hot days, when the plants are showing signs of temporary wilting or when it is raining, the Bordeaux mixture should

not be sprayed, particularly on newly emerged tender foliage. Exercise this caution particularly on nursery plants.

- After carrying out spray operations the appliances should be thoroughly washed with plenty of water to remove any copper deposits.
- The left over Bordeaux mixture should not be dumped in the field as this may prove toxic to the subsequent sowings. Because the copper ions build up in the soil, continuous use will cause heavy metal pollution. Copper also bioaccumulates in organisms. It is thus illegal to use in the United Kingdom as well as most European Union countries, with the exception of Belgium, Cyprus, Greece, Hungary, Italy, Malta, Portugal, Romania and Slovenia.

Bordeaux mixture has been found to be harmful to fish, livestock and due to potential buildup of copper in the soil to the earthworms.

Many other copper-based fungicides have been developed to capture the positive weathering and disease control features of Bordeaux without the challenges of preparing the material properly.

Copper-based active ingredients in other products include copper ammonium complex (Copper Count-N), copper hydroxide (Champion, Kocide, Nu-Cop, etc.), copper oxide (Nordox), copper oxychloride, (C-O-C-S), and copper sulfate (Cuprofix Disperss, etc).

B. Copper Ion

The active ingredient in all copper-based formulations is the positively charged copper ion (Cu^{+2}). Many organisms are sensitive to very small amounts of copper ion, such as bacteria, fungi (including pathogens like downy mildews) and especially aquatic organisms such as algae or water molds. Copper-based products have broad-spectrum activity against microorganisms due to copper's interaction with nucleic acids, interference with energy transport, and disruption of enzyme activity and integrity of cell membranes.

The recommended amounts of copper toxic to plant pathogens, are not toxic to plants or humans. Copper at moderate to high doses may be toxic to plants. Other forms of copper used for plant disease management, such as copper hydroxide, copper oxide, copper oxychloride and copper octanoate, are formulated to produce low doses of copper to reduce toxicity to plants. The goal of most copper-based products is to tie up or fix much of the free copper ions so it is not phytotoxic to plants but allow just enough of the copper ions to be released to inhibit disease causing pathogens.

Different copper formulations results in different amounts of copper ion released. Most products express the amount of copper they contain in terms of copper metallic equivalents. Unfortunately the copper equivalent does not directly relate to the amount of copper ion released. Acidic conditions result in a higher concentration of copper ion. When copper-based pesticides are tank mixed with acidic compounds more copper ion may be released, which can lead to phytotoxicity.

C. Fixed Copper Sulfate

Several products contain a fixed copper sulfate and may list the active ingredient as basic copper sulfate (Cuprofix) or copper sulfate pentahydrate (CS 2005, Mastercop, Instill, Phyton 27). These products can be used on a wide variety of crops to manage bacterial and fungal diseases. Phytotoxicity can still occur with some crops such as fruit marking of cherry and russeting of some pears. To prevent fruit marking of cherry, applications are only allowed during and prior to bloom or after harvest. Good drying conditions are important to avoid fruit russet risk. Read label warnings carefully to avoid damage to crops.

8.1.2. Copper Hydroxide and Copperoxychloride (0.25%)

A. Copper Hydroxide

Chemical analysis of Bordeaux mixture indicate that one of the products produced is copper hydroxide and therefore, now-a-days many products contain the active ingredient copper hydroxide (e.g. Champion, Kocide, Nu-Cop, Previsto and many others). These formulated materials are as effective as Bordeaux for disease management without the mixing problems.

Formulations of copper hydroxide can vary considerably in the amount of free copper ion found in solution and the degree of effective disease control. Use of a liquid formulation of copper hydroxide (Kocide LF) results in fewer free copper ions than dry formulations (Kocide DF or 101). In general, dry formulations of copper-based pesticides have resulted in better fungal and bacterial disease control than liquid formulations in trials conducted in western Oregon. A formulation of copper hydroxide suspended in an alginate matrix (Previsto) has been effective against several diseases such as fire blight of pome fruit while using less overall copper. Pome fruit russeting can occur with applications of copper-based bactericides, but less russet was reported with Previsto used in arid production areas.

B. Copper Oxychloride and Combinations

Copper oxychloride is a green to blue-green compound used for some of the bacterial disease control. The product C-O-C-S is composed of both copper oxychloride and basic copper sulfate. Similarly Badge is composed of both copper oxychloride and copper hydroxide. Both of these products have the same warnings about pH and copper sensitive crops.

In lilac tissue culture crop, C-O-C-S resulted in 80% control of copper sensitive isolates of *Pseudomonas syringae* but only 27% control of copper resistant bacteria.

The bactericidal potential of Copper oxychloride on tomato leaf spot pathogen Xanthomonas campestris pv.vesicatoria (table.9) indicated its bactericidal potential

Table 9: In vitro Bactericidal potential of Copperoxychloride against Xanthomonas.

Sr.No	Fungicide	Concentration (%)	Bactericidal potential	MIC
1.	Copper oxychloride	0.2	–	
		0.1	–	
		0.05	+	0.1%
		0.025	+	
		0.01	+	

+ = Bacterial growth; – = inhibition of bacterial growth
Source: Ajayasree et.al, 2018. J. Advanced Research in Biotechnology. 3(2):1-5.

8.1.3. Fungicides having Bactericidal Properties

8.1.3.1. Dithiocarbamates

Copper di-ethyl-dithiocarbamate, cadmium di-ethyl-dithiocarbamate, lead di-ethyl-dithiocarbamate, nickel di-ethyl-dithiocarbamate and zinc di-ethyl-dithiocarbamate is reported to have the fungicidal and bactericidal activity. Minimum Inhibitory Concentration (bactericidal) for bacterial species was 6.25–25.00 μg mL^{-1}. The dithiocarbamate fungicide Mancozeb was effective to inhibit the bacterial growth of plant pathogen *Xanthomonas.* The bactericidal potential of di-thiocarbamate group of fungicides (table.10) on tomato leaf spot pathogen *Xanthomonas campestris pv.vesicatoria* indicated its bactericidal potential at 0.05 percent.

Table. 10: Bactericidal potential (in vitro) of Mancozeb on bacterial leaf spot pathogen *Xanthomonas.*

Fungicide	Concentration (%)	Bactericidal potential	MIC
Mancozeb	0.2	–	
	0.1	–	
	0.05	–	0.05%
	0.025	+	
	0.01	+	

+ = Bacterial growth; – = inhibition of bacterial growth
Source: Ajayasree et.al, 2018. J. Advanced Research in Biotechnology. 3(2):1-5.

Antibacterial Activities of Novel Dithiocarbamate-Containing 4*H*-Chromen-4-one derivatives against three bacterial plant pathogens viz. *Xanthomonas oryzae* pv *oryzae* (*X. oryzae* pv *oryzae*.), *Ralstonia solanacearum* (*R. solanacearum*), and *Xanthomonas axonopodis* pv *citri* (*X. axonopodis* pv *citri.*) showed that most of the target compounds displayed good inhibitory effects against *X. oryzae* pv *oryzae* and *X. axonopodis* pv *citri*. Remarkably, compound E6 showed the best *in vitro* antibacterial activity against *X. axonopodis* pv *citri* with an EC_{50} value of 0.11 μg/mL, which was better than those of thiodiazole copper (59.97 μg/mL) and bismerthiazol (48.93 μg/mL). Compound E14 exhibited the best *in vitro* antibacterial activity against *X. oryzae* pv *oryzae* with an EC_{50} value of 1.58 μg/mL, which was better than those of thiodiazole copper (83.04 μg/mL) and bismerthiazol (56.05 μg/mL). Scanning electron microscopy analysis demonstrated that compounds E6 and E14 caused the rupture or deformation of the cell membranes for *X. axonopodis* pv *citri* and *X. oryzae* pv *oryzae* respectively. *In vivo* antibacterial activity test and the defensive enzymes activity test results indicated that the compound E14 could reduce *X. oryzae* pv *oryzae* more effectively than thiodiazole-copper or bismerthiazol.

Pyrrolidine dithiocarbamate (PDTC), an antioxidant with a metal-chelating activity, has an antimicrobial activity against various bacteria. The antibacterial activity of PDTC and other compounds showed that the bacterial growth was inhibited by PDTC, where a wide range of sensitivity was demonstrated among the tested bacteria. The antibacterial activity of PDTC was reduced by the addition of copper chloride; but in contrast it was enhanced considerably by zinc chloride. Two different zinc chelators, Ca-saturated EDTA (Ca-EDTA) and *N, N, N', N'*-tetrakis (2-pyridylmethyl) ethylenediamine, blocked the antibacterial activity of PDTC, whereas Zn-EDTA failed to reduce the activity of PDTC. These results demonstrate that PDTC possesses an antibacterial activity, for which zinc is required.

Similarly antibacterial activity of a series of phosphanegold(I) dithiocarbamates, ($R_3PAu[S_2CN(iPr)CH_2CH_2OH]$ where R = Ph (**2**), Cy (**3**) and Et (**4**), against 25 strains of Gram-positive and Gram-negative bacteria were determined through the disk diffusion method, the determination of minimum inhibitory concentration (MIC) and minimum bactericidal concentration (MBC) and by time-kill assay, compounds 2 and 3 have been shown to be specifically active against the tested Gram-positive bacteria, with MIC values ranging from 7.81 to 125 μg/ml. Compound 4 has a broad-spectrum activity against 24 strains of Gram-positive and Gram-negative bacteria, with MIC values ranging from 0.98 to 1,000 μg/ml. In time-kill studies, the bacteriostatic and bactericidal activities of the tested compounds towards susceptible strains were similar to their characteristics determined by MBC/MIC ratios. In the time-kill assay, 2 and 3 showed only bactericidal activity towards the susceptible strains tested, whereas 4 revealed varying degrees of bactericidal and bacteriostatic activities, indicating different antibacterial mechanisms are involved. The antibacterial activity of these compound against plant pathogenic bacteria needs confirmation for their use in bacterial plant disease control.

8.1.3.2. Zineb

The bactericidal activity of fungicides Zineb (table.11) against X.c.pv. vesicatoria indicated its bactericidal potential.

Table 11: Bactericidal potential of Zineb (in vitro) against bacterial leaf spot pathogen

Fungicide	Concentration (%)	Bactericidal potential	MIC
Zineb	0.2	–	
	0.1	-	
	0.05	-	0.025
	0.025	-	
	0.01	+	

+ = Bacterial growth; – = inhibition of bacterial growth
Source: Ajayasree et.al, 2018. J. Advanced Research in Biotechnology. 3(2):1-5.

8.1.3.3. Carbendazim (0.15%)

The report on the efficacy of 5 fungicides (Carbendazim, Bordeaux mixture, mancozeb, cuprasol and thiophanate-methyl) and 1 antibiotic (plantomycin) on the control of anthracnose-bacterial leaf spot of *Piper betle* caused by *Colletotrichum capsici* and *Xanthomonas campestris* pv *betlicola* revealed that the best control was achieved by spray treatment with carbendazim + plantomycin or Bordeaux mixture + plantomycin indicating the efficacy of carbendazim in the management of bacterial plant disease.

8.1.4. Stable Bleaching Powder

Stable bleaching powder is generally used in the management of soil borne bacterial plant pathogens. Bacterial wilt and brown rot caused by Ralstonia solanacearum is one of the most dreaded disease causing 30 to 70% yield loss in India. Soil application of bleaching powder at the rate of 12.5 kg/h, irrespective of other treatments, significantly reduced wilting of plants in comparison to control.

Bacterial wilt of potato caused by *Ralstonia solanacearum* race 3, biovar II was managed under glasshouse and field conditions in Nepal, by amending infested soils with stable bleaching powder (SBP) with a mixture of urea and lime (urea-lime). The use of SBP at the rate of 25 kg/ha was more effective and suitable. Alternatively, soil amendment with 428 kg/ha urea and 5 ton/ ha lime can be used to effectively control bacterial wilt disease where SBP is not a vialable. Similarly, in management of bacterial wilt incidence in tomato caused by Ralstonia solanacearum race 1 biovar 3, minimum disease incidence was found in calcium chloride (0.1%) treatment followed by bleaching powder (0.025%) after 9th day of inoculation. The bacterial wilt disease was reduced by 50-52.5% by applying these chemicals. Population of R. solanacearum was drastically reduced by 45.9 % in rhizosphere of tomato in bleaching powder treated soil as compared to control.

Use of stable bleaching powder (Klorocin) in paddy field controls the bacterial blight disease. Application at nursery stage and 10 days after transplanting @ 12.5 kg/ha was most satisfactory. Application of SBP before uprooting (nursery treatment) helps to stop the entry and multiplication of the bacteria in the youmg seedling from soil and irrigation water source, by increasing the initial concentration of chlorine at the primary entry point of the pathogen caused through root injury due to uprooting, and transplanting. Since the initial attack is from soil inoculum, a chemical suitable for soil application (SBP) is more effective for the control of this bacteria, than the antibiotic which do not persist longer in soil and degrade much faster. Antibiotics are also too expensive for soil application. Persistance of chlorine in soil and its direct effect on soil borne inoculum suggest the SBP is better suited for soil application to control X.c.pv.oryzae. Such an application during earlier stages of crop growth, when the plant is more vulnerable to the pathogen, gives better results.

In most of the cases where initial source of inoculum is from soil, may be due to soil borne nature of bacteria, leaf fall of infected bacterial leaves on to the soil and their degradation to release the bacterial pathogen as soil borne inoculum, the application of stable bleaching powder in soil helps to reduce the inoculum in the management of the bacterial diseases.

8.1.5. Synthetic Organic Bactericide

8.1.5.1. Cinoxacin

Cinoxacin (compound 64716) is a synthetic organic acid with antibacterial activity against most aerobic gram-negative bacilli. It is a new synthetic compound similar chemically and in antimicrobial activity to oxolonic acid and nalidixic acid. It is most effective against all species of *Enterobacteriaceae (some species are soil borne plant pathogenic bacteria)*. Relative to nalidixic acid, cinoxacin has slightly greater inhibitory and bactericidal activity. The drugs is more active in an acid than an alkaline medium.

Minimal inhibitory concentrations of cinoxacin (agar-dilution method) were determined for 419 strains. *Escherichia coli* was the most susceptible group of organisms. The majority of *Klebsiella* sp., *Enterobacter* sp (both are soil borne plant pathogenic bacteria)., *Proteus* sp., and *Serratia marcescens* were inhibited by 8 μg of cinoxacin per ml. *Pseudomonas aeruginosa* and all gram-positive isolates tested were resistant to 64 μg or less of cinoxacin per ml. Zones of inhibition using a 30-μg disk correlated well with agar-dilution minimal inhibitory concentrations ($r = -0.9$). Cinoxacin was bactericidal when tested with inocula of 5×10^6 organisms per ml. The in vitro properties of this agent were similar to those of nalidixic acid.

8.1.5.2. Parabens (esters of para-hydroxybenzoic acid)

The potassium salts of the methyl and ethyl parabens were microbiocidal against the fungus *Aspergillus niger* and some bacteria, whereas the potassium salts of propyl and butyl parabens and their respective parabens were not microbiocidal. The bactericidal activity.of parabens and potassium salts of parabens was enhanced against E. cloacae (a plant pathogenic bacteria) and S. aureus.

8.1.5.3. Silver Nano Particles

The development of resistant strains of bacteria to current antibiotics has become a serious problem in human health and also in plant health. Therefore, there is a strong incentive to develop new bactericides. Silver has long been known to exhibit a strong toxicity to a wide range of micro-organisms; and therefore, silver-based compounds have been used extensively in many bactericidal applications. Experimental evidence suggests that DNA loses its replication ability once the bacteria have been treated with silver ions. Other studies have shown evidence of structural changes in the cell membrane as well as the formation of small electron-dense granules formed by silver and sulfur. Silver ions have been demonstrated to be useful and effective in bactericidal applications and due to the unique properties of nanoparticles, the development of new bactericides like silver nano particles are a promising option as bactericide.

AgNPs interact with a wide range of molecular processes within microorganisms resulting in a range of effects from inhibition of growth, loss of infectivity to cell death which depends on shape, size, concentrations of AgNPs and the sensitivity of the microbial species to silver. Several studies have reported that the positive charge on the Ag^+ ion is crucial for its antimicrobial activity through the electrostatic attraction between the negatively charged cell membrane of the microorganism and the positively charged nanoparticles. In contrast, the antimicrobial activity of AgNPs on Gram-negative bacteria depends on the concentration of AgNPs and is closely associated with the formation of pits in the cell wall of bacteria; consequently, AgNPs accumulated in the bacterial membrane disturbing the membrane permeability, resulting in cell death. Another mechanism proposed is the inhibition of cell wall synthesis, protein synthesis mediated by the 30s ribosomal subunit, and nucleic acid synthesis.

Studies have shown that the gram-negative (G–ve) bacteria are more liable to AgNPs than the gram-positive (G+ve) bacteria. The cell wall of G-ve bacteria is more tapered as compared to G+ve strains. The thick cell wall decrease the diffusion of the AgNPs into the cellular environment. The diverse antibacterial activities of AgNPs over the G–ve and G+ve bacteria suggested that the uptake of AgNPs is essential for effective antibacterial or antimicrobial action. The antibacterial effects of AgNPs is also influenced by the panicles properties such as shape, size, surface charge, doses and diffusion state. It is usually recognized that AgNPs smaller than 10 nm could directly modify cell penetrability, go into the bacterial cells and initiate cell lysis. Apart from these, some other mechanisms are also reported, including interaction of Ag^+ ions with the biological macromolecules (enzymes and DNA), based on the mechanism of electron-discharge or free radical generation. As per the reported studies, AgNPs triggers the inhibition of both protein synthesis and cell wall synthesis, evidenced by the amassing of envelope protein precursor or disruption of the outer cellular membrane, leading to ATP leakage.

Commercial silver nanoparticles are not yet available for controlling bacterial plant diseases and further studies of AgNPs on plant pathogenic bacteria will pave way for its use in the control of bacterial plant diseases.

8.1.5.4. Triclosan

Triclosan is a widely accepted broad spectrum antimicrobial agent proven to be effective against many gram-positive and gram-negative bacteria. Triclosan acts by blocking bacterial fatty acid biosynthesis. The addition of Gantrez co-polymer has been shown to enhance the antimicrobial activity of triclosan. The minimum bactericidal concentration (MBC) of triclosan ranged from 12-94 µg/mL. The MBC of triclosan with Gantrez ranged from <0.3-10.4 µg/mL. The addition of Gantrez enhanced the bactericidal activity of triclosan. Both triclosan and triclosan with Gantrez demonstrated bactericidal activity against some of the endodontic pathogens. The studies on plant pathogenic bacteria by this bactericide is lacking.

8.1.5.5. Bactericin-100

The commercial bactericide Bactericin-100 was effective at a concentration of 50 ppm and above (table.12) against bacterial leaf spot pathogen of tomato while it was not effective at 25 ppm. The MIC for Bactericin-100 for this pathogen was 50 ppm whereas for Bordeaux mixture it was 0.1 percent.

Table 12: The bactericidal potential of bactericin-100 on tomato leaf spot pathogen *Xanthomonas campestris pv vesicatoria*

Bactericides	Concentration	Bactericidal potential	MIC
Bactericin-100	500	–	
	250	–	
	100	–	
	75	–	50 ppm
	50	–	
	25	+	

+ = Bacterial growth; – = inhibition of bacterial growth
Source: Ajayasree et.al, 2018. J. Advanced Research in Biotechnology.

8.1.5.6. Bronopol (2 bromo, 2 nitro-1, 3-propanediol)

Bronopol, or 2-Bromo-2-nitro-1, 3-propanediol, is an organic compound with wide-spectrum of antimicrobial properties. First synthesized in 1897, bronopol was primarily used as a preservative for pharmaceuticals and was registered in the United States in 1984 for use in industrial bactericides, slimicides and preservatives.

Bronopol is used as a microbicide or microbiostat in various commercial and industrial applications, including oil field systems, air washer systems, air conditioning or humidifying systems, cooling water systems, papermills, absorbent clays, metal working fluids, printing inks, paints, adhesives and consumer products. Compared to other aliphatic halogen-nitro compounds, bronopol is more stable to hydrolysis in aqueous media under normal conditions.

The inhibitory activity against various bacteria, including *Pseudomonas aeruginosa*, was demonstrated *in vitro.* The agent is largely available commercially as an antibacterial for a variety of industrial purposes while it is predominantly available for purchase as a pet animal litter antibacterial at the domestic consumer level. Nevertheless, ongoing contemporary re-evaluations of bronopol use in large markets such as Canada now place various compositional and product restrictions on the use of the agent in cosmetic products [L873] and in other products where it may not primarily be used in the role of a non-medicinal preservative antimicrobial [L874].

It is proposed that bronopol generates biocide-induced bacteriostasis followed by a growth at an inhibited rate in bacteria, via two distinct reactions between bronopol and essential thiols within the bacterial cell. Under aerobic conditions, bronopol catalyzes the oxidation of thiol groups, such as cysteine, to disulfides. This reaction is accompanied by rapid consumption of oxygen, where oxygen acts as the final oxidant. During the conversion of cysteine to cystine, radical anion intermediates such as superoxide and peroxide are formed from bronopol to exert a direct bactericidal activity. The oxidation of excess thiols alters the redox state to create anoxic conditions, leading to a second reaction involving the oxidation of intracellular thiols such as glutathione to its disulfide. The resulting effects are inhibition of enzyme function, and reduced growth rate following the bacteriostatic period. Under the anoxic conditions, the reaction between thiol and bronopol decelerates without the involvement of oxygen and the consumption of bronopol predominates. Bronopol is ultimately removed from the reaction via consumption and resumption of bacterial growth occurs.

Commercially the product Bronopol is also known as Bromopal / Bactinashak / Bactronol.The field evaluation of different antibiotics/antibactericides during 2004 and 2005 revealed that application of Bromopal at 500 ppm + Copper oxychloride at 2,000 ppm, followed by Zinc sulphate 1.0 g + Magnesium sulphate 1.0 g + Boron 1.0 g + Lime 1.0 g/lit. sprayed one day after spray of streptocycline 500 ppm + Copper oxychloride 2,000 ppm was found effective in controlling the bacterial blight of pomegranate disease with recording of 19.08 percent and 19.65 percent incidence as compared to control (78.65 percent incidence) after 5th sprays at 10 days intervals. The plant that received the spray of Bromopol at 500 ppm + Copper oxychloride at 2,000 ppm recorded maximum yield (of 8.03 t/ha) followed by Streptocycline at 500 ppm + Copper oxychloride at 2,000 ppm (7.74 t/ha).

8.2. Antibiotics for Management of Plant Pathogenic Bacteria

Antibiotic is defined as a chemical substance produced by one micro-organism which in low concentration can inhibit or even kill other micro-organism.

Because of their specificity of action against plant pathogens, relatively low phytotoxicity, absorption through foliage and systemic translocation and activity in low concentration, the use of antibiotic is becoming very popular and very effectively used in managing several plant diseases. They can be grouped as antibacterial antibiotics and antifungal antibiotics. Most antibiotics are products of several actinomycetes and a few are from fungi and bacteria.

8.2.1. Antibacterial antibiotics

1. Streptomycin sulphate

Streptomycin is an antibacterial antibiotic produced by streptomyces griseus. **Streptomycin is actually streptomycin sulphate and is sold as Agrimycin-100, Streptomycin sulphate, Plantomycin, Streptocycline, Paushamycin, Phytostrip, Agristrep and Embamycin.**

Agrimycin - 100 contains 15 per cent streptomycin sulphate + 1.5 percent terramycin(Oxy tetracycline).

Agristerp contains 37 percent streptomycin sulphate.

Phytomycin contains 20 percent streptomycin.

Streptocycline and **paushamycin** contains 9 parts of streptomycin and 1 part of tetracycline hydrochloride.

This group of antibiotics act against a broad range of bacterial pathogens causing blights, wilt, rots etc. This antibiotic is used at concentrations of 100-500 ppm. Some important diseases controlled are blight of apple and pear *(Erwinia amylovora)*, Citrus canker *(Xanthomonas campestris pv. citri)*, Cotton black arm *(X.c.* pv. *malvacearum)*, bacterial leaf spot of tomato *(Pseudomonas solanacearum)*, wild fire of tobacco *(Pseudomonas tabaci)* and soft rot of vegetables *(Erwinia carotovora)*.

In addition, it is used as a dip for potato seed pieces against various bacterial rots and as a disinfectant in bacterial pathogens of beans, cotton, crucifers, cereals and vegetables.

Although it is an antibacterial antibiotic, it is also effective against some diseases caused by Oomycetous fungi, especially foot-rot and leaf rot of betelvine caused by *Phytophthora parasitica var. piperina.*

2. Tetracyclines

Antibiotics belonging to this group are produced by many species of Streptomyces. **This group includes Terramycin or Oxymicin (Oxytetracycline).**

All these antibiotics are bacteriostatic, bactericidal and mycoplasmastatic. **These are very effective against seed-borne bacteria (table.13).**

This group of antibiotic is very effective in managing MLO diseases of a wide range of crops. These are mostly used as combination products with Streptomycin sulphate in controlling a wide range of bacterial diseases.

Oxytetracyclines are effectively used as soil drench or as root dip controlling crown gall diseases in rosaceous plants caused by *Agrobacterium tumefaciens.*

Table 13: The bactericidal potential of antibiotics belonging to streptomycin sulphate and tetracycline group on tomato leaf spot pathogen *Xanthomonas campestris* pv. *Vesicatoria*

Sr.No	Antibiotics	Concentration (ppm)	Bactericidal potential	MIC
1.	Streptomycin sulphate	1000	–	1000 ppm
		500	+	
		250	+	
		100	+	
		75	+	
		50	+	
		25	+	
2.	Streptocycline	500	–	25 ppm
	(Streptomycin sulphate)	250	–	
		100	–	
		75	–	
		50	–	
		25	–	
3.	Terramycin	500	–	75 ppm
	(Tetracycline)	250	–	
		100	–	
		75	–	
		50	+	
		25	+	

+ = Bacterial growth; – = inhibition of bacterial growth
Source: Ajayasree et.al, 2018. J. Advanced Research in Biotechnology.

The different antibiotics of these group differ in their MIC in the control of a particular bacterial plant pathogen.

8.2.2. Antifungal antibiotics

1. Aureofungin

It is a heptane antibiotic produced in sub-merged culture of *Streptomyces cinnamomeum* var. *terricola.*

It is absorbed and translocated to other parts of the plants when applied as spray or given to roots as drench.

It is sold as Aureofungin-Sol. Containing 33.3% Aureofungin and normally sprays at 50-100 ppm.

The diseases controlled are citrus gummosis caused by several species of Phytophthora, powdery mildew of apple caused by *Podosphaera leucotricha* and apple scab (Venturia inaequalis), groundnut tikka leaf spot, downy mildew, powdery mildew and anthracnose of grapes, potato early and late blight. As post harvest treatment it effectively check *Diplodia* rot of mango, *Alternaria* rot of tomato, *Pythium* rot of cucurbits and *Penicillium* rot of apples and citrus. As a trunk application/root feeding, 2 g of aureofunginsol+ 1g of copper sulphate in 100 ml of water effectively reduce Thanjavur wilt of coconut.

2. Griseofulvin

This antifungal antibiotic was first discovered to be produced by *Penicillium griseofulvum* and now by several species of *Penicillium,* viz., *P.patulum, P.nigricans, P.urticae, and P.raciborskii.*

It is commercially available as Griseofulvin, Fulvicin and Grisovin. It is highly toxic to powdery mildew of beans and roses, downy mildew of cucumber. It is also used to control *Alternaria solani* in tomato, *Sclerotinia fructigena* in apple and *Botrytis cinerea* in lettuce.

3. Cycloheximide

It is obtained as a by-product in streptomycin manufacture. It is produced by different species of *Streptomyces,* including *S.griseus* and S. *nouresi.*

It is commercially available as Actidione, Actidione PM, Actidione RZ and Actispray.

It is active against a wide range of fungi and yeast. Its use is limited because it is extremely phytotoxic. It is effective against powdery mildew of beans *(Erysiphe polygoni),* Bunt of wheat (*Tilletia* spp.) brown rot of peach *(Sclerotinia fructicola)* and post harvest rots of fruits caused by *Rhizopus* and *Botrytis* spp.

4. Blasticidin

It is a product of *Streptomyces griseochromogenes* and specifically used against blast disease of rice caused by *Pyricularia oryzae*. **It is commercially sold as Bla-s.**

5. Antimycin

It is produced by several species of *Streptomyces,* especially *S. griseus* and *S. Kitasawensis.* It is effectively used against early blight of tomato, rice blast and seedling blight of oats. **It is commercially sold as Antimycin.**

6. Kasugamycin

It is obtained from *Streptomyces kasugaensis.* It is also very specific antibiotic against rice blast disease. **It is commercially available as Kasumin.**

7. Thiolutin

It is produced by *Streptomyces albus* and effectively used to control late blight of potato and downy mildew of cruciferous vegetables.

8. Endomycin

It is a product of *Streptomyces endus* and effectively used against leaf rust of wheat and fruit rot of strawberry *(Botrytis cinerea).*

9. Bulbiformin

It is produced by a bacterium, *Bacillus subtills* and is very effectively used against wilt diseases, particularaly redgram wilt.

10. Nystatin

It is also produced by *Streptomyces noursei.* It is successfully used against anthracnose disease of banana and beans. It also checks downy mildew of cucuribits.

As a post harvest dip, it effectively reduces brown rot of peach and anthracnose of banana in stroage rooms. **It is commercially marketed as Mycostain and Fungicidin.**

11. Eurocidin

It is a pentaene antibiotic produced by *Streptomyces anandii* and called as pentaene G-8. It is effectively used against diseases caused by several species of *Colletotrichum* and *Helminthosporium.*

9

Chemical Molecules for Management of Viruses and Phytoplasmas

9.1. Management of Plant Viruses

Plant viruses cause severe crop losses worldwide. Conventional control strategies, such as cultural methods and biocide applications have limited success at mitigating the impact of plant viruses.

Planting resistant cultivars is the most effective and economical way to control plant virus diseases. Natural sources of resistance have been exploited extensively to develop virus-resistant plants by conventional breeding. Non-conventional methods have also been used successfully to confer virus resistance by transferring primarily virus-derived genes, including viral coat protein, replicase, movement protein, defective interfering RNA, non-coding RNA sequences, and protease, into susceptible plants. Non-viral genes (R genes, microRNAs, ribosome-inactivating proteins, protease inhibitors, dsRNAse, RNA modifying enzymes, and scFvs) have also been used successfully to engineer resistance to viruses in plants. Very few genetically engineered (GE) virus resistant (VR) crops have been released for cultivation and none is available yet in developing countries. However, a number of economically important GEVR crops, transformed with viral genes are of great interest in developing countries. The major issues confronting the production and deregulation of GEVR crops in developing countries are primarily socio-economic and related to intellectual property rights, biosafety regulatory frameworks, expenditure to generate GE crops and opposition by non-governmental activists.

Thermotherapy or heat therapy is the oldest method used for the elimination of viruses, viroids, and phytoplasmas from vegetatively propagated plants and has been used since the end of the nineteenth century. The first reference to the use of heat therapy of plants dates back to 1869 when Scottish gardeners immersed bulbs in hot water before planting. In 1936, Kunkel reported the use of dry heat or hot water treatment for the cure of peach yellows diseases

(Nyland & Goheen, 1969). Walkey and Cooper (1975) have also studied the effect of temperature on virus eradication.

Thermotherapy has been used for virus elimination in grapevines (Leonhardt, et.al,1998), stone fruits (Manganaris, et.al,2003; Mink,et.al., 1998), citrus (Muhammad, et.al, 2005), pome fruits (Howell, et.al.,1998; Tan, et.al.,2010), potatoes (Salazar, 1996), and strawberries and others (Nyland & Goheen, 1969; Panattoni,et.al., 2013). The infected plant is incubated in a controlled temperature cabinet at 30–40 °C for a period of 6–12 weeks. Elimination of viruses from the whole plant is usually impossible without causing severe damage to or even death of the plant. In most cases, after thermotherapy, cuttings or bud grafts are removed from the heat-treated plant and propagated *in vivo* or *in vitro* using meristem tips to produce new VF plants.

Alternatively, thermotherapy can be applied *in vitro* with the advantage of reducing procedure time. Shoots from infected plants are cultured *in vitro* and then incubated in a controlled temperature cabinet at 30–40 °C for several days. *In vitro* thermotherapy has been used successfully to eliminate *Apple chlorotic leaf spot virus* (ACLSV) from apricot (*Prunus* armeniaca) shoot cultures; *Prunus necrotic ring spot virus* (PNRSV) and ACLSV from peach (*Prunus persica*) shoot cultures; and Prune *dwarf* virus (PDV) and ACLSV from sour cherry (*Prunus avium*) shoot cultures (Gella & Errea, 1998).

Virus control in berry crops starts with the development of plants free of targeted pathogens, usually viruses, viroids, phytoplasmas, and systemic bacteria, through a combination of testing and therapy. These then become the top-tier plants in certification programs and are the source from which all certified plants are produced, usually after multiple cycles of propagation. In certification schemes, efforts are made to produce plants free of the targeted pathogens to provide plants of high health status to berry growers. This is achieved using a systems approach to manage virus vectors. Once planted in fruit production fields, virus control shifts to disease control where efforts are focused on controlling viruses or virus complexes that result in disease. In fruiting fields, infection with a virus that does not cause disease is of little concern to growers. Virus control is based on the use of resistance and tolerance, vector management, and isolation.

Currently, different inducers derived from various origins are being used to reduce the virus concentration. Preliminary screening of the inducers against tobamoviruses was conducted by using the local lesion assay (*Nicotiana glutinosa*). The results showed that all the inducers used for screening were effective in reducing the number of local lesions formed by the challenge inoculation of tobamoviruses. Both spray and seed treatment of inducers against the tobamoviruses reduced the concentration of viruses in the seedlings as evident from the results of Indirect Enzyme linked immunosorbent assay. Among the inducers used for the induction of resistance against tobamoviruses, *Bougainvillea spectabilis* extract was found to be most effective. The inducer-treated seeds also showed enhancement of seed germination and seedling vigor.

The antiviral benzimidazoles and their derivatives, ribavirin and plant hormones, comparing results obtained in animals and plants were appropriate. It also shows the symptom suppression in plants by benzimidazole derivatives and of the search for endogenous antiviral mechanisms in plants.

9.2. Management of Phytoplasma Diseases

Several approaches were recommended for the management of phytoplasma diseases including their insect vectors. However, unfortunately, not even a single effective control measure has been identified so far to completely control the phytoplasma infection. Control of insect vectors through pesticides is a plausible way to limit the spreading of phytoplasma diseases. However, complete elimination of vectors is not achievable despite heavy dosage application of chemicals (Firrao et al., 2007).

Use of clean propagating material which is infection free and vector free should be used. A novel and simple method of elimination of phytoplasma and insect eggs from grapevine cane is by treatment by heat (hot water treatment).

Rouging of symptomatic disease plant helps to reduce the disease inoculum and thereby the spread and intensity of the disease in the field.The severity of brinjal little leaf disease was reduced by rouging symptomatic plants and spraying insecticides (Sohi et al., 1974).

Spraying of diseased plants with antibiotic tetracycline is another way to revive the infected plants as a temporary management of phytoplasma diseases. The effect of tetracycline in suppressing the phytoplasma disease symptoms are temporary, and the symptoms reappear after some weeks/months or when exposed to insect vectors again. For example the effect of tetracycline was reported to be only a temporary remission of symptoms in brinjal infected with little leaf disease and could not eliminate the pathogen completely in the host plant (Raychaudhuri et al., 1970). Moreover, flowers and fruits were not observed in any of the brinjal cultivars treated with antibiotics (Upadhyay, 2016). Furthermore, antibiotics are too costly and their application is prohibited in several countries.

Vegetation composition surrounding a field/orchard/vineyard has a profound effect on the presence and dispersal of phytoplasma vector, therefore habitat management is important. Managing the habitat by application of various organic and synthetic mulches is important. Synthetis mulches such as plastic sheets can physically prevent the movement of vector into the soil. The use of reflective mulches for repelling the vector from the plant to avoid the transmission of phytoplasma is adopted. Control of aster yellow phytoplasma on carrot using aluminium foil mulch is an example of mulch use in management of phytoplasmas.

Barrier sprays is another method where kaoline is used. Kaoline is a non-abrasive, fine grained aluminosilicate mineral applied as a particle film. Kaoline can act in a couple of ways i.e by killing an insect by suffocating it or coating the plant and obstructing feeding and oviposition sites. The efficacy of kaoline is greatly hindered by water, in dry areas it may be very effective in controlling pests, but in areas with overhead irrigation or heavy rains it is washed away.

Varma et al. (1975) demonstrated that spraying gibberellic acid on infected brinjal plants induced symptom recovery and an increasing recovery rate was reported with gibberellic acid treatment followed by ledermycin.

Several bio-agents are also utilized for plant disease management. Treatment of tomato plants infected with "stolbur" phytoplasma with arbuscular mycorrhizal (AM) fungi led to reduced symptom expression and degeneration of phytoplasma cells (Lingua et al., 2002).

The most common vectors of phytoplasmas i.e. leafhopper and planthoppers are attacked by a range of predators. Spiders are very important predators of both adult and nymphs, especially in grassland ecosystem while miridae (Hemiptera) may be significant egg predator..

Genetic/molecular manipulation of plant species also helps in the management of vectors and phytoplasma diseases. One form of manipulation involves the modification of plant lectins in host plants, which affect various physiological functions of vector including blocking the absorption of free amino acids and sugars. Two plant lectins showed efficacy in phytoplasma vectors. The snowdrop lectin (Galanthus nivalis agglutinin, GNA) and Allium sativum leaf lectin (ASAL) have been shown to cause mortality in phytoplasma vector in rice.

Systemic acquired resistance (SAR) is a plant defence mechanism that is activated by certain chemicals against a particular pest / disease pathogen.

The potential of benzothiadiazole (BTH) to elicit SAR reaction and protect the model plant Arabidopsis thaliana from phytoplasma transmitted by Colladonus montanus has been demonstrated in laboratory.Treatment of plants with a host defense inducer salicylic acid on potato plant infected with the purple top disease found a significant reduction in their symptom expression. Wu et al. (2012) showed that pre-treatment with two applications of salicylic acid 2 and 4 days before phytoplasma inoculation by insects, significantly reduced purple top disease symptoms in tomato plants. Pre-treatment of salicylic acid also caused the upregulation of the expression of three defense-related genes which included LeWRKY1, LeMPK3 and LePRP1, 3 days after phytoplasma inoculation.

Developing resistant cultivars to either phytoplasmas or their insect vectors would be a long-lasting tool for the control of phytoplasma diseases. Limited work was accomplished on the development of resistant genotypes in vegetable crops. Chakrabarti and Choudhury (1975) reported that the wild relatives of brinjal, *Solanum integrifolium* and *S. gilo* were resistant to the little leaf disease.

Other management strategies such as rouging of infected plants, adjustment of date of sowing, use of clean propagating material, rotation with non-host crops, and removal of weeds coupled with vector control are effective methods for the containment of phytoplasma-associated diseases. Hence, an integrated approach may be the most viable and sustainable option by integrating components of cultural, physical, biological, resistance and chemical applications.

Phytoplasmas are normally controlled by the breeding and planting of disease-resistant crop varieties (perhaps the most economically viable option) and by the control of insect vectors.

Tissue culture can be used to produce healthy clones of phytoplasma-infected plants. Cryotherapy (i.e., the freezing of plant samples in liquid nitrogen) prior to tissue culture increases the probability of producing healthy plants in this manner.

Plantibodies targeting phytoplasmas have also been developed but its use is not yet commercialised.

10

Chemical Molecules as Nematicides for Nematode Management

10.1. Plant Parasitic Nematodes

Nematodes are multicellular, unsegmented, bilaterally symmetrical roundworms belonging to pseudocoelomates class of creatures. Many of them are parasites of plants. Different crops species are susceptible to several species of plant parasitic nematodes which limits the yield and quality of the food harvest. Root-knot nematodes (*Meloidogyne* spp.), foliar nematodes (*Aphelenchoides* spp.), root-lesion nematodes (*Pratylenchus* spp.) reniform nematodes (*Rotylenchulus* spp.) and other ectoparasitic nematodes are the important nematodes species to infect the crop plants.

These nematodes live in the soil and feed on the root system and damage the root system thoroughly. Besides damaging directly, plant parasitic nematodes also can enhance damage caused by other soil borne fungal and bacterial pathogens. These can also transmit the plant pathogens in the soil.

10.1.1. Symptoms of Nematode Infection and Nature of Damage

The nematode infested crop plant exhibit the symptoms on above ground plant parts of various intensities which includes of stunting, chlorosis (yellowing) and tend to wilt under dry conditions. Nematodes infestation are usually first detected in localized areas within a field. Gradually, the area of infected plants expands in size and the entire planting can eventually be affected. Infection can reduce crop yeild and quality. Different types of nematodes infects different crop plant with different types of symptoms.

The symptoms of root-knot nematodes include formation of distinctive swelling called root galls (root knots) on the roots of affected plants. Root-knot galls may vary in size and shape. On heavily infected plants, galls tend to fuse together so that large areas or entire root may be swollen. Root injuries from other nematodes include root necrosis resulting in severe root damage and subsequent dwarfing of plants. Fibrous or feeder roots are mostly attacked which may reduce the absorption ability of roots and other physiological functions of the plant. Root growth slows and secondary root development is limited.

Some nematode parasites on young foliage and flower stalk (Foliar nematode, *Aphelenchoides* spp. in tuberose). The infected flower stalk initially appears rough, stalk become crinkled, stunted and finally distorted and in severe cases flower buds failed to bloom. Brown streaks appear on the leaf bracts and petals and subsequently develop rusty brown spots. The severely infected flower stalk becomes rotten and brittle over drying. The number of flower stalk is also reduced and small crinkled and distorted flowers are produced which are not acceptable in the market.

10.1.2. How to Diagnose the Nematode Infection in Field?

The symptoms and damage on crop plants due to nematodes infection has often been confused with nutritional deficiencies problem and occasionally due to fungi or bacterial infection. Characteristic foliage symptoms such as yellowing of leaves, stunted growth, patches of poor crop growth in field and wilting of such plants in the day sunshine and recovering during night are some of the clues of nematode infection in the crop plants. If such plants are uprooted, they show characteristic below ground symptoms like galling and/ or necrotic on roots and reduced root length. Diagnosing the damage on flower crops can best be done by periodic field observations and examinations of roots in conjunction with testing the soil and plant sample for nematode extraction.

10.2. Management of Nematode Diseases

Several management practices can be used to combat nematode infection. These include cultural, resistant varieties, crop rotations, biological controls and application of nematicides. Damage by plant parasitic nematodes can be minimized by adequate water and fertilizer. Plant parasitic nematodes reduce the plant root system's ability to take up water and nutrients, especially when nematode population density is high at planting. Adequate water and fertilizer does not reduce the nematode density but reduces plant stress and thus the symptoms of nematode damage. Anything that moves soil can spread plant parasitic nematodes within fields and to other fields. The effective management of plant-parasitic nematodes has relied upon the application of chemical nematicides as a short-term control means by suppressing nematode population densities in soil to levels below known economic damage thresholds. The reduction of nematode population densities to below threshold levels will reduce their damage to roots and increase yield in infested fields. Chemical nematicides have been used singly or in combination with other nematode control practices since the late 19th century.

Nematicides are generally soil fumigants in chemicals forms that, when injected in to the soil, emit toxic fumes that penetrate air spaces in soil in sufficient concentration to kill nematodes. These must be sealed into the soil with water or a plastic tarp to ensure that a lethal concentration and exposure time are reached. Because fumigants are harmful to all living plants, a period of 2 weeks to 2 months must be allowed between soil treatment and planting of crop in order to avoid crop damage. Several non-fumigant nematicides are also available for several vegetable crops.

The use of nematicides is globally valued at about $1 billion annually, with 48% of this market being spent for the control of root-knot nematodes.

10.2.1. Types of Nematicides

Nematicides are chemically synthesized products that kill nematodes or adversely affect nematodes. These products can have nematicidal or nemastatic effects against nematodes. Nematicidal compounds are highly toxic and kill exposed nematodes, whereas nemastatic compounds do not kill nematodes but impede nematode movement toward host plant roots or delay nematode egg hatching. Nematicides are grouped into fumigant and non-fumigant nematicides based on their volatility in soil.

10.2.1.1. Fumigant Nematicides

Fumigant nematicides are toxic compounds with narrow- to broad-spectrum effects. In addition to killing plant-parasitic nematodes, they may also be effective against numerous types of soil borne pests and pathogens. Fumigants are used to disinfest soil and reduce the risk of yield loss caused by soil borne pests in high-value crops such as vegetables. When applied to the soil, fumigant compounds reach target organisms in the form of gas that moves through the open spaces between soil particles or by dissolving into the film of water that surrounds soil particles. Treating soil with fumigant nematicides can effectively control plant-parasitic nematodes over a range of soil types, but they are generally more effective in coarse-textured (sandy) soils than in fine-textured (clay) soils. The gaseous dispersal of fumigants may also be constrained in soils with water-filled pore space. Additionally, the volatility of fumigants is affected by soil temperature. Volatility is higher in warm soils and lower in cool soils. For example, Telone II (1, 3-dichloropropene) has promising nematicidal efficacy when injected into soil at soil temperatures ranging from 4.4 to 26.6 ^{0}C. Additionally adequate soil moisture is necessary for proper movement of the gases in the soil. Excessive soil moisture, therefore, will reduce the efficacy of the fumigant nematicides. For decades, soil fumigation has played a crucial role in controlling nematodes in the production of vegetable crops. Prior to 2005, chemical control options for soilborne nematodes, pathogens, and weeds were rather limited to methyl bromide, which provided a high level of control of various pathogens and pests. Following the methyl bromide phaseout initiated in January 2005 under the provisions of the U.S. Clean Air Act and the Montreal Protocol due to its ozone depletion effects, attention has focused on the application of other chemistries for the short-term management of nematodes. Managing the root-knot nematode (Meloidogyne spp.) is very important to vegetable production. The root-knot nematode is the most critical yield-limiting nematode pathogen to multiple vegetable crops and they are widely distributed in most of the vegetable-growing regions in the united state. More than 50% of all vegetable fields have at least one Meloidogyne species above the known damage threshold level. Sandy soils, which are common in southern Georgia, favor the establishment and development of root-knot nematodes, but they also are well aerated and drained, which increases the efficacy of soil fumigants. In fact, preplant soil fumigation of infested soil is the dominant approach for managing root-knot nematodes in vegetable crops. Fumigant nematicides are applied to the soil with fumigation equipment that injects the material into soil and covers the soil with a polyethylene mulch as a single operation. Transplanting vegetable seedlings into treated beds usually occurs two to three weeks after fumigation, depending on the fumigant type and rate, soil moisture, and soil temperature, when nematodes have been killed and all residues of fumigant nematicides have adequately dispersed. Chemical fumigants are highly effective in reducing nematode populations in soil. However, major drawbacks of these compounds are that they are difficult to apply, have a long re-entry period compared with non-fumigants, require buffer zones, and are rather expensive. It is also important to know that soil fumigation will not completely kill all plant-parasitic nematodes when applied inaccurately. The aim of soil treatments with fumigants is to reduce population densities of nematodes to allow for successful plant establishment and growth. These generally are systemic compounds that also may provide good insect control.A number of different factors affect the performance of these products, including soil temperature, soil moisture, soil tilth, organic matter, soil type, and time of application.Consult the product label for specific details on safe handling and application methods. Several fumigant nematicides (table 14) as mentioned below are available for use against plant-parasitic nematodes. The most common fumigants used to control nematodes in vegetable crops are Telone II, chloropicrin, Vapam (metam sodium), and Paladin (dimethyl disulfide or DMDS)

Table 14: Fumigant Nematicide available in the market

Trade name	Active Ingredient	Nature of activity	Manufacturer
1. Telone II	1,3,dichloropropene (1,3-D)	Nematicide	Corteva Agriscience, Dow DuPont
2. Chloe-O-pic	96.5-99%Chloropicrin	Nematicide/Fungicide	Corteva Agriscience,Dow DuPont
3. Telon C-17	73% 1,3-D; 17% Chloropicrin	Nematicide/Fungicide	Corteva Agriscience,Dow DuPont
4. Telon C-35	65% 1,3-D; 35% Chloropicrin	Nematicide/Fungicide	Corteva Agriscience,Dow DuPont
5. Telon EC	1,3-D	Nematicide	Corteva Agriscience,Dow DuPont
6. In Line	61% 1,3-D; 33% Chloropicrin	Nematicide/Fungicide	Corteva Agriscience,Dow DuPont
7. Pic clor-60	40% 1,3-D; 60% Chloropicrin	Nematicide/Fungicide	Corteva Agriscience,Dow DuPont
8. Vapam	Metam sodium	Broad Spectrum	AMVAC Chemical Corporation
9. K-Pam	Metam potassium	Broad Spectrum	AMVAC Chemical Corporation
10. Paladin	Dimethyl disulfide	Broad Spectrum	Arkema Inc
11. Dominus	Allyl isothiocyanate	Broad Spectrum	Isagro ,USA

The details about the important fumigant nematicidal products are given below.

a. Telone II

The nematicidal properties of Telone II were discovered in 1943, but its commercial application to control root diseases caused by nematodes increased much later, when other fumigants such as ethylene dibromide and methyl bromide were taken off the market. Indeed, Telone II became more popular for nematode control after methyl bromide began to be phased out in 2005 and vegetable producers had to apply alternatives. Telone II has been demonstrated nationally and globally to increase crop yields in nematode-infested fields at ranges relatively comparable to methyl bromide. Currently, it is the most commonly used fumigant in plasticulture and bare ground production systems in the Southeastern U.S. With the limited use of Telone II in the near future due to potential environmental concerns, there will be few alternatives available to control nematodes of economic importance.

b. Chloropicrin

Chloropicrin has been used as preplant soil fumigant for control of nematodes and soilborne diseases in various crops since 1919 in the U.S. It is manufactured in liquid form with a moderate vapor pressure and it disperses quickly through soil when applied below the soil surface under a tarp. The nematicidal effect of chloropicrin is not the primary reason for its application, as it also has fungicidal, insecticidal, and herbicidal activity, making it a broad-spectrum biocide. Chloropicrin is often used in combination with 1, 3-D to fumigate soil, and formulations of chloropicrin that include 1,3-D (e.g., Telone C-17 and Telone C-35) increase the nematicidal effect of chloropicrin. Such mixtures control both nematodes and other pathogens better than either product alone.

c. Vapam

Vapam (a methyl isothiocyanate generator) is a liquid product that has a slow nematicidal effect compared with Telone II. This product can be applied by drenching, spraying, or injection procedures. Vapam lacks consistent performance in the control of nematodes, therefore growers do not extensively use it as a single management option. However, research suggests that effective control of root-knot nematodes can be achieved when Vapam is combined with non-fumigant nematicides. This nematicide also has some toxicity effect on weeds at higher rates.

d. Paladin

Paladin has been recently marketed in the U.S. as a soil fumigant to control nematodes, weeds, and soilborne plant pathogens. As an emulsified compound, Paladin may be broadcasted, drenched, or drip-applied prior to planting for control of soilborne pests in vegetables crops including tomatoes, cucurbits, squash, peppers, eggplants, and melons.

The Other fumigant nematicide available (table.15) for the management of nematode diseases are:

Table 15: Other fumigant nematicides

Chemical	Application	Waiting period for planting	Crops	Comments
Methyl-bromide	Preplant,Tarped or mulched for 24 to 48 hrs	10 to 14 days	Most vegetables	Formulation with 98% methylbromide and 2% chloropicrin are appropriate for nematode control
SMDC (Sodium methyl-dithiocarbamate)	Preplant,Tarped. Dont enter within 48 hrs	14 to 21 days after treatment	General use fumigant	More effective when used with considerable water

10.2.1.2. Non-fumigant Nematicides

Non-fumigant nematicides are nonvolatile toxic chemicals that can be applied prior to planting, at planting, or after planting through soil drenching, drip irrigation, or spraying onto the crop foliage to reduce population densities of nematodes and protect crops from damage. These products are grouped into two categories: contact (killing nematodes in soil by direct exposure) or systemic (killing nematodes while they feed from plant roots) nematicides. When applied to the soil, non-fumigant compounds are dispersed by movement in soil water. In contrast to fumigant nematicides, the efficacy of non-fumigants does not depend on soil temperature. A major drawback of these compounds is that their efficacy is generally less than that of fumigants such as Telone II for nematode control. Quick leaching of the active ingredients of non-fumigants, particularly in sandy soils, may also reduce their effectiveness. The removal of many older non-fumigant nematicides from the market due to toxicity and environmental concerns has encouraged the development of a new generation of chemical compounds that address these concerns and offer effective nematode management.

A non-fumigant nematicides (table.16) as mentioned below are commercially available for use (in vegetable crops).

Table. 16: Non Fumigant nematicides.

Trade Name	Active Ingredient	Nature of activity	Manufacturer
1. Vydate	Oxamyl	Nematicide/insecticide	Corteva Agriscience, Dow DuPont
2. Nimitz	Fluensulfone	Nematicide	Adama
3. Velum prime	Fluopyram	Nematicide/Fungicide	Bayer Crop Science
4. Mocap 15 G	Ethoprop	Nematicide/Insecticide	AMVAC Chemical Corporation
5. Mocap EC	Ethoprop	Nematicide/Insecticide	AMVAC Chemical Corporation
6. Movento	Spirotetramat	Nematicide/Insecticide	Bayer Crop Science
7. Counter 20 G	Terbufos	Nematicide/Insecticide	AMVAC Chemical Corporation
8. Salibro	Auazaindolizine	Nematicide	Corteva Agriscience, Dow DuPont

The details of non fumigant nematicides are given below.

a. Vydate

Vydate® is a carbamate (a pesticide derived from carbamic acid) that has nematicidal and insecticidal properties against plant-parasitic nematodes and soil insects. Vydate is a systemic nematicide (any compound that can be taken up by the plant through roots, leaves or both, and is effective against nematodes inside the plant), and is used in liquid form, as the granular form is no longer labeled in the U.S. due to concerns regarding its consumption by birds. This nematicide may be applied before planting, at planting, and after planting through soil drench, in-furrow, drip injection, broadcast, or foliar spray. Vydate is effective against nematodes in carrots, cucumbers, tomatoes, eggplants, squash, peppers, watermelon, and cantaloupes.

b. Nimitz

Nimitz® is a systemic fluoroalkenyle compound that recently received U.S. Environmental Protection Agency (EPA) registration for use in vegetable crops. In vegetable production systems, Nimitz has received much interest as a methyl bromide alternative in part because it causes mortality of target nematodes within 24 to 48 hours of product application. This nematicide controls several types of plant-parasitic nematodes (e.g., root-knot, stubby root, sting, lesion, and needle nematodes) resulting in improved fruit quality and yield of several vegetable crops, including cucumbers, eggplants, tomatoes, and peppers. Nimitz can be broadcasted or applied through drip irrigation seven days before seeding or transplanting vegetables. Nimitz has shown the greatest efficacy in the control of plant-parasitic nematodes when used preplant. When applied on plant foliage, the active ingredient of Nimitz moves from the application point down into the roots, where it might affect parasitic nematodes. To prevent phytotoxicity and subsequent crop losses, foliar application of Nimitz is not recommended.

c. Velum Prime

Velum Prime® is manufactured as a liquid formulation and is one of the few available non-fumigant nematicides with systemic properties labeled for control of multiple species of plant-parasitic nematodes in the Southern U.S. This nematicide can be broadcasted or applied through soil drench and drip irrigation at planting of vegetables. Velum Prime is registered for vegetable crops including potatoes and sweet potatoes; cucurbit vegetables such as cucumbers, pickling cucumbers, squash, watermelons, and cantaloupes; fruiting vegetables such as tomatoes, eggplants, okra, and peppers; and brassica vegetables such as cabbage and broccoli.

d. Mocap

Mocap® is a granular (15G) or liquid (EC) systemic insecticide/nematicide containing ethoprophos, which is used to reduce plant-parasitic nematode damage in some vegetable crops, including potatoes, sweet potatoes, cucumbers, corn (sweet and field), and cabbage. The nematicidal activity of Mocap can be achieved before or at planting by band or broadcast application.

e. Movento

Movento® is currently registered as a liquid formulation with potential to control both plant-parasitic nematodes and insect pests. It can be applied as a soil or plant foliage spray or through chemigation. Once applied on the plant foliage, the systemic activity of the product will allow it to penetrate the leaf surface and translocate down within plant tissues toward the roots, where most parasitic species of nematodes feed. This nematicide has shown some beneficial activity when used against root-knot nematodes in carrots and sweet potatoes.

f. Counter

Counter® 20G is a granular systemic insecticide and nematicide that is registered for the control of plant-parasitic nematodes in sweet corn. This nematicide can be applied in-furrow or banded at planting using the Lock'n Load handling system because of its high toxicity.

g. Fluazaindolizine

Fluazaindolizine is a new effective and selective nematicide for the control of nematodes, but it has not yet been registered for use in the U.S. Over several years, fluazaindolizine has been tested in the U.S. in field trials against nematodes including root-knot nematodes, and it has shown good nematicidal effect when used at planting. If registered, this chemical could be a promising product to control nematodes in vegetable crops, and because it does not affect beneficial organisms in the soil, it could be considered for integrated pest management programs of specialty crops.

10.3. Biological Control of Plant Parasitic Nematodes

The microorganisms that parasitize or prey on nematodes or reduce nematode populations by their antagonistic effect or behaviour, the fungi hold important positions and some of them have shown great potential as biocontrol agents (table 17). Fungi continuously destroy nematodes in virtually all soils because of their constant association with nematodes in the rhizosphere. A large number of fungi are known to trap or prey on nematodes but the most important genera include *Paecilomyces, Verticillium, Hirsutella, Nematophthora, Arthrobotrys, Drechmeria, Fusarium* and *Monacrosporium*.

10.3.1. Table 17: Fungal species used in the management of nematodes

Type of Fungal agent	Mode of action	Advantages and Limitations
Facultative parasites Trapping fungi	Traps produced on modified mycelium give rise to infective trophic hyphae	***Advantages:*** easily produced *in vitro;* some species rhizosphere competent; wide host range. **Limitations:** difficult to ensure trapping activity coincides with activity of infective stages of cyst and root-knot nematodes; often do not produce resting structures and so may be difficult to formulate.
Paecilomyces lilacinus	*Hyphal penetration*	***Advantages:*** easily produced *in vitro;* rhizosphere competent; attacks the eggs of several nematode species; treatment of planting material (e.g. seed tubers) can be effective. ***Limitations:*** requires high soil temperatures; has given variable control in range of conditions; large numbers of propagules (10^6/g soil) required for nematode control; some isolates are pathogenic to humans.

Type of Fungal agent	Mode of action	Advantages and Limitations
Verticillium chlamydosporium	*Hyphal penetration*	***Advantages:*** easily produced *in vitro;* some isolates rhizosphere competent, and virulent (10^3 propagules/g soil required for nematode control); resistant resting spores produced; survives throughout growing season in soil. ***Limitations:*** seed treatments ineffective; efficacy dependent on nematode species, density and plant host.
Obligate parasites *Pasteuria* spp.	Adhesive spores	***Advantages:*** most isolates highly virulent; infective spores resistant to drying; good shelf-life; reduce infectivity of nematodes as well as fecundity. ***Limitations:*** very difficult to culture in *vitro;* isolates are highly specific; no proliferation in soil in absence of nematodes.
Hirsutella spp.	Adhesive spores	***Advantages:*** relatively easy to culture in *vitro;* attack infective nematodes in soil. ***Limitations:*** poor saprophytic competitor; limited spread in soil.
Rhizosphere bacteria	Toxins or modification of root exudates	***Advantages:*** easy to culture *in vitro;* can be applied as seed treatments; reduce plant damage. ***Limitations:*** effective for a relatively short period; activity affected by crop cultivar and nematode species; little effect on nematode multiplication.
Endophytic fungi (non-pathogenic root-infecting fungi and mycorrhizae)	Competition in roots and modification of root exudates	***Advantages:***include agents with potential to control migratory endoparasitic nematodes in roots; may improve plant growth even in absence of nematodes; reduce damage caused by wide range of nematodes and limit their multiplication; can be mass produced and formulated; could be applied to seeds or transplant material; may reduce fungal root rots. ***Limitations***: non-mycorrhizal fungi may be difficult to register as closely related to plant pathogens; efficacy affected by plant cultivar and other crops in rotation such as crucifers can reduce survival.

Sources: Brown and Kerry, 1987; Stirling and West, 1991; Sikora, 1992.

Similarly some of the bacterial species (table.18) gives promising control of plant parasitic nematodes.

10.3.2. Table.18: Bacterial groups with pathogenic activity against nematodes

Nematophagous bacterial group	Genus and species	Target nematodes	Pathogenic effects on nematodes	Action mode
Parasitic bacteria	**Four species:** Pasteuria. Penetrans, P.thornei; P.nishizawae; Candidatus.	323 nematode species of 116 genera.	Major economic important plant-parasitic nematodes have been observed to be parasitized by Pasteuria	Parasitism
Opportunistic parasitic bacteria	Bacillus nematocida (Bacillus sp. B16); Brevibacillus laterosporus. Bacillus sp. RH219 etc.	Panagrellus redivius and Bursaphelenchus xylophilus.	Br. laterosporus strain G4 could penetrate the nematode (P. redivius and B. xylophilus) cuticles and eventually digest the target organism in the laboratory.	Parasitism, production of enzymes and toxin.
Rhizobacteria	Distribution in more than 29 genera. Bacillus (more than 15 species) and Pseudomonas (more than 11 species) are two of the most dominant populations.	Reduce nematode populations in soil	Different rhizobacteria showed different degrees of suppression on nematodes in various conditions. Three commercial bionematicides from bacteria all belong to this group.	Interfering with recognition, production of toxin, nutrient competition, plant-growth promotion; induction of systemic resistance.
Parasporal crystal-forming bacteria	Bacillus thuringiensis (Cry5, Cry6, Cry12, Cry13, Cry14, Cry21).	Trichostrongylus colubriformis; Caenorhabditis elegans; and Nippostrongylus brasiliensis.	These Cry proteins showed toxicity to larval stages of free-living and parasitic nematodes.	Cry proteins caused damage to the intestines of nematodes.
Endophytic bacteria	The majority of rhizobacteria can also be identified as endophytic bacteria.	Root-knot nematode and root-lesion nematode etc.	Suppress root-knot nematodes and root-lesion nematode etc.	Rhizobacteria and endophytic bacteria use some of the same mechanisms.
Symbiontic bacteria of entomopathogenic nematodes	Two genera: Xenorhabdus. and Photorhabdus	Bursaphelenchus xylophilus; M. incognita and their eggs	Toxic to juveniles of root-knot and pine-wood nematodes and inhibit egg hatch.	Toxin production (ammonia, indole and stilbene derivative).

10.3.3. Advantage of using Bio-control Agents

- Cost effective, reusable, farmer & eco-friendly products
- Inherently harmless in comparison with chemical pesticides
- As a component of IPM, can greatly decrease the use of chemical pesticides
- Increase the yield of the crop & lead to sustainable productivity in a long run
- They have specific mode of action on pathogens
- Helps in developing disease suppressive soils

10.3.4. Enhancement of Biological Suppressiveness

- Incorporation of crop residues in the soil
- Addition of organic amendment such as manures and compost
    - Stimulate natural antagonist/ natural enemies of plant parasitic nematodes
    - It improves soil nutrients, soil physical conditions, and crop health

11

Chemical Molecules for Management of Algal Infection

The algae associated with the plants belongs to the Order *Trentepohliales* (*Ulvophyceae*, Chlorophyta). These are filamentous green algae parasitic to higher plants, generally causing minor diseases characterized by negligible leaf spots, but on certain crops, in moist environments, these algal diseases can cause economic injury to plant leaves, fruits, and stems (Nelson, 2008).

11.1. Algal Species Involved in Plant Infection

The species of the algal genus Cephaleuros particularly *Cephaleuros virescens* and *Cephaleuros parasiticus* are the main cause of algal diseases of crop plants. Besides the genus *Cephaleuros*, the algal genus *Phycopeltis, Printzina*, and *Trentepohlia* have also been associated with the plant habitat (Brook, et.al, 2015).

C. virescens is the most frequently reported algal pathogen of higher plants worldwide and has the broadest host range among Cephaleuros species. Frequent rains and warm weather are favorable conditions for these pathogens. Poor plant nutrition, poor soil drainage, and stagnant air are predisposing factors for the infection by the algae in the host crop.

11.2. Symptoms of Algal Infections

The symptoms and crop damage can vary greatly depending on the combination of *Cephaleuros* species, hosts and environments. Spots caused by *C. parasiticus* are a top-down, intercellular, full-thickness necrosis, destroying both upper and lower epidermal cell layers and all intervening tissues. Then, a bright but relatively sparse orange algal bloom bursts forth from the undersides of leaf lesions, rather than from the upper side of leaves as is usually the case with *C. virescens*, the more commonly occurring species in the genus.

The distinctive and much more common spots caused by *C. virescens* on many hosts appear on the upper leaf surface as raised yet flattened, textured, burnt-orange to brown or rust-colored, circular areas up to about 2 cm in diameter, having indistinct, filamentous margins and fuzzy surface topographies. Yet, this species is generally a relatively harmless, subcuticular leaf parasite.

A number of different algal species on a leaf is common in shaded, moist environments. Most of these algae are not parasitic on plants and can be easily rub off with fingers, but their presence on the leaf and plant surface hinder the process of photosynthesis and plant respiration, weakening the plant host and decrease the yield potential and quality of the plant produce.

11.3. Important Hosts Crops

Species of *Cephaleuros* are very common on the leaves, twigs and trunks of economically important tropical trees, shrubs and crop plants such as **tea** (*Camellia sinensis*), **kava** (*Piper methysticum*), **pepper** (*Piper nigrum*), **magnolia** (*Magnolia grandiflora*), **coffee** (*Coffea arabica*), **oil palm** (*Elaeis guineensis*), **avocado** (*Persea americana*), **vanilla** (*Vanilla planifolia*), **mango** (*Mangifera indica*), **breadfruit** (Artocarpus altilis), **guava** (*Psidium guajava*), **coconut** (*Cocos nucifera*), **cacao** (*Theobroma cacao*), as well some **citrus** (*Citrus* spp.) cultivars and several forest tree species.

11.4. Crop Damage

Generally, on most hosts, leaf infection is of little direct economic importance and is confined to low-hanging branches near ground level. However, where leaves are the main harvest from the crop, e.g. tea leaves from tea plantation crop, the leaf infection due to algal pathogen deteriorate the quality and yield of the produce with severe economic losses. On more susceptible hosts such as guava there can be significant tissue necrosis, extensive defoliation, economic injury to fruit quality, crop loss, and reduced photosynthesis and plant vigor.

We may separate *Cephaleuros* species and the damage they cause into two groups on the basis of their mode of parasitism: subcuticular vs. intercellular. *C. virescens* is a subcuticular parasite, while *C. parasiticus* is an intercellular parasite. The intercellular mode of parasitism permits more tissue damage than subcuticular parasitism. Symptoms include rough circular greenish-grey or rusty-red areas and shoot dieback. The damage to guava fruits is limited to the skin of the fruit and there is no soft rot. The fruit spots are slightly sunken and cracked with a dark, somewhat corky aspect. When severe, the fruits appear scabbed. A recent survey of *Cephaleuros* spp. host range in American Samoa found it to be extremely broad, affecting plants in dozens of families.

11.5. Management of *Cephaleuros* Diseases

When the disease is mild, remove and destroy leaves with spots as well as diseased branches. Also, rake up and destroy affected leaves that have fallen to the ground. Since algal leaf spot is most often a problem for plants that are under stress, an important step in management is to improve growing conditions. Make sure that plants are receiving the recommended amounts of sunlight, water, and fertilizer. If needed, improve air circulation around affected plants by removing or thinning out nearby vegetation. Keep weeds around affected plants under control (this will reduce relative humidity in the plant canopy and reduce plant stress); prevent climbing and vine-like weeds from establishing on the host plant. Try to select a more tolerant variety (for example, of guava) where the crop is grown in a wet environment. For highly susceptible crops, select a planting location with moderate rainfall, well drained soil and good air circulation. Manage plant spacing and thinning to improve aeration and light exposure. Avoid overhead irrigation whenever possible.

When algal leaf spot is severe, copper fungicide (Table 19) sprays are recommended. Sprays may need to be applied every 2 weeks, if wet conditions persist. However, when the affected plant is too tall to apply sprays safely, treatment may not be feasible for the homeowner and may require professional assistance.

Table 19: Copper Fungicides for Algal disease Control.

Fungicide Active Ingredient	Examples of Brand Names & Products
1.Copper Ammonium Complex	Monterey Liqui-Cop Fungicide Concentrate; & RTS
	Southern Ag Liquid Copper Fungicide Concentrate
2.Copper Sulfate	Bonide Copper Fungicide (wettable powder)
3.Copper Salts of Fatty Acids	Bonide Liquid Copper Fungicide Concentrate
	Camelot O Fungicide/ Bactericide Concentrate
	Natural Guard Copper Soap Fungicide Concentrate
	Monterey Liquid Copper Fungicide RTU
	Espoma Organic Copper Soap RTU

RTS = Ready to Spray (a hose-end sprayer bottle); RTU = Ready to Use (a pre-mixed sprayer bottle).

Please note that some copper fungicides may injure plants if applied incorrectly. Always read the entire label and follow all directions when applying a pesticide.

Use chemicals only according to directions on the label; the Bordeaux mixture is commonly recommended chemical preparation for control of algal infection.

12

Management of Protozoan Diseases

Protozoan parasites are Trypanosomatids of the class Kinetoplastida predominately restricted to invertebrate hosts (i.e. possess a monoxenous life-cycle). However, several genera are pathogenic to humans, animals and plants, and have an invertebrate vector that facilitates their transmission (i.e. possess a dixenous life-cycle).

Phytomonas is one such dixenous genus of protozoan parasite that includes plant pathogens of different hosts transmitted by phytophagous insects. Over 100 years after trypanosomatids were first discovered in plant tissues, *Phytomonas* parasites have now been isolated across the globe from members of 24 different plant families, however; to date only two species of Phytomonas are definitively known to cause plant disease.

12.1. The Genus *Phytomonas*

Phytomonas, as members of the family Trypanosomatidae, have structures that are characteristic of the family, including the flagellum-associated kinetoplast, subpellicular microtubules, the paraxial rod, and glycosomes (Souza, 1991). Within a host plant, *Phytomonas* exhibits a fusiform structure twisted 2-5 times along the longitudinal axis (Dollet, 1984). Within the plant, the organisms can be in several flagellated stages: mostly promastigote with some paramastigotes in the phloem and lacticiferous tubes, and amastigote form in the latex (Souza, 1991). However, most of the species are mainly observed in the promastigote form, with an elongated body and a single 10-15 μm long flagellum emerging from the anterior flagellar pocket (Souza, 1991).

This main form of their morphology is said to be consistent with insect parasites in their sister group *Leishmania (Jaskowska et.al, 2015)*. Species range from 10 to 20 μm in length and have widths close to 1.5 μm (Souza, 1991). It appears that reproduction of cells occurs only in the promastigote stage (Dollet, 1984). When viewed as live samples under the light microscope, *Phytomonas* can be seen to be incredibly active (Dollet, 1984).Different species that infect different host plants have differences in external morphology, for example, oil palm (*Elaeis guineensis*)-infecting *Phytomonas* exhibit fewer twists than those in coconut (*Cocos nucifera*).

Like other trypanosomatids, the cell surface of *Phytomonas* species can be divided into the three regions i.e. body surface, flagellar surface, and the flagellar pocket. Also like other trypanosomatids, *Phytomonas* does not have a cell wall, but instead are protected from host responses and environmental conditions by membrane-anchored proteins and glycoinositol phospholipids (Jaskowska et.al, 2015). The pellicular cell membrane is also lined with microtubules that run along the longitudinal axis of the organism, with a single row of four microtubules in the flagellar pocket. A paraxial rod also runs parallel to the axoneme of the single flagellum on one side, giving the flagella increased thickness, robustness, and strength (Dollet, 1984).

These organisms also have glycosomes, which are specialized peroxisomes. Depending on the species examined, these glycosomes may take the form of two rows separated by filamentous fibres. Some isolates of the genus also have a contractile vacuole located at the anterior end, near the flagellar pocket. This vacuole varies in size, but can be as large as 2 μm. It has been found that the endoplasmic reticulum (ER) of some species form subpellicular sheets that run parallel to the longitudinal axis. Depending on the species and isolates in question, the ER may also have ribosomes in paracrystalline array and incredibly thin intracisternal space. The cytoplasm is considered to be ribosome-rich. In close association with the flagellum is the kinetoplast. The kinetoplast DNA networks of *Phyotomonas* species have been found to vary in isolates extracted from various insect and plant hosts, ranging from a loose appearance to compact networks reminiscent of those of other trypanosomatids (Souza, 1991).

The genus p*hytomonas* contains promastigote trypanosomatid flagellates that are parasites and have a life cycle completed in two hosts, a plant and an insect. In plants, some *Phytomonas* species live in the phloem sieve tubes of non-laticiferous plants, such as coconut and oil palms, red ginger and coffee, and are definitely pathogenic. Others live in the latex-containing cells of laticiferous plants and are not considered to be pathogenic, although one (*P. franτai*) has been associated with the empty root and subsequent decline of cassava. Still other trypanosomatid flagellates, some of them *Phytomonas* species and some that may not even belong to *Phytomonas*, parasitize and cause damage only to the fruit and seed of several plants.

The reproduction in the protozoan occurs during the promastigote stage. Elongation of the kinetoplast occurs first, and is followed by splitting of the anterior end of the cell. Because the single flagellum morphology of trypanosomes arose from the loss of one flagellum in the flagellar pocket, this splitting results in one piece of anterior cell having one flagellum and another that does not. A new flagellum will later grow on the part that lacked a flagellum (Dollet, 1984). Further longitudinal division distributes the nuclei between the forming daughter cells.

Reproduction and multiplication of Phytomonas generally stops after entering the insect host. Within the first week inside the host, the parasites elongate to form "giant" versions of themselves inside the pylorus of the host. After 12 days, cells migrate to the salivary glands via the haemolymph. It is here, in the salivary glands, where multiplication resumes again, producing regular sized forms for infection of plant hosts. Depending on species and isolate, these promastigote forms may or may not be attached to the cell walls of the salivary glands with their flagella (Dollet, 1984), par se, in *Phytomonas serpens*, cells do attach to the interior of salivary glands, which indicates a true developmental stage within the insect host in some species (Jaskowska et.al, 2015).

12.2. Major Protozoan Diseases

The diseases which are presently known of protozoan etiology includes Phloem necrosis of coffee, Hartrot of Coconut palm and Martichez of Oil Palm.

a. Phloem necrosis of Coffee

Phloem necrosis of coffee (*Coffea liberica*) was described by Stahel in 1917 and its association with flagellated *Phytomonas leptovasorum* stahel was established (Vermeulen, 1963). The disease is characterized by deposition of callose in the sieve tubes and necrosis (Dollet, 1984). The insect vector of the parasite has not been confirmed, but Stahel mainly suspected *Lincus spathuliger* as a culprit. Vermeulen (2005) also suspected a Hemipteran insect, due to the presence of flagellate protists in the midgut of certain bugs often found on coffee plant roots. While in the plant host, these parasites inhabit only the phloem, a characteristic that it shares with *P. staheli*.

Despite being first characterized in *C. liberica*, the disease can also be caused by *P. leptovasorum* in *C. arabica, C. excelsa, C. stenophylla*, and *C. abeocutae*. The disease caused by this phytomonad can be either acute or chronic, with the chronic form being much more frequent (95%). The two forms differ in the amount of foliage loss and killing time, but both cases involve the browning and death of the roots. The symptoms of phloem necrosis disease include yellowing, drooping leaves. Advanced infections result in root die-back and can eventually lead to tree fatality. More mature trees appear to be more susceptible to phloem necrosis disease, although younger trees can be infected through root grafts.

b. Hartrot of Coconut

Hartrot of coconut (*Cocus nucifera*) also known as fatal wilt, bronze leaf wilt, lethal yellowing, and Coronie wilt is caused by *Phytomonas staheli* (Jaskowska et.al, 2015). This is an acute wilt, and begins at the leaves of the tree, moving down the plant towards the root. As such, the first symptoms to appear are yellowing of the leaf tips, followed by yellowing of new leaves. As unripe fruits begin to fall prematurely, rotting of the root begins. After ten weeks, the plant dies, and a foul odor arises due to secondary infection by bacteria and fungi. In the beginning, it was believed that the disease was only present in the east coast of South America, but in fact occurred in Colombia and Ecuador as well (Dollet, 1984). The disease is transmitted through its vector Lincus sp (Louise et.al, 1986).

c. Marchitez of Oil palm

Marchitez of oil palm (*Elaeis guineensis) is also caused by Phytomonas staheli*, known as *marchitez sorpresiva* (sudden wilt/ Sudden wither of Palm) is a deadly disease in Central and South America. The pathogen also causes slow wilt disease (Jaskowska et.al, 2015). The protozoa has been found in association with sudden wilt in the phloem of roots, meristem zone, spear base, and stalks of inflorescence. Palms have been attacked from the age of one year. Several insects have been suggested to be the vectors of the protozoa, namely the root miner, *Sagalassa valida,* and the bugs *Myndus crudus* and *Lincus lethifer*. The alternate host plant of the bugs is a grass, *Panicum maximum.*

As in hartrot of coconut, disease symptoms first appear as yellowing of leaf tips. However, in oil palm, the lower leaves are targeted first, with upper leaves being invaded next. Roots also deteriorate, starting from the growing tips and moving toward primary roots. As in coconut hartrot, fruits are also lost. Within 3–5 weeks, the plant dies and leaves turn into dried grey masses. The characteristic symptoms are (1) sudden rotting of all developing bunches, (2) reddish discoloration on top of petioles, and (3) rapid drying of fronds from lower older fronds upwards. The affected palms die in 2 to 3 weeks due to the advanced rotting in the root system.

The results of several papers reported sudden wither to be potentially very serious in oil palm (Giovanni et.al, 2013). In young fields of 2 to 5 years old, 15–20% mortality has been reported. In the same estate, the sudden wilt destroyed two-thirds of the estate when the palms were 12 years old. Therefore, the effect of damage caused by sudden wither on Fresh Fruit bunches yield of the estate is very serious. *Phytomonas staheli* can spread between trees quickly, and can infect up to 30 trees within a few weeks. The disease has great impact across the north of Latin America.

12.3. Economic Significance

The diseases caused by phytomonas particularly hartrot of coconut, wilt of palm and coffee phloem necrosis are problematic in areas of South America where they threaten the economies of developing countries. Both diseases of coconut and palm caused by *Phytomonas staheli are* acute lethal wilts that begin in the leaves and progress to the rotting of the spear and root, although this rotting may be the result of secondary infection by bacteria and other organisms. Phloem necrosis of coffee caused by *Phytomonas leptovasorum* is an important disease in coffee plantation in several countries. A third species, *Phytomonas françai*, inhabits the latex ducts of cassava (*Manihot esculenta*) and has been linked with yield loss diseases known as "*chochamento de raizes*," or empty root syndrome. This disease is characterised by poor root development and chlorosis of the leaves.

However, there are only two proven pathogenic species, and those species pose some economic risk in South America. *P. leptovasorum* (the aetiological agent of coffee phloem necrosis) has been isolated in Suriname (Vermeulen, 2005) and Brazil (Bezerra and Figueiredo, 1982). With the rapid expansion of coffee plantations across South America and the resulting change in the distribution of crop pathogens and pests, coffee phloem necrosis poses a potential risk to these expanding coffee economies. For example, Brazil is the world's largest exporter of green coffee beans trading more than 1,700,000 tonnes in 2011 at a value of 8,000 million dollars (http://faostat.fao.org/). *P. staheli* (the aetiological agent of wilt diseases of oil and coconut palm) has previously been isolated from Ecuador, Suriname, Venezuela, Brazil, Costa Rica, Honduras, and Colombia (Randolph and Gerardo, 1982). In these regions, oil palm is a major commercial crop with 397,000 tonnes of palm oil exported by Colombia and Ecuador alone in 2011 at a value of 493 million dollars (http://faostat.fao.org/).

12.4. Management Approaches

No effective or practical control measures exist for diseases that affect roots and stems.

There is no known cure, and early eradication of infected plants is important. It is often difficult to recommend control due to the difficulty in providing the correct diagnosis of these diseases.

Measures to control the suspected vectors are used to break disease transmission. Application of insecticide around the base of palms is recommended to control the root miner, *Sagalassa valida,* which is considered a vector of the protozoa.

In areas where *Lincus lethifer* is considered the vector, application of insecticide in combination with the spraying of herbicide to eradicate *P. maximum* is practiced.

It has been reported that some oil palms remained healthy and survived severe infestation of sudden wilt in the plantations. Selection of resistant planting material is therefore possible.

Triazolo-pyrimidine derivatives (Magan et.al, 2004) and botanical extract of piperaceae plants (Cincini et.al, 2020) have shown some effect against the plant pathogenic protozoan, but are under commercial formulation stages.

13

Methods of Application of Chemical Molecules As Plant Medicines

13.1. Seed Treatment

Seed is a basic input to raise the crop plants. Healthy seeds produces the healthy crop. However, the seed carries different pathogens either externally or sometimes internally in the seed to affect the seed emergence and healthy seedling growth. Therefore, the seed treatment with fungicides/bactericides, based on the infection of the pathogen, is highly essential to counter the attack and infection of these pathogens to the germinating seedlings. In addition, when the seed is sown, it is also vulnerable to attack by many common soil-borne pathogens, particularly fungal or bacterial, leading to either seed rot, seedling mortality or produce seedling blight and subsequent infection at a later stage.

Seed treatment is probably the effective and economical method of seed borne disease control and is being advocated as a regular practice in crop protection against soil and seed-borne pathogens. Seed treatment is therapeutic when it kills pathogens that infect embroys, cotyledons or endosperms under the seed coat; eradicative when it kills pathogens that contaminate seed surfaces and protective when it prevents penetration of soilborne pathogens into the seedling.

There are various types of seed treatment and broadly they may be divided into three categories i.e. (a) Mechanical, (b) Chemical and (c) Physical

A. Mechanical Method

When the seeds are infected with some pathogen, there may be alteration in size, shape, weight of seeds besides discolouration of such infected seeds. Due to this it is possible to detect the infected seeds and separate them from the healthy ones. In the case of ergot diseases of cumbu, rye and sorghum, the fungal sclerotia are usually larger in size and lighter than healthy grains, so by sieving or flotation, the infected grains may be easily separated. Such mechanical separation eleminates the infected grains and may be easily separated to a larger extent. This method is also highly useful to separate infected grains in the case of "tundu" disease of wheat. To achieve this, dissolve 2kg of common salt in 10 litres of water (20% solution). Drop the seeds into the salt solution and stir well. Remove the ergot affected seeds and sclerotia which float on the water surface. Wash the seeds in fresh water 2 or 3 times to remove the salts on the seeds. Dry the seeds in shade and use for sowing.

B. Chemical methods

Use of fungicides/bactericides or antibiotics as seed treatment is one of the most efficient and economical methods of chemical disease control. On the basis of their tenacity and action, the seed dressing chemicals may be grouped as

(a) **Seed disinfectant**, which disinfect the seed but may not remain active for a long period after the seed has been sown and

(b) **Seed protectants**, which disinfect the seed surface and stick to the seed surface for sometime after the seed has been sown, thus giving temporary protection to the young seedlings against soil borne fungi. Now, the systemic fungicides are impregnated into the seeds to eliminate the deep seated infection in the seeds.

The seed dressing chemicals may be applied by (i) Dry treatment (ii) Wet treatment and (iii) Slurry.

(i) Dry Seed Treatment

In this method, the fungicide/bactericide adheres in a fine form on the surface of the seeds. A calculated quantity of fungicide is applied and mixed with seed using machinery specially designed for the purpose. The fungicides may be mixed with the seeds of small lots using simple Rotary seed Dresser (Seed treating drum) or of large seed lots at seed processing plants using Grain treating machines.

Normally at field level, dry seed treatment is carried out in dry rotary seed treating drums which ensure proper coating of the chemical on the surface of seeds. The dry dressing method is used in pulses, cotton and oil seeds with the antagonistic fungus like *Trichoderma viride* by mixing the formulation at the rate of 4g/kg of the seed, e.g. dry seed treatment in paddy where a required amount of fungicide is mixed with required quantity of seeds in a seed treating drum or polythene lined gunny bags, so as to provide uniform coating of the fungicide over the seeds. Any one of the fungicide viz. **Thiram** or **Captan** or **Carboxin** or **Tricyclazole** may be used for seed treatment at the rate of 2g/kg. Treat the seeds atleast 24 hours prior to soaking for sprouting.

(ii) Wet seed treatment

This method involves preparing fungicide/bactericide suspension in water, often at field rates and then dipping the seeds or seedlings or propagative materials for a specified time. The seeds cannot be stored and the treatment has to be done before sowing. This treatment is usually applied for treating vegetatively propagative materials like cuttings, tubers, corms, setts rhizomes, bulbs etc., which are not amenable to dry or slurry treatment.

a. **Seed dip / Seed soaking:** For certain crops, seed soaking is essential. Seeds treated by this methods have to be properly dried after treatment. The fungicide/bactericide adheres as a thin film over the seed surface which gives protection against invasion by soil-borne pathogens, eg. Seed dip treatment in paddy. To perform this, prepare the fungicidal solution by mixing any of the fungicides viz., **carbendazim** or **pyroquilon** or **tricyclazole** at the rate of 2g/litre of water and soak the seeds in the solution for 2 hrs. Drain the solution and keep the seeds for sprouting or sowing. To perform the Seed dip treatment in Wheat, prepare 0.2% of carboxin (2g/litre of water) solution and soak the seeds for 6 hours. Drain the solution and dry the seeds properly before sowing. This effectively eliminates the loose smut pathogen, *Ustilago nuda tritici.*

b. **Seedling dip / root dip:** The seedlings of vegetables and fruits are normally dipped in 0.25% **copper oxychloride** or 0.1% **carbendazin** solution for 5 minutes to protect against seedling blight and rots.

c. **Rhizome dip:** The rhizomes of cardamom, ginger and turmeric are treated with 0.1% **emisan** solution for 20 minutes to eliminate rot causing pathogen present in the soil.

d. **Sett dip / Sucker dip:** The sets of sugarcane and tapioca are dipped in 0.1% **emisan** solution for 30 minutes. The suckers of pine apple may also be treated by this method to protect from soilborne diseases.

(iii) Slurry treatment (Seed pelleting)

In this method, fungicide is applied in the form of a thin paste (active material is dissolved in small quantity of water). The required quantity of the fungicide slurry is mixed with the specified quantity of the seed so that during the process of treatment slurry gets deposited on the surface of seeds in the form of a thin paste which later dries up. Almost all the seed processing units have slurry treaters. In these slurry treaters, the requisite quantity of fungicides slurry is mixed with specified quantity of seed before the seed lot is bagged, eg. Seed pelleting in ragi. To achieve this, mix 2.5g of **carbendazim** in 40 ml of water and add 0.5g of gum to the fungicidal solution. Add 2 kg of seeds to this solution and mix thoroughly to ensure a uniform coating of the fungicide over the seed. Dry the seeds under the shade. Treat the seeds 24 hrs prior to sowing. The slurry treatment is more efficient than the rotary seed dressers.

(iv) Special method of seed treatment

This is used for the hard seed coated fuggy seeds, e.g. Acid delinting in cotton. This method is followed in cotton to kill the seed-borne fungi and bacteria particularly the Xanthomonas leaf spot pathogen. The seeds are treated with **concentrated sulphuric acid** @ 100 ml/kg of seed for 2-3 minutes. The seeds are then washed 2 or 3 times thoroughly with cold water and shade dried. After drying, they are again treated with **captan** or **thiram** @ 4g/kg before sowing.

13.2. Soil Treatment

It is well known that soil harbours a large number of plant pathogens including nematodes. The primary sources of many plant pathogens happens to be in soil where dead organic matter supports active or dormant stages of pathogens. In addition, seed treatment does not afford sufficient protection against seedling diseases and a treatment of soil around the seed is necessary to protect them. Soil treatment is largely curative in nature as it mainly aims at killing the pathogens in soil and making the soil safe for the growth of the plant. Chemical treatments of the soil is comparatively simple, especially when the soil is fallow as the chemical is volatile and disappears quickly either by volatilization or decomposition. Soil treating chemicals should be non-injurious to the plants in the soil adjacent to the area where treatment has been carried out because there may be standing crop in adjacent fields. The soil treatment can be achieve either by (i) Soil drenching, (ii) broadcasting, (iii) furrow application, (iv) fumigation, and (v) chemigation.

(i) **Soil drenching:** This method is followed for controlling damping off and root rot infections at the ground level. Requisite quantity of fungicide suspension is applied per unit area so that the fungicide reaches to a depth of atleast 10-15 cm, eg. **Emisan, PCNB, Carbendazim, Copper fungicides,** etc.

(ii) **Broadcasting:** It is followed in granular fungicides wherein the pellets are broadcasted near the plant.

(iii) **Furrow application:** It is done specifically in the control of some diseases where the direct application of the fungicides on the plant surface results in phytotoxicity. It is specifically practiced in the control of powdery mildew of tobacco where the sulphur dust is applied in the furrows.

(iv) **Fumigation:** Volatile toxicants (fumigants) such as **methyl bromide, chloropicrin, formaldehyde** and **vapam** are the best chemical sterilants for soil to kill fungi and nematodes as they penetrate the soil efficiently. Fumigations are normally done in nursery areas and in glass houses. The fumigant is applied to the soil and covered by thin polythene sheets for 5-7 days and removed. For example, Formaldehyde is applied at 400 ml/100 Sq.m. The treated soil is irrigated and used 1 or 2 weeks later. Vapam is normally sprinkled on the soil surface and covered. Volatile liquid fumigants are also injected to a depth of 15-20 cm, using sub-soil injectors.

(v) **Chemigation:** In this method, the fungicides are directly mixed in the irrigation water. It is normally adopted using sprinkler or drip irrigation system.

13.3. Foliar Application

A. Spraying

This is the most commonly followed method. Spraying of fungicides/ bactericides is done on leaves, stems and fruits. Wettable powders are most commonly used for preparing spray solutions. The most common diluent or carrier is water. The dispersion of the spray is usually achieved by its passage under pressure through nozzle of the sprayer. The amount of spray solution required for a hectare will depend on the nature of crops to be treated. For trees and shrubs more amount of spray solution is required than in the case of ground crops. Depending on the volume of fluid used for coverage, the sprays are categorized into high volume, medium volume, low volume, very high volume and ultra low volume. The different equipments used for spray application are: Foot-operated sprayer, rocking sprayer, knapsack sprayer, motorised knapsack sprayer (Power sprayer), tractor mounted sprayer, mist blower and aircraft or helicopter (aerial spray). It is advisible to spray the crops in the direction of the wind or when the wind is not blowing.

B. Dusting

Dusts are applied to all aerial parts of a plant as an alternative to spraying. Dry powders are used for covering host surface. Generally, dusting is practicable in calm weather and a better protective action is obtained if the dust is applied when the plant surface is wet with dew or rain drops. The equipments employed for the dusting operation are: Bellow duster, rotary duster, motorised knapsack duster and aircraft (aerial application). This is generally used to control fungal diseases like mildews.

13.4. Post-harvest Treatment

Fruits and vegetables are largely damaged after harvest by fungi and bacteria. Many fungicides/bactericides have been used as spray or dip or as fumigation. Post harvest fungicides are most frequently applied as aqueous suspensions or solutions. Dip application has the advantage of totally submerging the commodity so that maximum opportunity for penetration to the infection sites. This method is generally adopted for the products which has to be stored in the storage for a considerable period of time. Systemic fungicides, particularly **thiabendazole, carbendazim, metalaxyl, fosety-AI** have been found to be very effective against storage diseases. The waiting period before marketing the treated product is most important for avoiding the residual effect on the consumers. Please read the Label of the post harvest fungicides properly before its applications. In addition, **dithiocarbamates** and **antibiotics** are also applied to control the post-harvest diseases. Wrapping the harvested products with fungicide impregnated wax paper is the latest method available.

13.5. Special Method of Applications

1. Trunk Application / Trunk Injection

It is normally adopted in coconut gardens to control Thanjavur wilt caused by *Ganoderma lucidum*. In the infected plant, a downward hole is made to a depth of 3-4 inches at an angle of 45 degree at the height of 3 inches from the ground level with the help of an auger. The solution containing 2g of **Aureofungin** and 1 g of **copper sulphate** in 100 ml of water is taken in a saline bottle and the bottle is tied with the tree. The hose is inserted into the hole and the stopper is adjusted to allow the solution in drops. After the treatment, the hole is covered with clay.

2. Root Feeding

Root feeding is also adopted for the control of Thanjavur wilt of coconut instead of trunk application. The root region is exposed; actively growing young root is selected and given a slanting cut at the tip. The root is inserted into a polythene bag containing 100 ml of the fungicidal solution. The mouth of the bag is tied tightly with the root.

3. Pseudostem Injection

This method is very effective in controlling the aphid vector *(Pentalonia nigronervosa)* of bunchy top disease of bannana. The banana injector is used for injecting the insecticide. Banana injector is nothing but an Aspee baby sprayer of 500 ml capacity, in which the nozzle is replaced by leurlock system and aspirator needle No. 16. The tip of the needle is closed and two small holes are made in opposite direction. It is for free flow of fluid and the lock system prevents the needle from dropping from the sprayer. One ml of **monocrotophos** mixed with water at 1:4 ratio is injected into the pseudostem of 3 months old crop and repeated twice at monthly intervals. The same injector can also be used to kill the bunchy top infected plants by injecting 2 ml of 2, 4-D (**Femoxone**) mixed in water at 1:8 ratio.

4. Corm Injection

It is an effective method used to control Panama will of banana caused by *Fusarium oxysporum* f. sp. *cubense.* Capsule applicator is used for this purpose. It is nothing but an iron rod of 7 mm thickness to which a handle is attached at one end. The length of the rod is 45 cm and an iron plate is fixed at a distance of 7 cm from the tip. The corm is exposed by removing the soil and a hole is made at 45 degree angle to a depth of 5 cm. One or two gelatin capsules containing 50-60 mg of **carbendazim** is pushed in slowly and covered with soil. Instead of capsule, 3 ml of 2% carbendazim solution can also be injected into the hole.

5. Paring and Pralinage

It is used to control *Fusarium* wilt and burrowing nematode (*Radopholus similis)* of banana. The roots as well as a small portion of corm is removed or chopped off with a sharp knife and the sucker is dipped in 0.1% **carbendazim** solution for 5 minutes. Then, the sucker is dipped in **clay slurry** and **furadan granules** are sprinkled over the corm @ 40 g/corm.

14

Precautions in Selection of a Proper Plant Health Medicine

Proper selection of plant health medicines like fungicide, bactericide, antibiotics, algaecide, viricide, nematicide, weedicide, plant harmones and auxins and its application at the correct dose and the proper time are highly essential for the management of plant health. The basic requirement of an application method is that it delivers the chemical to the site where the active compound will prevent the pathogen/pest damaging the plant. This is mostly achieved by spray, fog, smoke, aerosol, mist, dust, or granules applied to the growing plant or by seed or soil treatment. In addition, some trees and shrubs can be protected by injection of pesticide liquid into the trunk or by brushing wounds with pesticidal paints or slurries.

In the case of sprays, mists, aerosols and fogs, the chemical is in droplets of water of another fluid. In the case of smokers, the solid particles of the pesticides are carried by the air. In the case of dusts and granules, the pesticide is straightly mixed with an inert carrier, impregnated into and coated on the particles, which are applied mechanically.The object of spraying or dusting is to cover the entire susceptible surface of host with a thin covering of a suitable concentration of the pesticide before the pathogen/pest has come into contact with the host. However, these practices may not effectively eradicate the inoculums present on the surface of the seeds or deep-seated in the seed. So, the application of chemicals as seed dressing is highly essential. In addition, soil harbours several pathogens/pests which cause root diseases/root damages in several crop plants. So treatment of soil with chemicals is also highly useful in reducing the pest inoculums load present in the soil. The pesticidal application varies according to the nature of the host part infected and nature of survival and spread of the pathogen/pest.

Although pesticides play an important role in crop production, intensive use can have adverse effects on crops, useful antagonistic organisms, and predators or parasites of noxious insects. However, the most serious problem is the development of pesticide/fungicide-resistant strains leading to failures of pest/disease control, and reduction in the number of available active ingredients.

14.1. Adverse Effects on Crops and Useful Organisms

Phytotoxic effects of fungicides/pesticides are rarely noticed in arable crops, but can be a problem in orchard and glasshouse crops. Sometimes these effects are only temporary and are overcome during the growing season. In some instances all cultivars of a specific crop may be affected, for example, inhibitors of sterol C 14 demethylation (DMIs), which reduce growth of barley and wheat when applied at high rates on seeds. This effect is mainly related to inhibition of gibberellin biosynthesis. Several pesticides/fungicides/ bactericides reduce saprophyte populations of microorganisms in the phyllosphere of crops. If such chemicals reduce such populations (e.g. yeasts) and have little or no effect on certain minor pathogens, infection by these pathogens may, in exceptional circumstances, increase when plant protection treatments are applied. One example is the role of phyllosphere yeasts on the incidence of Septoria nodorum in wheat. A reduction of the population of yeast may also affect the fermentation of grape juice. This result has been observed when multisite fungicides were applied to control Botrytis cinerea (e.g. captan, dichlofluanid, thiram). Consequently, these fungicides cannot be applied just before harvest. Recent investigations conducted with DMIs, applied against Uncinula necator, indicate that they can disturb fermentation by acting directly on the yeast or by modifying the sterol content of the grape. Fungicides may also effect antagonistic interactions between pathogens and the soil microflora. Many systemic fungicides reduce competition for nutrients by common soil fungi like Penicillium spp and Trichoderma spp, thereby increasing pathogens with a natural insensitivity to these compounds. Such effects may also be based on a reduction of antibiosis or parasitism. Differential sensitivity of pathogens may sometimes result in disease trading, which is a change of the dominant pathogen. For instance, use of benzimidazole fungicides to control common eyespot caused by Pseudocercosporella herpotrichoides resulted in a higher incidence of sharp eyespot caused by Rhizoctonia solani. Overall, side effects of fungicides on the soil microflora are economically important in only a limited number of pesticide-crop combinations. Development of integrated crop protection requires fungicides with few or no adverse effects on parasites and predators of noxious insects and mites. Several old fungicides such as sulphur, pyrazophos, chinomethionate, which are suitable against powdery mildew of fruits, are not recommended because they are more or less toxic to various Phytoseiidae and Coccinellidae. However, intensive use of this type of fungicide is not recommended because of the risk of resistance. Problems encountered from the adverse effects of fungicides are much less severe than those associated with the application of insecticides.

Section-II

Chemical Molecules to Control Plant Physiological Processes and Physiological Disorders

15

Plant Growth Regulators

Plant growth regulators or phytohormones are organic substances produced naturally in higher plants, controlling growth or other physiological functions at a site remote from its place of production and active in minute amounts. Thimmann in1948 proposed the term *Phytohormone* as these hormones are synthesized in plants (Thimmann, 1963). *Plant growth regulators* include auxins, gibberellins, cytokinins, ethylene, growth retardants and growth inhibitors. Auxins are the first hormones discovered in plants and later on gibberellins and cytokinins were discovered.

15.1. Use of Plant Growth Regulators

The use of plant growth regulators in agricultural production began in the United States in the 1930s. The first discovery and use of plant growth regulators was with acetylene and ethylene, which enhanced flower production in pineapple. Subsequently, use of plant growth regulators has grown exponentially to become a major component of agricultural commodity production.

Certain herbicides and insecticides that are not true plant growth regulators cause some plant-growth-regulating effects. For example, the widely used insecticide carbaryl is used to thin apple fruit from trees and to aid in encouraging annual bearing. According to the American Society for Horticultural Science, plant growth regulators fall into six major classes (table. 20) with the plant development function(s) and their practical uses.

Table 20: Plant growth regulator class, associated function(s) and practical uses

Classes	Funtions	Practical uses
Auxin	Shoot elongation	Thin tree fruit, increase rooting and flower formation
Gibberellins	Stimulate cell division and cell elongation	Increase stalk length, increase flower and fruit size
Cytokinin	Stimulate cell elongation	Prolong storage life of flowers and vegetables and stimulate bud initiation and root growth
Ethylene generator	Ripening	Induce uniform ripening in fruit and vegetables
Growth inhibitor	Stops growth	Promote flower production by shortening internodes
Growth retardant	Slows growth	Retard tobacco sucker growth

15.1.1. Auxins

Auxins are a group of phytohormones produced in the shoot and root apices and they migrate from the apex to the zone of elongation. Auxins promote the growth along the longitudinal axis of the plant and hence the name (auxeing: to grow). The term, auxin was introduced by Kogl and Haagen- Smit in 1931(Kogl and Haagen-Smit, 1933). Went (1928) isolated auxin from the Avena coleoptile tips by a method called *Avena coleoptile or curvature test* and concluded that no growth can occur without auxin. Auxins are widely distributed throughout the plant however, abundant in the growing tips such as coleoptile tip, buds, root tips and leaves. **Indole Acetic Acid (IAA) is the only naturally occurring auxin in plants. The synthetic auxins include, 2, 4-D: 2, 4 dichloro phenoxy acetic acid and 2, 4, 5 – Trichloro phenoxy acetic acid**

Auxin's role in/on plant

1. **Cell division and elongation:** The primary physiological effects of auxin are cell division and cell elongation in the shoots. It is important in the secondary growth of stem and differentiation of xylem and phloem tissues.

2. **Apical dominance:** In many plants, if the terminal bud is intact and growing, the growth of lateral buds just below it remains suppressed. Removal of the apical bud results in the rapid growth of lateral buds. This phenomenon in which the apical bud dominates over the lateral buds and does not allow the lateral buds to grow is known as *apical dominance*. It is pointed out that the apical dominance might be under the control of auxin produced at the terminal bud and which is transported downward through the stem to the lateral buds and hinders the growth. They

removed the apical bud and replaced it with *agar* block. This resulted in rapid growth of lateral buds. But when they replaced the apical bud with agar block containing auxin, the lateral buds remained suppressed and did not grow.

3. **Root initiation:** In contrast to stem, the higher concentration of auxin inhibits the elongation of roots but the number of lateral roots is considerably increased i.e., higher concentration of auxin induces more lateral branch roots. **Application of IAA in lanolin paste (lanolin is a soft fat prepared from wool and is good solvent for auxin) to the cut end of a young stem results in an early and extensive rooting. This fact is of great practical importance and has been widely utilized to promote root formation in economically useful plants which are propagated by cuttings.**

4. **Prevention of abscission:** Natural auxins prevent the formation of abscission layer which may otherwise result in the fall of leaves, flowers and fruits.

5. **Parthenocarpy:** Auxin can induce the formation of parthenocarpic fruits (fruit formation without pollination and fertilization). In parthenocarpic fruits, the concentration of auxin in the ovaries is higher than in the ovaries of plants which produce fruits only after fertilization. In the later cases, the concentration of the auxin in ovaries increases after pollination and fertilization.

6. **Respiration:** Auxin stimulates respiration and there is a correlation between auxin induced growth and respiration. Auxin may increase the rate of respiration indirectly through increased supply of ADP by rapidly utilizing ATP in the expanding cells.

7. **Callus formation:** Besides cell elongation, auxin may also be active in cell division. In many tissue cultures, where the callus growth is quite normal, the continued growth of such callus takes place only after the addition of auxin.

8. **Eradication of weeds:** Some synthetic auxins especially 2, 4- D and 2, 4, 5-T are useful in eradication of weeds at higher concentrations.

9. **Flowering and sex expression:** Auxins generally inhibit flowering but **in pine apple and lettuce it promotes uniform flowering.**

15.1.1.1. Distribution of Auxin in Plants

In plants, auxin (IAA) is synthesized in growing tips or meristematic regions from where it is transported to other plant parts. Hence, the highest concentration of IAA is found in growing shoot tips, young leaves and developing auxiliary shoots. In monocot seedling, the highest concentration of auxin is found in coleoptile tip which decreases progressively towards its base.

In dicot seedlings, the highest concentration is found in growing regions of shoot, young leaves and developing auxiliary shoots. Within the plants, auxin may present in two forms. i.e., *free auxins* and *bound auxins*. Free auxins are those which are easily extracted by various organic solvents such as diethyl ether. Bound auxins on the other hand, need more drastic methods such as hydrolysis, autolysis, enzymolysis etc. for extraction of auxin. Bound auxins occur in plants as complexes with carbohydrates such as glucose, arabinose or sugar alcohols or proteins or amino acids such as aspartate, glutamate or with inositol.

15.1.1.2. Mechanism of Action

IAA increases the plasticity of cell walls so that the cells stretch easily in response to turgor pressure. It has been suggested that IAA acts upon DNA to influence the production of mRNA. The mRNA codes for specific enzymes responsible for expansion of cell walls. Recent evidences indicate that IAA increases oxidative phosphorylation in respiration and enhanced oxygen uptake. The growth stimulation might be due to increased energy supply and it is also demonstrated that auxin induces production of ethylene in plants.

15.1.2. Gibberellins

A Japanese scientist Kurosawa found that the rice seedlings infected by the fungus *Gibberella fujikuroi* grow taller and turned very thin and pale. An active substance was isolated from the infected seedlings and named as Gibberellin.

15.1.2.1. Pathways of gibberellin's Biosynthesis in Plants.

The primary precursor for the formation of gibberellins is acetate.

Acetate + COA → Acetyl COA → Mevalonic acid → MA pyrophosphate →

Isopentanyl pyrophosphate → Geranyl pyrophosphate → GGPP → Kaurene → Gibberellins.

15.1.2.2. Role of gibberellins in/on plant

1. **Seed germination:** Certain light sensitive seeds eg. Lettuce and tobacco show poor germination in dark. Germination starts vigorously if these seeds are exposed to light or red light. This requirement of light is overcome if the seeds are treated with gibberellic acid in dark.

2. **Dormancy of buds:** In temperate regions the buds formed in autumn remain dormant until next spring due to severe cold. This dormancy of buds can be broken by gibberellin treatments. In potato also, there is a dormant period after harvest, but the application of gibberellin sprouts the dormant buds vigorously.

3. **Root growth:** Gibberellins have little or no effect on root growth. At higher concentration, some inhibition of root growth may occur. The initiation of roots is markedly inhibited by gibberellins in isolated cuttings.

4. **Elongation of internodes:** The most pronounced effect of gibberellins on the plant growth is the elongation of the internodes. Therefore in many plants such as dwarf pea, dwarf maize etc gibberellins overcome the genetic dwarfism.

5. **Bolting and flowering:** In many herbaceous plants, the early period of growth shows rosette habit with short stem and small leaves. Under short days, the rosette habit is retained while under long days bolting occurs i.e. the stem elongates rapidly and is converted into polar axis bearing flower primordia. This bolting can also be induced in such plants by the application of gibberellins even under non-inductive short days.

   In *Hyoscyamus niger* (a long day plant) gibberellin treatment causes bolting and flowering under non-inductive short days. While in long day plants the gibberellin treatment usually results in early flowering. In short day plants, its effects are quite variable. It may either have no effect or inhibit or may activate flowering.

6. **Parthenocarpy:** Germination of the pollen grains is stimulated by gibberellins; likewise, the growth of the fruit and the formation of parthenocarpic fruits can be induced by gibberellin treatment. In many cases, e.g.pome and stone fruits where auxins have failed to induce parthenocarpy, the gibberellins have proven to be successful. **Seedless and fleshly tomatoes and large sized seedless grapes are produced by gibberellin treatments on commercial scale.**

Function of gibberellins

One important function of gibberellins is to cause the synthesis of the enzyme α - amylase in the aleurone layer of the endosperm of cereal grains during germination. This enzyme brings about hydrolysis of starch to form simple sugars which are then translocated to growing embryo to provide energy source.

15.1.2.3. Distribution of Gibberellins in Plant

Gibberellins are found in all parts of higher plants including shoots, roots, leaves, flower, petals, anthers and seeds. In general, reproductive parts contain much higher concentrations of gibberellins than the vegetative parts. Immature seeds are especially rich in gibberellins (10-100 mg per g fresh weight). In plants, gibberellins occur in two forms i.e free gibberellins and bound gibberellins. Bound gibberellins usually occur as gibberellin – glycosides.

15.1.3. Cytokinins (Kinetin)

Kinetin was discovered by Skoog and Miller in 1950 from the tobacco pith callus and the chemical substance was identified as 6-furfuryl aminopurine. Because of its specific effect on *cytokinesis* (cell division), it was called as cytokinins or kinetin.

The term, cytokinin was proposed by Letham (1963). Fairley and Kingour in 1966 used the term, *phytokinins* for cytokinins because of their plant origin.

Chemically cytokinins are kinins and they are purine derivatives. Cytokinins, besides their main effect on cell division, also regulate growth and hence they are considered as natural plant growth hormones. Some of the very important and commonly known naturally occurring cytokinins are Coconut milk factor and Zeatin. Cytokinin was also identified as a constituent of t-RNA.

15.1.3.1. Naturally Occurring Cytokinins

Cytokinins can be extracted from coconut milk (liquid endosperm of coconut), tomato juice, flowers and fruits of *Pyrus malus*; fruits of *Pyrus communis* (Pear), *Prunus cerasiferae* (plum) and *Lycopersicum esculentum*; Cambial tissues of *Pinus radiata*, *Eucalyptus regnans* and *Nicotiana tabacum*; immature fruits of *Zea mays*, *Juglans* sp. and *Musa* sp; female gametophytes of *Ginkgo biloba*; fruitlets, embryo and endosperms of *Prunus persica*; seedling of *Pisum sativum*; root exudates of *Helianthus annus* and tumour tissues of tobacco.

According to Skoog and Armstrong (1970), at least seven well established types of cytokinins have been reported from the plants.

15.1.3.2. Role of Cytokinins in/on Plant

1. **Cell division**: The most important biological effect of kinetin on plants is to induce cell division especially in tobacco pith callus, carrot root tissue, soybean cotyledon, pea callus etc.
2. **Cell enlargement**: Like auxins and gibberellins, the kinetin may also induce cell enlargement. Significant cell enlargement has been observed

in the leaves of *Phaseolus vulgaris*, pumpkin cotyledons, tobacco pith culture, cortical cells of tobacco roots etc.

3. **Effect on apical dominance**: External application of cytokinin promotes the growth of lateral buds and hence counteracts the effect of apical dominance
4. **Dormancy of seeds**: Like gibberellins, the dormancy of certain light sensitive seeds such as lettuce and tobacco can also be broken by kinetin treatment.
5. **Delay of senescence (Richmand - Lang effect):** The senescence of leaves usually accompanies with loss of chlorophyll and rapid breakdown of proteins. Senescence can be postponed to several days by kinetin treatment by improving RNA synthesis followed by protein synthesis. Richmand and Lang (1957) while working on detached leaves of *Xanthium* found that kinetin was able to postpone the senescence for a number of days.
6. **Flower induction**: Cytokinins can be employed successfully to induce flowering in short day plants.
7. **Morphogenesis:** It has been shown that high auxin and low kinetin produced only roots whereas high kinetin and low auxin could promote formation of shoot buds.
8. **Accumulation and translocation of solutes:** Plants accumulate solutes very actively with the help of Cytokinin and also help in solute translocation in phloem.
9. **Protein synthesis:** Osborne (1962) demonstrated the increased rate of protein synthesis due to translocation by kinetin treatment.
10. **Other effects:** Cytokinins provide resistance to high temperature, cold and diseases in some plants. They also help in flowering by substituting the photoperiodic requirements. In some cases, they stimulate synthesis of several enzymes involved in photosynthesis.

15.1.3.3. Commercial Applications

Cytokinins have been used for increasing shelf life of fruits, quickening of root induction and producing efficient root system, increasing yield and oil contents of oil seeds like ground nut.

15.1.4. Ethylene

Ethylene is the only natural plant growth hormone exists in gaseous form.

15.1.4.1. Role in/on plant

The main role of ethylene is to hasten the ripening of fleshy fruits eg.Banana, apples, pears, tomatoes, citrus etc.

- It stimulates senescence and abscission of leaves
- **It is effective in inducing flowering in pine apple**
- It causes inhibition of root growth
- It stimulates the formation of adventitious roots
- **It stimulates fading of flowers**
- It stimulates epinasty of leaves.

15.1.5. Abscisic Acid

Addicott (1963) isolated a substance strongly antagonistic to growth, from young cotton fruits and named Abscissin II. Later on this name was changed to Abscisic acid. This substance also induces dormancy of buds therefore it also named as Dormin. **Abscisic acid (ABA) is a naturally occurring growth inhibitor.**

15.1.5.1. Physiological Effects

The two main physiological effects are

1. **Geotropism in roots**: Geotropic curvature of root is mainly due to translocation of ABA in basipetal direction towards the root tip.
2. **Stomatal closing**: ABA is synthesized and stored in mesophyll chloroplast. In respond to water stress, the permeability of chloroplast membrane is lost which resulted is diffusion of ABA out of chloroplast into the cytoplasm of the mesophyll cells. From mesophyll cells it diffuses into guard cells where it causes closing of stomata.

Other effects includes

i. Induces bud dormancy and seed dormancy

ii. Induces tuberisation

iii. Induces senescence of leaves, fruit ripening, abscission of leaves, flowers and fruits

iv. Increase the resistance of temperate zone plants to frost injury.

Different plant growth regulators registered and available in market are given in table 21,22 and 23.

15.1.6. Table 21: Plant Growth Regulators Registered for Use

Active ingredient	Registered crops and function	Trade name
Ancymido	Ornamental plants – **growth inhibitor**	Abide, A-Rest
Butralin	Tobacco – **shoot inhibitor**	Butralin
Alcohols	Tobacco – **shoot inhibitor**	Fair, Royaltac M, Sucker-Plucker, Off-Shoot, Contact-85
Chlormequat chloride	Ornamental flowers – **shoot inhibitor**	Citadel, Cycocel, E-Pro
Cytokinin	Ornamental plants, vegetables, fruits including citrus – hasten maturity, enhance fruit color, growth enhancer, enhance tuber color, improve tuber quality	Conklin, Culbac, Cytoplex, Early Harvest, Foli-Zyme, Goldengro, Happygro, Incite, Megagro, Ascend, Radiate, Stimulate, Suppress, Validate, X-Cyte
Daminozide	Ornamental plants – **growth inhibitor**	B-Nine, Compress, Dazide
Ethephon	Turfgrass – **reduces mowing frequency** Various fruits, vegetables, and nuts – **hastens ripening and maturity, enhance fruit color and floral stimulant** Cucurbits – **increases flowering** Ground covers – **inhibits flowering** Ornamental trees – inhibits fruiting Tobacco and cotton – **hastens maturity** Cereal grains and grasses grown for seed – **reduces lodging**	Boll Buster, BollD, Cerone, Cotton Quik, Ethrel, Finish, Flash, Florel, Mature, MFX, Prep, Proxy, Quali-Pro, SA-50, Setup, Super Boll, Whiteout
Flurprimidol	Ornamental woody plants and ground covers – **reduces pruning** Turfgrass – **reduces mowing frequentcy**	Cutless, Legacy, Mastiff, Topflor
Gibberellic acid	Small fruits, cucurbits – **increase fruit set** Citrus – **promote rind/peel integrity, prevent fruit drop** Rice, cotton – **growth enhancer**	Ascend, Cytoplex, Early Harvest, Falgro, Florgib, Foli-Zyme, GA3, GibGro, Green Sol, Incite, N-Large, PGR IV, Pro-Gibb, Release, Rouse, Ryzup, Stimulate

Active ingredient	Registered crops and function	Trade name
Gibberellin mixtures	Cut flowers – **plant preservative** Tree fruit – **increase fruit size, hasten maturity, shoot stimulant** Evergreen trees – **floral stimulant, stimulate germination**	BVB, Chrysal, Fascination, Procone
Indole-3-butyric acid (IBA)	Ornamental plants and trees – **root stimulant**	Numerous trade products
Maleic hydrazide, potassium salt	Tobacco – **shoot inhibitor** Stored bulbs – **sprout inhibitor**	Fair, Rite-Hite, Royal, Sucker Stuff,
Mefluidide	Ground covers, shrubs, ornamental trees – **reduces pruning** Turfgrass – **reduces mowing frequency**	Embark, Sta-Lo
Mepiquat chloride	Cotton - **growth inhibitor, enhance uniform fruit maturity**	Pix
Mepiquat pentaborate	Cotton – **growth inhibitor, enhance uniform fruit maturity**	Pentia
Naphthalene-acetic acid (NAA)	Ornamental plants – **stimulates rooting, increase vegetative Growth**	DipN Grow, Goldengro, Hi-Yield
1-Naphthaleneacetamide (NAD)	Woody ornamental cuttings – **rooting stimulant**	Rootone
n-Decanol	Tobacco – **shoot inhibitor**	Antac, FST-7, Royaltac
Paclobutrazol	Ornamental flowers – **promotes uniform flowering** Ornamental plants – **reduces internodal length** Ornamental trees – **growth inhibitor** Turfgrass – **increased plant thickness, growth inhibitor**	Bonzi, Cambistat, Cutdown, Downsize, Florazol, Paclo, Paczol, Piccolo, Profile, Shortstop, Trimmit, Turf Enhancer
Prohexadione calcium	Peanut, Tree fruit, turfgrass grown for seed – **growth inhibitor, reduce vegetative growth**	Apogee
Trinexapac-ethyl	Turfgrass – **reduces mowing frequency**	Armor Tech, Goldwing, Governor, Groom, Legacy, Primeraone, Primo, Provair, Solace,

15.1.6.1. Table.22: List of marketed plant growth regulators

Active ingredient	Formulation	Concentration	Product name	Company name	Crops
2,4-D (sodium salt)	SC	25 g/lkg	Deccomone Solution	Bayer	Citrus
2-(1-naphthyl)acetamide	WP	100 g/kg	Golden Thin	Applied Agricultural Products	Apple
2-(1-naphthyl) acetamide/2-(1-2-methylnaphthyl) acetic acid/2- (1-2- methylnaphthyl) acetamide/4-indol-3-ylbutyric acid	DP	0,67/0, 33/0,13/0,75 g/kg	Rootstim	Applied Agricultural Products	Ornamentals
2-(1-naphthyl)acetic acid	SL	45 g/l	Planofix	Bayer	Apple, Pear, Pineapple
3,5,6 TPA	TB	100 g/kg	Maxim	Hyper Agrochemicals	Citrus
4-indol-3-ylbutyric acid	DP	1 g/kg	Seradix B No 1	Bayer	Ornamentals
4-indol-3-ylbutyric acid	DP	3 g/kg	Seradix B No 2	Bayer	Ornamentals
4-indol-3-ylbutyric acid	SL	8 g/kg	Seradix B No 3	Bayer	Ornamentals
4-indol-3-ylbutyric acid	Powder	3 g/kg	Dynaroot B No 2	PBR Trading International	Flowers & Ornamentals (Plant cuttings)
4-indol-3-ylbutyric acid/1-naphthyl-acetic acid	SL	10/5 g/l	Dip & Root	PBR Trading International	Flowers & Ornamentals (Plantcuttings)
4-indol-3-ylbutyric acid/2(1-naphthyl)acetic acid	SL	10/5 g/l	Dip'n Grow	Fleuron	Ornamentals
6-benzyl adenine/gibberellins	SL	19/19 g/l	ELPPA	RT Chemicals	Apple, Plum
6-benzyl adenine/gibberellins	SL	19/19 g/l	Perlan	Universal Crop Protection	Apple ,Plum

Active ingredient	Formulation	Concentration	Product name	Company name	Crops
6-benzyl adenine/gibberellins	SL	19/19 g/l	Promalin	Valent BioSciences	Apple, Plum, Flowers Ornamentals (Protea)
alkoxylated fatty alkylamine polymer	EC	1000 g/l	Armothin	Universal Crop Protection	Plum, Nectarine, Peach
alkylamine polymer/ nitrogen(organic)	SL	471/534 g/l	Partner A Plus	Universal Crop Protection	Apple
Aminoethoxy vinylglycine hydrochloride	SP	150 g/kg	ReTain	Valent BioSciences	Apple, Nectarine
ammoniated nitrates	AL	170 g/l	GAN	Universal Crop Protection	Apple, Grape, Grape-seedless
auxins/cytokinins	SL	11/0,031 mg/l	Kelpak	Kelp Products	Deciduous Fruit trees & Grape Vineyards, Various Vegetables, Agricultural Crops
auxins/cytokinins	SL	2,2/0,0062 mg/l	Kelp	Wonder	Flowers, Ornamentals, Vegetables
auxins/cytokinins	SL	2,2/0,0062 mg/l	Kompel Kelpak	Starke Ayres	Flowers, Ornamentals Vegetables
auxins/cytokinins	SL	22/15 mg/l	Marinure Seaweed Foliar Spray	Canyon International	Potato, Wheat
auxins/cytokinins	SL	45/30 mg/l	Marinure DS Seaweed Foliar Spray	TL Organichem	Potato ,Wheat
calcium arsenate	WP	855 g/kg	Calcium Arsenate	Kynoch Agrochemicals	Citrus
calcium arsenate	WP	855 g/kg	Calcium Arsenate	Sunwood Chemicals	Citrus

Active ingredient	Formulation	Concentration	Product name	Company name	Crops
calcium arsenate	WP	855 g/kg	Calsenate	Unisun	Citrus
calcium arsenate	WP	855 g/l	Calcium Arsenate	Natural Plant Protection	Citrus
Carbaryl	SL	480 g/l	Sevin XLR Plus	Bayer	Apple
Carbaryl	WP	850 g/kg	Sevin 850 WP	Bayer	Apple
chlormequat chloride	SL	750 g/l	CeCeCe 750	BASF	Pear, Wheat
chlormequat chloride/ethephon	SL	300/150 g/l	Uprite	Bayer	Wheat
Chlorpropham	DP	10 g/kg	Neo-Stop	Natural Plant Protection	Potato
Chlorpropham	HN	500 g/l	Neostop L500 HN	Natural Plant Protection	Potato
chlorthal-dimethyl	SC	360 g/l	Erazor	Universal Crop Protection	Tobacco
Cloprop	SL	80 g/l	Swelpine	Applied Agricultural Products	Pineapple
Cyanamide	SL	520 g/l	Deurbraak	RT Chemicals (Erintrade)	Apple, Apricot, Grape, Grape-seedless, Kiwi, Pear, Plum
Cyanamide	SL	520 g/l	Dorcy	Almond Agro Chemicals	Apple, Apricot, Pear, Plum Grape, Kiwi, Grape (seedless)
Cyanamide	SL	520 g/l	Dormex	Degussa Africa	Apple, Apricot, Plum, Grape, Kiwi, Grape-seedless, Pear Cherry
decan-1-ol	EC	690 g/l	Decanol EC	Universal Crop Protection	Tobacco
decan-1-ol	EC	690 g/l	Fair Tac	Natural Plant Protection	Tobacco

Active ingredient	Formulation	Concentration	Product name	Company name	Crops
decan-1-ol	EC	690 g/l	Royaltac	Crompton Chemical	Tobacco
decan-1-ol/maleic hydrazide	EC	462/71 g/l	FST-7	Natural Plant Protection	Tobacco
decan-1-ol/octan-1-ol	EC	400/300 g/l	C8-10 EC	Universal Crop Protection	Tobacco
decan-1-ol/octan-1-ol	EC	400/300 g/l	Fair 85	Natural Plant Protection	Tobacco
Dichlorprop	EC	50 g/l	Citrimax	Universal Crop Protection	Citrus
dichlorprop (2-butoxyethyl ester)	EC	50 g/l	Corasil E	Nufarm Agriculture	Citrus
Dimethipin	SC	250 g/l	Harvade	Crompton Chemical	Cotton
Dinocap	EC	350 g/l	Karathane 350 EC	Dow AgroSciences	Apple, Pear
diquat dibromide	SL	200 g/l	Midstream	Syngenta	Potato, Sunflower
diquat dibromide	SL	200 g/l	Reglone	Syngenta	Sunflower
diquat dibromide/paraquat dichloride	SL	80/120 g/l	Preeglone	Syngenta	Hops
Ethephon	SL	480 g/l	Effon 480 SL	Plaaskem (Gouws and Scheepers)	Apple, Cherry, Citrus, Cotton , Grape, Maize Peach, Plum, Pineapple Sugarcane
Ethephon	SL	480 g/l	Ethapon SL	Applied Agricultural Products	Apple, Cherry, Citrus, Cotton, Grape, Maize Peach, Plum, Pineapple Sugar cane, Tobacco
Ethephon	SL	480 g/l	Ethephon	GAP Chemicals	Pineapple, Sugarcane
Ethephon	SL	480 g/l	Ethephon 480 SL	Villa Crop Protection	Sugar cane, Pineapple, Wheat, Cotton Apple, Peach, Plum, Grape Cherry

Active ingredient	Formulation	Concentration	Product name	Company name	Crops
Ethephon	SL	480 g/l	Ethephon 480 SL	Universal Crop Protection	Apple, Cherry, Cotton Grape, Peach, Pineapple Plum, Sugarcane Wheat
Ethephon	SL	480 g/l	Ethrel	Bayer	Apple, Cherry, Citrus Cotton, Grape, Maize Peach, Pineapple, Plum Prune, Sugarcane, Tobacco
Ethephon	SL	480 g/l	Ripen-It	RT Chemicals	Citrus, Pineapple, Sugarcane
fluazifop-p-butyl	EC	125 g/l	Fusilade Super	Syngenta	Grasses (chemical mowing), Sugarcane
fluazifop-p-butyl	EC	150 g/l	Fusilade Forte	Syngenta	Grasses (chemical mowing), Sugarcane
Forchlorfenuron	EC	10 g/l	Sitofex	Degussa Africa	Grape, Kiwi
Gibberellins	SG	400 g/kg	ProGibb 40 %	Valent BioSciences	Citrus, Grape, Pear, Potato Mango,Hop
Gibberellins	SL	32 g/l	Falgro 4 SL	Universal Crop Protection	Citrus, Grape, Grape-seedless, Potato, Pear, Mango , Hops
Gibberellins	SL	32 g/l	Gibb Liquid 4	RT Chemicals	Potato, Citrus, Grape
Gibberellins	SL	32 g/l	ProGibb 4%	Valent BioSciences	Grape, Pear, Citrus, Potato Mango, Hops, Grape-seedless, (Sultana)
Gibberellins	SL	32 g/l	Supagibb 4 SL	Plaaskem (Gouws and Scheepers)	Grape, Mango, Citrus Potato

Active ingredient	Formulation	Concentration	Product name	Company name	Crops
Gibberellins	SP	200 g/kg	Falgro 20 SP	Universal Crop Protection	Citrus, Grape, Grape-seedless, Potato, Pear, Mango Hops
Gibberellins	TB	100 g/kg	Gibb Tab	RT Chemicals	Potato, Citrus, Grape
Gibberellins	TB	133,3 g/kg	Falgro Tablets	Universal Crop Protection	Citrus, Grape, Grape-seedless, Potato, Pear Mango, Hops
gibberellins + fertilizer	SL	0,003 g/l	Margaret Roberts Organic Supercharger	Ball Straathoff	Lawn Flowers Ornamentals
gibberellins/trace elements	SL	0,003 g/l	Nitrosol	Fleuron	Ornamentals Flowers
haloxyfop-P-methyl	EC	108 g/l	Gallant Super	Dow AgroSciences	Sugarcane
indolylacetic acid/fertilizer	SL	0,05 g/l	3:2:1 (22) Supranure Plus	Wonder	Ornamentals Potplants
maleic hydrazide	SL	250 g/l	Royal MH-30 SL	Crompton Chemical	Onion
maleic hydrazide	WG	800 g/kg	Fazor WSG	Crompton Chemical	Onions
mepiquat chloride	SL	50 g/l	Pix	BASF	Cotton
Methylcyclopropene	SP	33 g/kg	SmartFresh	Rohm & Haas	Apple, Avocado
Methylcyclopropene	TB	6,3 g/kg	SmartFresh SmartTabs	Rohm & Haas	Avocado, Kiwi, Plum Tomato
mineral oil	EC	869,3 g/l	Budbreak	BASF	Apple, Pear, Plum
mineral oil	EC	880 g/l	BP Ampron	BP SA	Apple, Pear, Plum Dormex Apple Apricot
mineral oil (heavy grade)	EC	864 g/l	86% Winter Oil	Cropchem	Apple, Pear, Apricot and Plum

Active ingredient	Formulation	Concentration	Product name	Company name	Crops
mineral oil (light)	EC	790 g/l	Wenfinex	Wenkem	Apple
mineral oil (medium grade)	EC	720g/l	Pestridol	Applied Agricultural Products	Apple, Pear, Plum, Apricot
n-decanol	EC	690 g/l	Antak	Sunwood Chemicals	Tobacco
n-decanol	EC	690 g/l	Deca Suckeride 690 EC	Dow AgroSciences	Tobacco
n-decanol	SL	790 g/l	Suckeride	Cropchem	Tobacco
paraquat dichloride	SL	145 g/l	Skoffel	Universal Crop Protection	Potato
paraquat dichloride	SL	200 g/l	Agroquat	Natural Plant Protection	Potato
paraquat dichloride	SL	200 g/l	Avi Parakwat	Avima	Potato, Sugarcane
paraquat dichloride	SL	200 g/l	Gramoxone	Syngenta	Hops, Potato, Sugarcane
paraquat dichloride	SL	200 g/l	Paragone SL	Plaaskem (Gouws and Scheepers)	Cotton, Potato
paraquat dichloride	SL	200 g/l	Paraquat	WPK Landbou	Potato
paraquat dichloride	SL	200 g/l	Paraquat	Cropchem	Potato
paraquat dichloride	SL	200 g/l	Paraquat SL	Kempton Chemicals	Potato
paraquat dichloride	SL	200 g/l	Skoffel Super	Universal Crop Protection	Potato
paraquat dichloride	Sl	200 g/l	Agriquat	Applied Agricultural Products	Potato
Pendimethalin	EC	330 g/l	Accotab	BASF	Tobacco
Pendimethalin	EC	330 g/l	Bacstop 330 EC	Dow AgroSciences	Tobacco
Pendimethalin	EC	330 g/l	Topout	BASF	Tobacco
prohexadione-calcium	WG	100 g/kg	Regalis	BASF	Apple

Active ingredient	Formulation	Concentration	Product name	Company name	Crops
sodium chlorate	SL	400 g/l	Bladval Ultra	Plaaskem (Gouws and Scheepers)	Cotton
sodium chlorate	SL	400 g/l	Zed	Plaaskem (Gouws and Scheepers)	Potato
Thidiazuron	EC	3 g/l	Lift	Almond Agro Chemicals	Apple, Pear
thidiazuron/diuron	SC	120/60 g/l	Dropp Ultra	Bayer	Cotton
trinexapac-ethyl	ME	120,9 g/l	Primo Maxx 120,9 ME	Syngenta	Lawn
Uniconazole	SC	50 g/l	Sunny 50 SC	Philagro	Avocado

(*Source*: *http://www.nda.agric.za/doaDev/sideMenu/ActNo36_1947/AR/PGRs.htm*)

15.1.7. Growth retardants

These chemicals reduce shoot growth and increase root growth and thus enable the plants to resist drought. They may also induce stomatal closure.

There are number of synthesized compounds which prevent the gibberellins from exhibiting their usual responses in plants such as cell enlargement or stem elongation, and therefore they are called as anti gibberellins or growth retardants. The synthesized growth retardants are

1. Cycocel (2- chloroethyl trimethyl ammonium chloride (CCC)
2. Phosphon D – (2, 4 – dichlorobenzyl – tributyl phosphonium chloride)
3. AMO – 161
4. Morphactins
5. Maleic hydrazide

15.1.7.1. Table.23: List of marketed plant growth Retardant

Active ingredient	Formulation	Concentration	Product name	Company name	Crops
glyphosate-isopropylamine	SL	360 g/l	Bounty SL	Meridian Agritech	Sugarcane, Grasses
glyphosate-isopropylamine	SL	360 g/l	Glyphogan 360 SL	Makhteshim-Agan	Grasses (chemical mowing)
glyphosate-isopropylamine	SL	360 g/l	Glyphosate 360 Acid	Monsanto	Sugarcane, Grasses (chemical mowing)
glyphosate-isopropylamine	SL	360 g/l	Mamba 360 SL	Dow AgroSciences	Sugarcane, Grasses (chemical mowing)
glyphosate-isopropylamine	SL	360 g/l	Roundup	Monsanto	Grasses (chemical mowing), Sugarcane
glyphosate-isopropylamine	SL	360 g/l	Roundup Ultra	Monsanto	Sugarcane, Grasses (chemical mowing)
glyphosate-isopropylamine	SL	360 g/l	Scat 360 SL	Universal Crop Protection	Grasses (chemical mowing)
glyphosate-trimesium	SL	480 g/l	Touchdown Plus	Syngenta	Grasses (chemical mowing)
Paclobutrazol	SC	250 g/l	Avocet	Universal Crop Protection	Avocado
Paclobutrazol	SC	250 g/l	Cultar	Syngenta	Avo, Litchi, Mango, Macadamia, Nectarine, Peach Pecan, Plum
Paclobutrazole	SC	250 g/l	Paclobutrazole 250	RT Chemicals	Avocado
Daminozide	SP	850 g/kg	B-Nine SP	Crompton Chemical	Flowers Ornamentals

16

Anti-Transpirant Chemicals to Mitigate Water-Stress

Anti-transpirants are the materials or chemicals which decrease the water loss from plant leaves by reducing the size and number of stomata.

Anti-transpirants are usually foliar sprays, although they may sometimes be used more conveniently as dips for immersing the above ground plant parts. Foliar sprays may reduce transpiration in three different ways:

1. Reflecting materials reduce the absorption of radiant energy and thereby reduce leaf temperatures and transpiration rates;
2. Emulsions of wax, latex or plastics dry on the foliage to form thin transparent films which hinder the escape of water vapour from the leaves; and
3. Certain chemical compounds can prevent stomata from opening fully (by affecting the guard cells around the stomatal pore), thus decreasing the loss of water vapour from the leaf.

16.1. Types of Anti-transpirants

16.1.1. Metabolic inhibitors

These reduce the stomatal opening and increase the leaf resistance to water vapour diffusion without affecting carbon dioxide uptake e.g. **phenyl mercury acetate, abscisic acid (ABA),** and **aspirin.**

1.1.2. Film forming anti-transpirants

These form a colorless film on the leaf surface that allows diffusion of gases but not of water vapour e.g. **silicone oil, phenyl mercuric acetate and waxes.**

Nearly 99 per cent of the water absorbed by the plant is lost in transpiration. Antitranspirants are of four types.

16.1.2.1. Stomatal closing type

Most of the transpiration occur through the stomata on the leaf surface. Some fungicides like **phenyl mercuric acetate (PMA)** and **herbicides like Atrazine** in low concentration serve as anti transpirants by inducing stomatal closing. These might reduce the photosynthesis. PMA was found to decrease transpiration than photosynthesis.These antitranspirants reduce the closure of stomata and reduce the rate of transpiration. Since the stomata are made to close, the rate of CO_2 diffusion into the leaf is also reduced leading to low photosynthetic rates.eg. Phenyl Mercuric Acetate (PMA), Abscisic Acid (ABA) and high CO_2 concentration. CO_2 is an effective antitranspirant. A little rise in CO_2 concentration from the natural 0.03% to 0.05% induces partial closure of stomata. Its higher concentration cannot be used as this result in complete closure of stomata affecting adversely, the photosynthesis and respiration.

PMA is widely used chemical for stomatal closer. It has the ability to close the stomata but it has a disadvantage that it is toxic to fruits and vegetables. This chemical also inhibits phosphorylation and hence damages the foliage by blocking photosynthesis.

Herbicides such as triazine, atrazine, simazine (which are the inhibitors of ETC at QA and QB sites) at the lower concentration can also be used as antitranspiratns.

16.1.2.2. Film forming type

Plastic and waxy material form a thin film on the leaf surface and result into physical barrier and inhibits the loss of water vapour from the leaf. But they allow CO_2 to pass into the leaf through lower epidermis.eg. Waxes, Plastic films, Silicone oils, ethyl alcohol, Tag 9; S 789 foliate.

16.1.2.3. Reflectance type

The principle of using this type of chemicals is to increase the light reflection by the leaves, thus decreasing the leaf temperature or head load on the leaf. The water loss is reduced without affecting the CO_2 assimilation. These materials form a coating on the leaves and increase the leaf reflectance (albedo). By reflecting the radiation, vapour pressure gradient and thus reduce transpiration. **Application of 5 percent Kaolinite (kaoline) spray reduces transpiration losses, eg. Diatomaceous earth product (Celite), hydrated lime/Lime water (Lime wash), calcium carbonate, magnesium carbonate, zincs sulphate etc.**

Anti-transpiratnts are also useful for reducing transplantation shock of nursery plants (Horticultural plants).

Different anti-transpirants are

1. Metabolic inhibitor like phenyl mercuric acetate, some alkanyl succinic acids.
2. Growth retardant such as ABA, Cycocel.
3. Herbicides, fungicides
4. Salicylic acid.
5. Colourless plastics, silicon oil, wax or plastic.
6. Light reflecting materials (e.g. Kaolin) emulsions or white wash.

17

Chemicals to Induce Early Flowering

The use of flower inducers stimulates flower growth during the resting year so the tree produces annually. Flower inducers can also increase the number of viable blooms during the natural fruit producing year, for an increase in production. These can also be used on varieties that provide a crop every year, to boost harvest. The inducers work best on healthy, mature trees that are under 10 years of age. The nutritional factors also plays an important role in flowering process in plants. The flower inducers includes various chemicals (KNO_3, CaN), Pesticides and fungicides, fertilizers, growth hormones (GA3, Salicyclic acid, Ethylene) etc.

17.1. Potassium Nitrate

Potassium nitrate (KNO_3) is a flower inducer that contains potash and nitrogen, needed for flower growth. KNO_3 caused a slight flowering promotion in October and a heavier one in December, on unpruned trees.

The interaction of pruning x KNO_3 on flowering was reported to be significant (Rojas and Leal, 1997). **Potassium nitrate** (KNO_3) has been shown to stimulate flowering in sufficiently mature stems. (Ramirez and Devenport, 2010)

November and December foliar sprays of **ammonium nitrate** (NH_4NO_3) have been successful to promote early flowering of several mango cultivars in Nayarit, Mexico (Salazar-Garcia *et al;* 1999).

Yeshita *et.al.,* (2004) evaluated effects of application of potassium nitrate alone and in combination with urea at different concentrations on flowering, fruit set and fruit quality of "Tommy Atkins" mango grown in the rift valley of Ethiopia and concluded that potassium nitrate concentrations especially in combination with urea (5 litre solution of 4% KNO_3+0.5 g urea tree^{-1} and 5 litres of 4% KNO_3+1 g urea tree^{-1}) produced better results for most of the flowering and yield parameters. There was a non-significant difference for the qualitative parameters between the treated and non-treated trees. The supplementation of nitrogen through the spraying of KNO_3 and urea is believed to be the reason for the greater flowering and yield of the sprayed to the unsprayed trees.

17.2. Calcium Nitrate

Calcium nitrate (CaN) contains 45 percent calcium to induce flowering in mango trees. This chemical treatment is a lower cost option than KNO_3 for stimulating flower growth.

17.3. Fertilizer

Nitrogenous fertilizer are useful for citrus fruit bearing trees to stimulate flower production and healthy tree growth. Organic fertilizers have a lower risk of burning the shallow root system. Apply the fertilizer early in the season, before midsummer, for best results with inducing flowers. Sandy soils drain quickly and might require an extra half dose of fertilizer with each treatment if new growth is slow. Flower inducers sprayed onto the tree are needed in addition to a soil applied fertilizer.

17.4. Pesticide Including Herbicides

A pesticide and fungicide treatment about two weeks after flowers emerge prevents pest damage to the delicate buds. Flowers take about 20 days to fully open and are susceptible to pests during this stage.

17.5. Plant Nutritients

Nitrogen and carbohydrate reserves: It has been found that higher starch, total carbohydrate and C/N ratio in the shoots favoured flower formation. Induction of flowering in the 'off' years and increased flowering in 'on ' years could be achieved by ringing of branches, a practice known to increase C/N ratio of shoots. The higher C/N ratio is reported to be the driving force for flowering in mango cultivars like Dashehri, Langra, Alphonso, etc. Recent investigations have revealed a possible relationship between fruit bud differentiation and amino acid, RNA, DNA, phenols, and proteins. Higher levels of reducing sugar and calcium and low levels of insoluble carbohydrates, non-reducing sugars, magnesium and potassium at the time of flower bud differentiation in Alphonso mango have been reported.

Phosphorous content: The investigations carried out at the Indian agricultural research institute (IARI), New Delhi on Dashehri cultivar indicated a high level of phosphorous in leaves and flowering shoots to be favourable for flower bud differentiation. The flower bud formation in mango is governed by proper adjustment of a number of factors including high level of phosphorous, specially the organic fractions.

17.6. Hormones

Endogenous gibberellins may be the natural factors involved in juvenility and the non-flowering condition in mango (Chacko *et al.,* 1976). **Ethephon** and **KNO_3 reportedly stimulate mango** flowering in the tropics (Davenport and Nunez-elisea, 1990). Flowering intensity was negatively correlated with GA_3-like substances but not correlated with ABA like substances (Pongsomboon *et al.,* 1997). PBZ application to increase number of flowering shoots due to lower vegetative growth and higher reserves in tree, stimulate production of flowering shoots during weakly inductive conditions (Burondkar and Gunjate, 1993).

17.6.1. Gibberellins

The main effect of gibberellins in plants is to cause stem elongation and flowering. These are also prominently involved in mobilization of endosperm reserves during early embryo growth and seed germination.

The flower inducing activity of exogenously applied gibberellins is clear in rosette forming long day plants such as carrot. From this fact, gibberellins have been considered as florigen. This conclusion was not accepted because gibberellins do not induce flowering in short day plants. However, there is no reason to assume that florigen must be the same in different plant species.

17.6.2. Abscisic Acid

Abscisic acid has a dual function in flowering of Pharbitis nil. It inhibits flowering when given before or early in the dark period, and promotes it when given after or later in the dark period. Together with the estimation of endogenous levels and the effect of inhibitors of biosynthesis, abscisic acid has been concluded to inhibit the time measuring process and promote florigen transport and flower evocation.

17.6.3. Ethylene

Ethylene induces flowering in pineapple and plant species belonging to the Annonaceae and the bulbious plants. Ethylene is commercially used as a flower inducing agent in pineapple cultivation. On the other hand, ethylene inhibits flowering in many other plant species. It is possible that ethylene is a florigen of the Annonaceae. However, it is not clear whether ethylene synthesis is induced under inductive photoperiods, or whether ethylene is actually involved in the regulation of flowering.

17.6.4. Chemicals other than Plant Hormones

The endogenous level of polyamines, especially putrescine is known to increase in photoinduced leaves of Sinapis alba. Difluoro methylornitine, an inhibitor of putrescine biosynthesis inhibits flowering, and this inhibiting effect was nullified if putrescine is applied simultaneously. However, polyamines alone do not influence flowering.

Benzoic acid, nicotinic acid, nicotinamide and L-pipecolic acid have been isolated and identified as flower inducing substances from the acetone extract of fronds of Lemna gibba. Differences in their levels between flowering and vegetative plants could not be detected, and thier role as an endogenous regulator of flowering has not been proven.

Asparagus grows vegetatively for 2 to 3 years after the seed germination prior to flowering. The treatment of seeds with **triazole** compounds or carbamate compounds induced flowering in about 1 month after seed germination. These compounds are artificially synthesized chemicals and therefore may not be involved in the natural regulation of flowering.

18

Chemicals to Control Fruit Dropping

Fruit drop just before harvest is a serious problem for various crop varieties. The severity of fruit drop is influenced by several factors but plant stress and premature ethylene production is at the basis of true physiological drop. Stress factors such as late season heat, primary and secondary pest infestation, and nutrient imbalance or deficiency, and heavy summer pruning, all can contribute to drop.

Abscission is shedding or dropping of plant parts such as leaves and fruit can be delayed or hastened by chemical treatment. From the grower's point of view, the control of this drop is of extreme importance.The practical effective delay of abscission by chemical means was discovered about 10 years ago. Growth regulators can be used to thin or to set crop, to remove the crop or to hold it on until it is ready to harvest, to increase fruit colour and to modify maturity.

Fruit drop also can be reduced by Chemical thinning which helps to reduce the crop load which in turn promotes increased fruit size. It is also well known that excessive crop leads to biannual bearing. This can result in economic losses both the on year when excessive cropping reduces fruit size and the off year when yields can be light or non-existent. Reducing the cropload by means of chemical thinning, hand thinning or the combination of both will help to reduce these problems. Various chemical thinners used are Naphthalene acetamide, Naphthalene Acetic Acid, Carbaryl, 6-benzyladenine, 6-benzylaminopurine, Ammonium-Thiosulphate etc.

The plant growth regulators such as indole or naphthalene type-such as indole-acetic acid, naphthalene-acetic acid and naphthalene acetamide, and 2, 4-D; 2, 4, 5-T etc are used to reduce fruit drop.

18.1. Growth Regulators to Control Fruit Drop

NAA (naphthalene acetic acid): is an auxin-type growth regulator recommended for use at 10 to 20 ppm.There are now several manufacturers of NAA. The optimum conditions for maximum absorption are at or near 21°C-24°C and high humidity. Absorption is less on foliage injured by insects, diseases or frost, and at temperatures below16°C. The addition of a non-ionic type spreader sticker improves absorption of NAA under less than ideal weather conditions. NAA inhibits fruit abscission but the fruit continues to mature at an accelerated rate. The higher the concentration used and the greater the number of applications, the greater the ripening effect.

2, 4-D: By delaying abscission layer formation, 2, 4-D sprays may actually be used on citrus to store citrus fruit on the trees since these fruits do not break down on ripening as fast as apples and pears. In oranges it was found that 2, 4-D prevented flower abscission for several months but did not ultimately increase fruit-set since only the usual small percentage of mature fruits developed from these flowers.

The best known plant growth regulator is 2, 4-D which is now widely used as a weed killer at a concentration of about 1000 ppm. This chemical, applied at a relatively low concentration-between 5 and 10 ppm effectively delays pre-harvest drop of Winesap and Stayman-Winesap apples.

19

Chemical Nutrient Deficiencies/ Toxicities in Plants and Their Rectification

Sixteen chemical elements are known to be important for a plant growth, their metabolic activities and survival. These can be divided in to two main groups

1. **Non mineral nutrients:** These are carbon, hydrogen and oxygen. These are found in air and water and derived from these sources by the plants.
2. **Mineral nutrients:** There are thirteen mineral nutrients required by the plants. These are divided into two groups based on their quantitative requirement by the plant.

19.1. Major/Macronutrients

These are required in a comparatively large quantities. These are the essential plant nutrients. These are further divided into two groups

19.1.1. Primary nutrients

Nitrogen, Phosphoros and potassium are the primary nutrients.

19.1.2. Secondary nutrients

Calcium, magnesium and Sulfur are the secondary nutrients.

19.2. Micronutrients

These are required in small quantities by the plant and useful in plant metabolism and physiological activities. **These are Boron, Copper, Iron, Manganese, Zinc, Molybednum and Chloride.**

Though these micronutrients are required in small quantities, the deficiencies or excess amount of these mineral nutrients in the plant hamper their metabolic activities and growth, and are expressed with various symptoms on the plants which must be rectified for proper plant growth, vitality and yield.

Symptoms caused by nutrient deficiencies are generally grouped into five categories:

1) Stunted growth,
2) Chlorosis,
3) Interveinal chlorosis,
4) Purplish-red coloring and
5) Necrosis.

Stunting is a common symptom for many deficient nutrients due to their varied roles in the plant.

19.3. Representative Symptoms of Nutrient Deficiencies/Toxicity in Plants

1. Nitrogen

Nitrogen is important in the growth and development of vital plant tissues and cells. It gives the plant their green colour. Its presence is required in plant protein synthesis and in purine and pyrimidines of DNA and RNA.

Deficiency symptoms: The visual symptoms of nitrogen deficiency are relatively easy to detect in some plant species. Symptoms include stunted poor plant growth, and pale green or yellow leaves (photo.5) because they are unable to make sufficient chlorophyll. Upper leaves remains light green whereas lower leaves turn yellow. The chlorophyll content of the plant leaves is reduced. Flowering and fruiting in the plant is also reduced. The deficiency appears on the older leaves first.

Photo 5: Nitrogen deficiency symptoms

Toxicity symptoms: The foliage and leaves become much dark green. Leaf tip may turn down (photo.6). The claw leaves eventually start turning yellow with spots. Excess amount may cause yellowing of plant with blighted appearance and death.

Photo. 6: Nitrogen toxicity symptoms

2. Phosphorus

Phosphorus is responsible for regulation of protein synthesis, increase stalk and stem strength, increase flower formation and seed production, and increase resistant to plant diseases. It increases N fixing capacity in legume.

Deficiency symptoms: Crops usually display no obvious symptoms of phosphorus deficiency other than a general stunting of the plant during early growth. By the time a visual deficiency is recognized, it may be too late to correct in annual crops. Some crops, such as corn, tend to show an abnormal discoloration when phosphorus is deficient.

In general the plants remain stunted. Older leaves often show dark dull green colour. Leaves are darker than normal. Stem and leaf stalk may turn purple (photo.7). Plant maturity is often delayed. Deficiency symptoms generally appear first on older leaves.

Photo 7: Phophorous deficiency symptoms

Toxicity symptoms: Excess phosphorus can interfere with availability of Cu and Zn, and induce symptoms of these micronutrient deficiency.

3. Potassium

Potassium control the opening and closing of stomata. It activates enzymes responsible for specific functions. Facilitate protein and starch synthesis in plants. It plays a role in osmoregulation of water and other salts in plant tissues and cells.

Deficiency symptoms: Stunted growth with slow or poor developed roots and stem. Chlorosis with yellowing of leaves which starts at leaf tips and leaf edges (photo.8), usually in younger leaves. Dead or yellow patches develop on younger leaves. Plant growth, root and fruit development are usually reduced in potassium-deficient plants.

Photo 8.: Potassium deficiency symptoms

Toxicity symptoms: Leaf tip and marginal necrosis (photo.9).

Photo 9: Potassium toxicity symptoms

4. Calcium

Calcium stimulates the absorption of ammonium. In the form of calcium pectate, it is responsible for holding together the cell walls of plants.It is important in guard cell physiology, specially in stomato closure. It activates certain enzymes and send signals that co-ordinate certain cellular activity.

Deficiency symptoms: New Young leaves and fruits display calcium deficiencies first. Calcium deficiency symptoms appear initially as localized tissue necrosis leading to stunted plant growth, necrotic leaf margins on young leaves or curling of the leaves, and eventual death of terminal buds and root tips. Generally, the new growth and rapidly growing tissues of the plant are affected first. Blossom end rot of tomatoes is a classic case of calcium deficiency (photo.10).

Photo 10: Calcium deficiency symptoms

Toxicity symptoms: High level of calcium can compete with magnesium and potassium uptake causing their deficiencies.

5. Magnesium

It plays a role in chlorophyll formation, synthesis of amino acid and cell protein, and in vitamin A and C concentration. It makes the plant resistance to unfavourable factors.

Deficiency symptoms: Magnesium deficient plants begin to degrade the chlorophyll in the old leaves, thereby causing interveinal chlorosis, or yellowing between leaf veins, sometimes with reddish brown tints while veins remain green giving the leaves a marbled appearance (photo.11).The deficiency causes early leaf fall.

Photo 11: Magnesium deficiency symptoms

6. Sulfur

It is essential for growth and development of all crop, formation of chlorophyll, production of protein, synthesis of oils in oilseeds and activation of enzymes.

Deficiency symptoms: A sulfur deficient plant show yellowing or pale green coloring throughout the plant. Younger leaves suffer from chlorosis with their tips becoming necrotic. Overall plant development and growth will be stunted without enough sulfur in the soil. Leaf tips may yellow and curl downward. Severely affected plants show yellowing/ whitening of entire leaf including veins (photo.12).

Photo 12: Sulfur deficiency symptoms

Toxicity symptoms: Leaf size will be reduced, overall growth stunted, yellowing or scorching of leaf edges.

7. Boron

Deficiency symptoms: symptoms of boron deficiency are expressed at growing tips of the root or shoot, and generally include stunting and distortion of the growing tip that can lead to tip death, brittle foliage, and yellowing of lower leaf tips.It also causes limited budding, bud break, distorted shoot and fruit growth (photo.13), short internodes, hallow stem and increased branching.

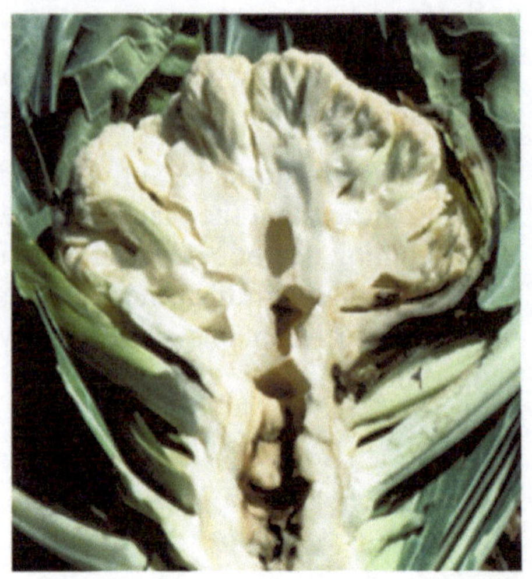

Photo 13: Boron deficiency symptoms

Toxicity symptoms: Chlorotic leaf tip, leaf necrosis, leaf falling, even plant death.

8. Chloride

Chloride (Cl) is needed in small quantities and aids in plant metabolism, photosynthesis, osmosis and ionic balance within the cell. Participates in several physiological metabolism processes, plant growth and development, osmotic and stomata regulation, evolution of oxygen in photosynthesis, disease resistance and tolerance, important in opening and closing of stomata. Chloride function in cat ion balance and transport within the plant.

Chloride deficiency can occur if there is consistently less than 2 ppm chloride in the growing medium.

Defiencies symptoms: Chloride deficiency symptoms include wilting due to a restricted and highly branched root system, often with stubby tips, and Leaf mottling and leaflet blade tip wilting with chlorosis. Chlorotic blotches with necrotic spots located between veins or on the margin of the younger leaves are also observed (photo.14). In advance stages Cl deficiency could cause plant wilting.

Photo 14: Chloride deficiency symptoms

Toxicity symptoms: Chloride toxicity starts as premature yellowing of leaves, then marginal or tip necrosis of older leaves (photo.15).

Photo 15: Chloride toxicity symptoms

9. Copper

Copper play part in several enzyme productions and is key to the formation of chlorophyll. Normal range is 0.05 to 0.5 ppm. Activate some enzymes in plants which are involved in lignin synthesis and is essential in several enzyme systems. Require in the process of photosynthesis, respiration and assist in plant metabolism of carbohydrates and proteins, serve flavor and color in vegetables and flowers.

Deficiency symptoms: Deficiency symptoms of copper are dieback of stems and twigs, yellowing of leaves, stunted growth and pale green leaves that wither easily. Chlorosis of either whole leaf or between the veins of newer leaves occurs. Small necrotic spots may form with the chlorotic areas of the leaf, especially on the leaf margin. Leaves become small in size, curl upward (photo.16), develop necrotic leaf edges and wilt. Apical meristem become necrotic and die .The growth of lateral branches is affected.

Photo 16: Copper deficiency symptoms

Copper deficiencies are mainly reported on sandy soils which are low in organic matter. Copper uptake decreases as soil pH increases.

Toxicity symptoms: Excess copper can restrict root growth by burning the root tips. Compete with plant uptake of Fe and sometimes Mo and Zn. Reduced branching and eventually plant die.

10. Iron

Iron is very important to health and plant growth. Essential for chlorophyll formation, components of cytochromes, essential for N fixation, assist in nitrate and sulphate reduction and energy production within the plant.

Deficiency symptoms: The most notable symptom of iron deficiency is chlorosis or yellowing between the veins of the youngest leaves i.e.interveinal chlorosis of new leaves (photo.17).Younger leaves are yellow and white with green veins.Older leaves are normal.

Iron deficiency in plant is also known as "lime-induced chlorosis". When soil iron concentration is high, but can not become available for absorption if soil pH is higher than 6.5, shows iron deficiency.

Photo 17: Iron deficiency symptoms

Fe chlorosis develops when soil pH is high.

11. Manganese (Mn)

Required for chlorophyll synthesis and oxygen evolution during photosynthesis, activate some enzyme systems contributors to various biological systems including photosynthesis, respiration and nitrogen assimilation, involved in pollen germination, pollen tube growth, root cell elongation, resistance to root pathogens.

Deficiency symptoms: Manganese deficiency symptoms, which often look like those of iron deficiency, appear as interveinal chlorosis (yellow leaves with green veins) on the young leaves (photo.18), and sometimes tan, sunken spots that appear in the chlorotic areas between the veins or elongated holes between veins. Plant growth may also be reduced and stunted with leaf shading, and failure to mature.

Photo.18: Manganese deficiency symptoms

Mn deficiency can occure when the pH of the growing medium exceed 6.5. The deficiency is most common on alkaline soils (high pH) particularly if the irrigation water contains high levels of bicarbonate.

Toxicity symptoms: Burning of tips and margin of older leaves with reddish brown spot occurs in the form of patches on older leaves (photo.19). Mn toxicity occur when fertilizer application rate is excessive, and the pH of growing media is below 5.5.

Photo.19: Manganese toxicity symptoms

12. Molybdenum

Require in smallest quantity by plants i.e. 0.3 to 1.5ppm. Component of two enzymes that convert nitrate in to nitrite and then in to ammonia before it is used to synthesize amino acid within the plant. Needed to symbiotic N fixing bacteria in legumes to fix atmospheric nitrogen.

Deficiency symptoms: Resembles to N deficiency symptoms. Deficiency symptoms show up on older and middle leaves as thin chlorotic leaf margin around the leaf perimeter and then the margin become necrotic (photo.20), whole leaf turn pale.

Photo 20: Molybdenum deficiency symptoms

13. Zinc

Helps plant to produce chlorophyll, an important component to photosynthesis. It is also a constituent of certain enzymes.

Deficiency symptoms: Chlorosis of leaves, often interveinal. In some species, young leaves are the most affected, but in others both old and new leaves are chlorotic. Necrotic spots develops on areas of chlorosis. Chlorotic areas may turn bronze coloured (photo.21); Growth suppression, reduce internode length, and rosetting also occur.

Photo 21: Zinc deficiency symptoms

Toxicity symptoms: Growth reduction, Leaf chlorosis, occur at low pH.

19.4. Chemicals Used to Rectify the Nutrient Deficiencies

Nitrogen deficiency: Use fertilizers having Nitrogen like nitrate fertilizers, Ammonical fertilizers, urea, ammonium sulfate, ammonium chloride, diammonium phosphate, Ammonical- nitrate fertilizers, Ammonium nitrate, Ammonium-sulphate nitrate, Calcium ammonium nitrate, Amide fertilizers etc.

Phosphorous deficiency: Use fertilizers having phosphorous like Diammonium phosphate, single super phosphate, triple super phosphate, Di-calcic phosphate, Rock phosphate, Bone meal etc.

Potassium deficiency: Use fertilizers having potash like Muriate of potash, potassium sulfate, potassium nitrate, sulfate potash magnesia, Potassium silicate, kainite etc.

Calcium deficiency: Use calcium sources like calcium nitrate, calcium ammonium nitrate etc.

Magnesium: Use magnesium sources like dolomitic limestone, sulfate of potash magnesia, magnesium nitrate fertilizers, magnesium sulphate etc.

Sulfur: Use sulfur sources like Ammonium sulfate, single superphosphate, gypsum, fertilizers containing elemental sulfur, liquid sulfur fertilizers etc.

Boron: Use boron sources like Borax, boric acid or solubor, boron-containing granular potash fertilizer called Aspire etc.

Chloride: Use chloride sources like murate of potash, calcium chloride, ammonium chloride and magnesium chloride etc.

Copper: Use copper sources like copper sulphate, multimicronutrient containing copper. Copper deficiency due to high soil pH should be treated by lowering soil pH with acidifiers such as sulfur or ammonium sulfate.

Iron: Use of iron sources like Ferrous sulfate or iron chelate compound, multimicronutrient containing iron. Two common iron chelate are Fe EDTA and Fe EDDHA. Iron sulphate and iron EDTA are only useful in soil up to pH 7.1, but they can be used as foliar spray.

Mangenese: Use of manganese sources like manganese sulphate or multimicronutrient containing manganese. Mangenese deficiency can be rectified by adjusting or lowering the soil pH either with ammonium sulfate or sulfur.

Molybdenum: Use Ammonium molybdate, Molybdenum trioxide, Sodium molybdate or multimicronutrient containing molybdate. Acidic soils are deficient in molybdenum which can be rectified with liming.

Zinc: Use micronutrient foliar spray that contains zinc. For soil application, use chelated zinc. Add compost or organic matter to sandy soils to help the soils to manage zinc better. Cut back on high phosphorous fertilizers because they reduce the amount of zinc available to the plants.

The applications of nutrients to rectify the deficiencies should be based on soil tests for the required amount of nutrients which differ from crop to crop.

Section-III

Pesticides for Management of Insect Pest

20

Classification of Insect Pest Based on Their Appearance and Feeding Habit (For Proper Insecticide Selection and Application)

20.1. Leaf and Foliage Eating Insect Pest

1. Heliothis

2. Semilooper

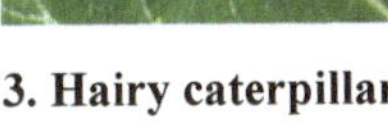

3. Hairy caterpillar

4. Leaf eating caterpillar

5. Plant beetles

6. Locust

7. Gaint Snail

20.2. Plant Sap Sucking Insect pests

1. Aphids

2. Jassids

3. White Flies

4. Wooly Aphid

5. Plant bugs

6. Leaf hopper

7. Thrips

8. Mites

9. Mealy bug

20.3. Bollworm and Fruit Borer

1. Bollworm

2. Pink bollworm

3. Fruit Borer

20.4. Shoot Borer, Stem borer and Cutworm

1. Shoot borer 2. Stem borer

3. Stem borer 4. Cutworm

20.5. Fruit flies and Friut Moths

1. Fruith Moth

2. Fruit flies

20.6. Bark Eating Caterpillar

1. Bark eating caterpillar

20.7. Leaf Tissue Scratching Insect Pest

1. Leaf miner

20.8. Web Forming Insect Pest

1. Leaf roller

2. Web forming caterpillar

3. Spider mite

20.9. Root Damaging Insect Pest

1. Root borer

2. Termite

3. Root aphid

4. Root Mealybugs

5. **Root weevils**

6. **Root Maggot**

7. **White grub**

8. **Fungus gnat larvae**

Photo 22: Insect pests of crop plants. Courtesy: Dr Borkar's Laboratoty, Nashik, India.

21

Insecticides for Control of Insect Pests

Medicines used for controlling insect pests are known as Pesicides which have assumed great importance in today's high yielding and intensive agricultural system as well as public health programme, all over the world. The term pesticide is an all-inclusive word meaning killer of pests and includes substances used for controlling, preventing, destroying, repelling or mitigating any pest.

Insecticides are organic or inorganic chemical substances or mixture of substances intended for preventing, destroying, repelling or mitigating the effect of any insect including crawling and flying insects; and these may occur naturally or is synthesized. Most of the synthesized insecticides by their nature are hazardous on health if not used properly. Insecticides vary in its hazardous nature and therefore, range from the extremely hazardous to those unlikely to produce any acute hazard. Some insecticides are insect repellants or insect growth regulators and used in agriculture, public health, horticulture and food storage. Insecticides have been used to boost food production to a considerable extent as these save the crop yields by controlling insect pest damages to the crop yield and similarly to control vectors of plant diseases as well as human diseases. However, these advantages that are of great economic benefits sometimes come with disadvantages when subjected to critical environmental and human health considerations. Many insecticides are newly synthesized whose health and environmental implications are unknown.

Insecticides of various chemistries like hydrocarbon oils (tar oils), arsenical compounds, organochlorine, organophosphorous compounds, carbonates, dinitrophenols, organic thiocynates, sulfur, sodium fluoride, pyrethroids, rotenone to nicotine, in solid or liquid preparation have been used in plant protection. Interestingly, most of these have been withdrawn due to the deleterious effect of the substances on human health, crop ecosystem and environment. Certain group of pesticides are effective against certain pests and not on others and therefore knowledge of pesticide groups are important for the crop pest management.

21.1. Pesticide Groups

The pesticides are generally classified into various groups based on pest against which the compounds are used, their chemical nature, mode of entry and mode of action.

21.1.1. Based on mode of entry

A) **Stomach poison:** The insecticide applied in/on the leaves and other parts of the plant when ingested, act in the digestive system of the insect to kill it. e. g. Malathion.

B) **Contact Poison:** The toxicant causes death of the pest species by means of contact of the toxicant to the insect. e. g. **Fenvalerate.**

C) **Fumigant:** The Toxicant in vapour form enter into the tracheal system (respiratory) of the insect through spiracles and causes death of the insect. e. g. **Aluminium phosphide**

D) **Systemic poison:** Insecticide when applied to plant or soil are absorbed by foliage, or roots and translocated through vascular system in the plant parts and the insect feeding on such plant die. e. g. **Dimethoate.**

21.1.2. Based on mode of action

A) **Physical poison:** Substances which exert a physical effect on insect to kill them. e. g.**Activated clay**.

B) **Protoplasmic poison:** Toxicant which precipitate cellular protein of the insect to cause death. e. g. **Arsenicals.**

C) **Respiratory poison:** Chemicals which inactivate respiratory system of the insect. e. g. **hydrogen cyanide**.

D) **Nerve poison:** Chemicals which knock down the nervous system of the insect to kill it. .**e.g.Malathion**.

E) **Chitin inhibition:** Chemicals which inhibit chitin synthesis in the cell wall of insect body. **e.g.Diflubenzuran**

21.1.3. Based on chemical nature

(Chemicals in under line letters are banned in India)

I. Inorganic pesticides

Inorganic chemicals used as insecticides e.g. **Arsenic, Fluorine, Sulphur, lime sulphur (Insecticides), Zinc phosphide (Rodenticide)**

II. Organic pesticides

- **Animal origin insecticides** –

  e. g. Nereistoxin extracted from marine annelids – commercially available as **cartap, padan.**

- **Plant origin insecticides** :

  Nicotine from tobacco plants,

  Pyrethrum from *Chrysanthemum* flowers,

  Rotenoids from roots of *Derris* and *Lonchocarpus*

  Neem – *azadirachtin,*

  Pongamia glabra,

  Garlic etc.

- **Synthetic organic compounds**: These organic chemicals are synthetically produced in laboratory.

i. **Chlorinated hydrocarbon or organochlorines**

   e. g. DDT, HCH, Endosulfan, Lindane, Dicofol

ii. **Cyclodienes**

   e. g. Chlordane, Heptachlor

iii. **Organophosphates: (Esters of phosphoric acid)**

   e. g. Dichlorvos, Monocrotophos, phosphamidon, Methyl parathion, Fenthion, Dimethoate, Malathion, Acephate, Chlorpyriphos

iv. **Carbamates: (Derivatives of Carbamic acid)**

   e. g. Carbaryl, Carbofuran, Carbosulfan

v. **Synthetic pyrethroid**

   (Synthetic analogues of pyrethrum)

   e.g.Allethrin, permethrin, Fenvalerate

vi. **Miscellaneous compounds**

   Neonicotinoids (Analogues of nicotine)

   e.g. Imidacloprid Spinosyns (Isolated from actinomycetes), Spinosad Avermectins (Isolated from bacteria), Avermectin, Vertimec Fumigants: e.g. Aluminium phosphide, Hydrogen cyanide, EDCT

21.2. Inorganic Pesticides

Several mineral compounds of arsenic, fluorine, copper, mercury, selenium, antimony, zinc, thalium etc., which do not contain carbon in the composition, constitute inorganic insecticides. Being crystalline and non-evaporative materials, these have long shelf lives.

Pesticides in this group, for example borates, silicates and sulfur, are minerals that are mined from the earth and ground into a fine powder. Some work as poisons and some work by physically interfering with the pest. Older "inorganics" included such highly toxic compounds as arsenic, copper, lead and tin salts. Current inorganic pesticides are relatively low in toxicity and have low environmental impact. Borate insecticides, such as Bora Care and Timbor have many uses in structural pest management and are very safe compared to older conventional pesticides.

Inorganic pesticides are simpler compounds. They have a crystalline, salt-like appearance, are environmentally stable, and usually dissolve readily in water. The earliest chemical pesticides were inorganic, and included substance such as sulfur and lime.

21.2.1. Arsenicals

These compounds have two forms

- **Trivalent arsenites**
- **Pentavalent arsenates**

Arsenicals are stomach poisons, effective only by oral ingestion.

i. **Arsenites**

   White tasteless, odourless compound.

   About 20 times as toxic as DDT.

   e. g. Calcium arsenites, Sodium arsenates.

Can be used in solids baits for locusts, snails and as ant paste.

Paris green

First commonly used arsenical for the control of Colorado beetle by 1865.

It is 3:1 or 2:1 mixture of copper meta-arsenites and copper acetate.

Highly phytotoxic.

Rapidly settles down from suspensions.

Poor adhesive quality.

Used as poison baits for leather jackets and slugs or as mosquito-larvicide.

No longer used in India

ii. Arsenates

Compared to arsenites, arsenates show less insecticidal activity.

Safer to use on plants being highly water insoluble.

Non-phytotoxic.

a. Lead arsenate

Very low phytotoxicity.

Gives uniform deposits on plants

Basic lead arsenate remain stable in the pH range of 6.5 to 11.

Safer to use on tender foliage.

Used as a dust, spray or poison bait ingredient to control chewing insects like locusts, grasshopper, Lemon butterfly etc.

b. Calcium arsenate

Comparatively cheaper.

More water soluble than lead arsenates.

Used to control Epilachna beetle on brinjal and potato.

c. Zinc arsenate

Used as wood impregnation against insect attack.

21.2.2. Fluorine Compounds

Fluorine has played a significant role in insect control since about 1896 when sodium fluoride and various iron fluorides were patented in England as insecticides. Sodium fluoride was used in the United States for cockroach control before 1900 and was introduced in 1915 for the control of poultry lice. However, the use of fluorine insecticides did not become general until the 1930′s when the disadvantages of arsenical residues on food crops became apparent and the inorganic fluorine compounds were introduced as safer substitutes. Systematic investigation of organofluorine insecticides began about 1935 in the I. G. Farben industrie and the fluoroalcohols and fluorophosphates (phosphorofluoridates) were intensively investigated largely through the research of Schrader (1952). During World War II fluoro-DDT or "Gix" was used for the control of insects of medical importance. More recently, fluoroacetamide and analogues have been used as systemic insecticides and a large variety of other fluorinated organic compounds have shown insecticidal activity. Sulfuryl fluoride has recently been marketed as a fumigant for household and structural pests.

i. **Sodium fluoride** (NaF):

   Contains about 45% fluorine.

   Used as dust mainly for the control of cockroaches and ants.

   Hazardous compound since 2-4 gram dose is fatal to humans.

ii. **Calcium and Magnesium fluosilicates**:

   Effective against chewing insects.

   e. g. Cryolite (Sodium fluoaluminate) – Safest insecticides.

   Used against household insects.

   The insects are killed by absorption of waxes and consequent loss of water from the insect cuticle

21.2.3. Boric acid

Effective against cockroaches and other crawling household pests.

Act as stomach poison as well as contact (cuticular) wax absorber.

21.2.4. Thalium sulphate or Thallus sulphate

White crystalline water soluble, colourless and tasteless powder for rodent control and ant baits.

21.3. Organic Pesticides

21.3.1. Synthetic pesticide

The use of synthetic pesticides in the US began in the 1930s and became widespread after World War II. By 1950, pesticide was found to increase farm yield far beyond pre-World War II levels. Farmers depend heavily on synthetic pesticides to control insects in their crops. Today, it is one of the most commonly used pesticide in controlling insects.

In 1997, US pesticide purchase accounted for nearly one-third of the world market. Annual pesticide expenditure totaled approximately $11.9 billion. Two-thirds of pesticide purchased is used in agriculture.

The use of synthetic pesticides in agriculture comes with a cost for the environment, and the health of animals and humans.

21.3.1.1. Danger of synthetic pesticide

The synthetic pesticide DDT was widely used in urban aerial sprays to control urban mosquito, gypsy moth, Japanese beetle and other insects in the 1940's. By 1972, DDT was banned from the United States due to widespread development of resistance to DDT and evidence that DDT use was increasing preterm births and also harming the environment. DDT was found to cause behavioral anomalies and eggshell thinning in populations of bald eagles and peregrine falcons. Although DDT is banned in the US and many other countries, DDT continues to be manufactured and applied in underdeveloped nations where some of the US food supply is grown.

Dursban, one of the most common pesticide used in households, schools, hospitals and agriculture was banned in 2000 by the USEPA due to unacceptable health risk, especially to children. Toxicology studies have found that exposures to Dursban early in life may affect the function of the nervous system later in life, with possilbe links to changes in normal learning and behavior. Yet, six manufacturers in the US are allowed to continue making the chemical for use on foreign crops.

21.3.2. Types of synthetic pesticide

There are many classes of synthetic pesticides. The main classes consist of organochlorines, organophosphates, carbamates, and pyrethroids. Exposure to pesticides can cause acute (short term) or chronic (long term) effects on animals and humans, especially in the reproductive, endocrine, and central nervous systems (table.24).

Table 24: Classes of Synthetic pesticide and their effect on human system

Class	Examples	Area of Effect
Organochlorines	DDT, toxaphene, dieldrin, aldrin	Reproductive, nervous, endocrine, and immune system
Organophosphates	Diazinon, glyphosate, malathion	Central nervous system
Carbamates	Carbofuran, aldicarb, carbaryl	Central nervous system
Pyrethroids	Fenpropanthrin, deltamethrin, cypermethrin	-

Today, there are more than 500 species of insects and mites that are resistant to some form of pesticides. As a result of the increasing resistance, countries have started to apply more products, combine pesticides, increase applications, or substitute with more toxic replacements.

Harm to Humans

There are three types of harmful effects caused by pesticides on humans: acute effects, delayed effects, and allergic effects.

Acute effects are injuries or illnesses that appear immediately after exposure. The effects are usually obvious and reversible if appropriate medical care is given right away.

Delayed effects are illnesses or injuries that do not appear immediately. These include cancer.

There are over 160 synthetic pesticides

21.3.2.1. Organochlorine Pesticides (OCPs)

These persistent, bio-accumulative pesticides include DDT, dieldrin, heptachlor and chlordane. In all, there are 13 pesticides on the list of scheduled wastes.Since they were first introduced into Australia in the mid-1940s, OCPs have been used in many commercial products, in different forms (eg powders and liquids) and in different types and sizes of containers. At one time up to 150 commercial products may have been registered in Australia, although they have been deregistered for several years now. Some of the more common product names include Hortico Dieldrin Dust, Mustex 25% DDT, Shell Dieldrex and Yates Garden Dust. They may be found in small to large containers, ranging from home garden packs (such as glass poison bottles or cardboard shaker packs) to large steel drums. Sizes of containers range from 50 ml to 200 litres. While most containers will have been clearly labelled, the original labels may have since deteriorated, or are no longer attached.

Originally, OCPs were widely and commonly used to protect crops, livestock, buildings and households from the damaging effects of insects. Deregistration has left Australia with stocks of unused and unwanted OCPs. Unable to be used, and with no easily accessible means of destruction, these pesticides are now being stored by individuals and government agencies, until a permanent solution to manage and destroy them is developed.

Why are OCPs harmful?

Generally, OCPs:

- Resist degradation by chemical, physical or biological means. They are persistent and have half-lives (the time taken for half of the quantity of pesticide to be degraded) ranging from months to years and in some cases decades;
- Are toxic to humans and other animals and are very highly toxic to most aquatic life. They can have serious short-term and long-term impacts at low concentrations. In addition, non-lethal effects such as immune system and reproductive damage of some of these pesticides may also be significant; and
- Build up in the fatty tissues of humans, plants and animals. Most of them are attracted to fatty tissues and organs and are accumulated significantly in animals such as fish. This means that animals high up the food chain such as birds of prey and humans can accumulate higher levels of the pesticides than animals lower down the food chain.

While these general observations are true, it is important to remember that OCPs are a diverse group of chemicals and their toxicity, their potential to build up in tissues and their persistence varies.

i. DDT Group

Chemically DDT is known as Dichloro Diphenyl Trichloroethane.

It belongs to chemical class of diphenyl aliphatics which means that it consist of an aliphatic or straight carbon chain with two phenyl rings.

DDT was first synthesized by Othmar Zeidler in 1874. However the insecticidal properties of DDT were first discovered in 1939 by Paul muller of the J.R. Geigy Company for which he obtained a Swiss patent in 1942 and the Nobel Prize in Medicine in 1948. The early dust formulation containing 5 or 10 % DDT proved to be very effective against a wide variety of insect pests of agricultural crops, in particular in controlling infestations of Colorado potato beetle which threatened the Swiss potato crop in 1939. The first full scale use of DDT was during World War II in Naples, Italy.

The control of insects affecting man and animal was revolutionized by the application of DDT. Aside from its favourable toxicity to a wide range of insects, its lack of color, odor, and skin irritancy made it an ideal chemical for the control of ectoparasites. It is proved to be an excellent agent against larvae of mosquitoes and blackflies, against adults of all the important biting flies (such as blackflies, tsetse flies, sandflies, horseflies, deerflies, etc.) Direct spraying of DDT on the dairy cattle was avoided due to its accumulation in body fats and milk. Also it is toxic to certain species of mites, ticks, and dipterous larvae.

Commercial preparations of DDT were completely safe to plant foliage with the exception of cucurbits. DDT could persist on the foliage for over a month since neither heat nor light destroy it, rain does not dissolve it, and due to its low pressure, its loss by evaporation is very slow. Its use on feed for dairy cattle was prohibited from early times due to its accumulation in cow's milk.

DDT showed great promise for the control of a wide variety of insects, including the Colorado potato beetle, European corn borer, various species of leafhopper, flea beetles, alfalfa weevil, pea aphids, Japanese beetles, codling moth, oriental fruit moth and many others.

Although DDT is effective against nymph and adults of the two spotted mite and the red spider mite, it does not kill their eggs and in the absence of natural enemies, the mite population explodes and becomes more abundant than before. Thus, DDT has been found to encourage the development of infestations of several mite species in orchards, greenhouses and crop plants.

A worldwide programme of malaria eradication was undertaken by World Health Organization in 1955 resulted in striking reductions of malaria cases in many areas of the world, particularly in Africa.

Many pesticide companies have been involved with DDT production. By 1991, only Enichem Synthesis (Italy), Hindustan Insecticides (India) and P.T Montrose Pesticido Nusantara (Indonesia) were listed by UN Environment Programme (UNEP) as basic producers. It is also produced in Mexico.

The various formulations of DDT under various trade names (table.25) were used, for agricultural pest control.

Table 25: Commercial insecticide of DDT ingredient and the pest controlled

Name of insecticide	DDT (Dichloro Diphenyl Trichloroethane)
Trade names	Zerdane Anofax Cezare Dinozide Guesapan Guesarol Gyron Ixodex Neocide Neocidol Zerdane Chlorophenothane, Dedelo, p,p'-DDT, Dinocide, Didimac, Digmar, ENT 1506, Genitox, Gexarex, Hildit, Kopsol, Neocid, Pentachlorin, Rukseam,
Active ingredient	p,p'-DDT
Formulations	Dustable powders, emusifiable concentrates
Pest controlled	Colorado potato beetle, European corn borer, various species of leafhopper, flea beetles, alfalfa weevil, pea aphids, Japanese beetles, codling moth, oriental fruit moth

However this insecticide is banned presently and its production is restricted.

DDT restrictions

Control actions to ban or severely restrict DDT have been taken by over 38 countries that began in the early 1970s. In at least 26 countries, DDT is completely banned, and in 12 others it is severely restricted. In these latter cases, it is permitted for use by government agencies for special programmes, usually involving vector control programmes.

Total ban

Canada (*1985*)*;* Chile (*1985*)*;* Cuba (*1970*)*;* Liechtenstein (*1986*)*;* Korea (*1986*)*;* Poland (*1976*)*;* Singapore (*1984*)*;* Switzerland (*1986*).

DDT was banned in the US in 1972, except for public health emergencies. Small amounts were used until 1980. In 1991, the US exported 92 tons of DDT.

ii. Methoxychlor

Methoxychlor is one of a few organochlorine pesticides that have seen an increase in use since the ban on DDT in 1972. This is due to its relatively low toxicity and relatively short persistence in biological systems. Methoxychlor is a general use pesticide.

Methoxychlor is a contact and stomach insecticide effective against a wide range of pests encountered in agriculture, households, and ornamental plantings. It is registered for use on fruits, vegetables, forage crops and on shade trees. Methoxychlor is also registered for veterinary use as a poison to kill parasites on dairy and beef cattle.

It is used to protect crops, ornamentals, livestock, and pets against fleas, mosquitoes, cockroaches, and other insects (table.26) with its various trade brands.

Table. 26: Commercial insecticide of Methoxychlor ingredient and the pest Controlled

Name of insecticide	Methoxychlor
Trade names	Marlate, Chemform Methoxy-DDT Dimethoxy-DT, DMDT, ENT 1716, Higalmetox, Methoxychlore, OMS 466 and Prentox.
Nature	Contact and stomach insecticide
Formulations	Wettable and dustable powder, emusifiable concentrates, granules, and as an aerosol.
Pest controlled	Colorado potato beetle, European corn borer, various species of leafhopper, flea beetles, alfalfa weevil, pea aphids, Japanese beetles, codling moth, oriental fruit moth

It was intended to be a replacement for DDT, but has since been banned based on its acute toxicity, bioaccumulation, and endocrine disruption activity.

The amount of methoxychlor in the environment changes seasonally due to its use in farming and foresting. It does not dissolve readily in water, so it is mixed with a petroleum-based fluid and sprayed, or used as a dust. Sprayed methoxychlor settles on the ground or in aquatic ecosystems, where it can be found in sediments. Its degradation may take many months. Methoxychlor is ingested and absorbed by living organisms, and it accumulates in the food chain. Some metabolites may have unwanted side effects.

Environmental Fate

Methoxychlor is very persistent in soil and its half-life is greater than six months. However, period may be as fast as one week in some instances. The chemical is tightly bound to soil and is insoluble in water, so it leaches slowly, if at all. Methoxychlor degrades much more rapidly in soil that has a supply of oxygen (aerobic) than in soil without oxygen (anaerobic). Any movement of the pesticide is expected to take place while attached to suspended soil particles in runoff. In the EPA pilot groundwater survey, methoxychlor was found in a number of wells in New Jersey (not quantified) and at extremely low concentrations (from 0.1 to 1.0 ppt) in water from the Niagara River, the James River, and a Lake Michigan tributary. Many other rivers tested throughout the United States did not contain methoxychlor.

In water the major products of breakdown in a neutral solution are anisoin, anisil, and DMDE. The half-life in distilled water is 37 days but in some river waters the half-life is as rapid as two to five hours. Methoxychlor evaporates very slowly, but the evaporation may contribute to the cycling of the product in the environment.

On mature soybean foliage, the washoff rate was 8% per cm of rainfall with a total of 3% washoff for a season. Dislodgeable residues account for less than 1% of the amount applied.

iii. Dicofol (Kelthane)

Dicofol is an organochlorine pesticide that is effective against mites, but controversial because of its environmental impact. Written as $C_{14}H_9Cl_5O$ or 2,2,2-trichloro-1,1-bis(4-chlorophenyl)ethano, this acaricide is used on a variety of crops including cotton, apples, citrus fruits, mint, strawberries, beans, peppers, tomatoes, walnuts, stone fruits, cucumbers, ornamentals, and other fruits and vegetables. Among the various mites that dicofol is effective against, it is often used on the red spider mite.

Dicofol is used to kill crop-feeding mite pests such as the red spider mite. It is a contact poison which kills the pest after being ingested and picked up from the surface of the crop. In many countries, dicofol is also used in combination with other pesticides such as the organophosphates parathion-methyl, and dimethoate.

Dicofol is an organochlorine acaricide (a chemical that kills mites) that is structurally similar to DDT. DDT is one of the intermediate products produced during dicofol manufacture. In 1986, the US Environmental Protection Agency (EPA) temporarily cancelled the use of dicofol because relatively high levels of DDT contamination were ending up in the final product. Modern processes can produce technical grade dicofol that contains less than 0.1% DDT.

There are a number of commercial formulations (table.27). In its purest form, dicofol is a white crystal, however in most applications it appears as an amber-colored, viscous liquid that smells like hay.

Table 27: Commercial insecticide of Dicofol ingredient and the pest controlled

Name of insecticide	Dicofol (Kelthane)	
Trade names	Acarin, Cekudifol, Decofol, Dicaron, Dicomite, Difol, Hilfol, Kelthane, Mitigan	
Nature	Acaricide	
Active ingredient	Dicofol 185g/Lt	
Formulations	EC	
Pest controlled	Red spider mite, Scarlet mite, Pink mite, Purple mite, yellow mite	
Crop	Pest	a.i (gm/ha)
Tea	Red spider mite, Scarlet mite, Pink mite, Purple mite, Yellow mite	230
Okra	Red Spider mite	250-500
Citrus	Red Spider mite	0.05%
Litchi	Red Spider mite	0.05%
Cotton	Red Spider mite	500-1000
Brinjal	Yellow mite	500-1000
Bottle & Bitter gourd	Red Spider mite	250-500

Uses: Dicofol is used to kill crop-feeding mite pests such as the red spider mite. It is a contact poison which kills the pest after being ingested and picked up from the surface of the crop. In many countries, dicofol is also used in combination with other pesticides such as the organophosphates parathion-methyl, and dimethoate. Application figures are not generally available as most countries do not record usage data according to particular active ingredients. The UK data show the average amount of dicofol active ingredient used between 1994 and 1997 was 1,143 kg per year. It is approved for use on apples, pears, blackcurrants, hops, strawberries, and protected crops of cucumbers, tomatoes and ornamentals.

In the US, dicofol is mostly used on cotton, apples and citrus crops. Other crops include strawberries, mint, beans, peppers, tomatoes, pecans, walnuts, stonefruit, cucurbits, and non-residential lawns/ornaments. During 1998 the manufacturers voluntarily cancelled all residential turf uses. Based on available US data from 1987 to 1996, total annual domestic agricultural usage of dicofol averaged 390,000 kg of active ingredient for 290,000 hectares treated. The largest markets for dicofol in the US are cotton (over 50%) and citrus (almost 30%). Most of the usage is in California and Florida.

Environmental impact

Dicofol was brought back after a new manufacturing process showed that the chemical had less than 0.1 percent DDT. It is now approved for many fruits, vegetables and field crops, but the U.S. Environmental Protection Agency (EPA) asked manufacturers to stop producing the chemical for lawn uses and they complied.

The EPA suggested people stop using dicofol on lawns because of the potential environmental repercussions. In 2006, the EPA created an amended Reregistration Eligibility Decision for the chemical which has since drastically cut down on its use. In 2011, the EPA requested voluntary cancellation of all dicofol uses in the U.S. by late 2016. Dicofol is the last organochlorine pesticide to go through these processes in the U.S., and its likely all due to the environmental concerns. As OSU explained, there are several impacts the chemical can have on different environments and organisms.

- Birds - Dicofol is a dangerous substance to most birds. It is responsible to eggshell thinning and reduced fertility. Quail, peasant, dove and screech owl are among the affected birds.
- Waterlife - This chemical is significantly more toxic to aquatic life. Small animals like shrimp are specifically at risk as well as algae, fish and invertebrates.
- Bees - Dicofol is not toxic to bees and there are no known negative side effects.
- Water - Dicofol absorbs into the sediment in most water, but it also degrades in UV light. It is insoluble in water and unlikely to leech through the soil in great depths. A more likely route to water pollution is through erosion or improper application.
- Land and vegetation - Dicofol has a 60-day half-life in soil and is degradable in UV light. It is found in soil but only on the surface. The chemical can remain in plants for up to two years, according to OSU.

- Humans - Long or short-term intake of residual dicofol is unlikely to present any danger to human health, according to the EPA.

According to the US Department of Agriculture, dicofol has a half-life in the environment for 60 days. Other US EPA laboratory and field studies show that dicofol has a half-life ranging from days to months. In ecological monitoring studies conducted in New York, Florida and California, dicofol dissipated from the soil surface with a half-life ranging from two to four months. The chemical is likely to be more persistent in acidic than neutral or alkaline soils or waters. Dicofol and DDT are similar chemical structures, and the US EPA considers that dicofol has to a lesser degree some ability to accumulate in the environment.

Acute toxicity

It is classified by the World Health Organisation as a Class III, 'slightly hazardous' pesticide. Like many insecticides and acaricides, dicofol is a nerve poison. The exact mode of action is not known, although in mammals it causes hyper stimulation of nerve transmission along nerve axons (cells). This effect is thought to be related to the inhibition of certain enzymes in the central nervous system. Symptoms of ingestion and/or respiratory exposure include nausea, dizziness, weakness and vomiting; dermal exposure may cause skin irritation or a rash; and eye contact may cause conjuctivitis. Poisoning may affect the liver, kidneys or the central nervous system. Very severe cases may result in convulsions, coma, or death from respiratory failure. Dicofol can be stored in fatty tissue. Intense activity or starvation may mobilise the chemical, resulting in the reappearance of toxic symptoms long after actual exposure.

Regulation: In 1990, the use of dicofol was suspended in Sweden for environmental reasons. In Switzerland its use is permitted for research purposes only. Throughout the European Union dicofol containing less than 78% pp'-dicofol or more than lg/kg (0.1%) of DDT or DDT related compounds cannot be used.

The 1998 US EPA review of dicofol recommended a number of changes in order to protect the environment and wildlife. Dicofol applications are limited to no more than one per year. In the UK, the maximum number of treatments permitted is two per year for apples and hops, and two per crop for strawberries, protected crops and tomatoes.

Nevertheless dicofol is one of the world's last organochlorines still in widespread use. It has perhaps survived because its environmental persistence is low relative to DDT, although this is not the case when compared with the vast majority of other pesticides on the market.

iv. Hexachlorocyclohexane (HCH)

a. β- Hexachlorocyclohexane (β-HCH): is an organochloride which is a byproduct of the production of the insecticide lindane (γ-HCH). It typically constitutes 5-14% of technical grade lindane.

This pesticide was widely used during the 1960s and 1970s, particularly on cotton plants. Although banned as a pesticide more than 30 years ago, traces of β--HCH can still be found in water and soil. Animal studies show that organochlorine pesticides, including β--HCH, are neurotoxic, cause oxidative stress, and damage the brain's dopaminergic system. Human studies show that exposure to β -HCH is linked to Parkinson's and Alzheimer's disease. β -HCH was present in elevated levels in some patients as early as 2009. It was manufactured by exhausting chlorination of benzene and for this reason was called erroneously β -BHC. This synonym still persists. As of 2009, the Stockholm Convention on Persistent Organic Pollutants classified α-HCH and β-HCH as Persistent organic pollutants (POPs), due to the chemical's ability to persist in the environment, bioaccumulative, biomagnifying, and long-range transport capacity.

b. Lindane: also known as γ- **hexachlorocyclohexane**, **gammaxene**, **Gammallin** and sometimes incorrectly called **benzene hexachloride** (BHC), is an organochlorine chemical variant of hexachlorocyclohexane that has been used both as an agricultural insecticide and as a pharmaceutical treatment for lice and scabies.

Lindane is a neurotoxin that interferes with GABA neurotransmitter function by interacting with the $GABA_{A\ receptor}$-chloride channel complex at the picrotoxin binding site. In humans, lindane affects the nervous system, liver and kidneys, and may well be a carcinogen. It is unclear whether lindane is an endocrine disruptor.

The chemical was originally synthesised in 1825 by Faraday. It is named after the Dutch chemist Teunis van der Linden (1884–1965), the first to isolate and describe γ-Hexachlorcyclohexane in 1912. Its pesticidal action was discovered only in 1942, after which lindane production, by Imperial Chemical industries Ltd (ICI), and use started up in the United Kingdom. It has been used to treat food crops and to forestry products, as a seed treatment, a soil treatment, and to treat livestock and pets. It has also been used as pharmaceutical treatment for lice and scabies, formulated as a shampoo or lotion. It is estimated that between 1950 and 2000, around 600,000 tonnes of lindane were produced globally, and the vast majority of which was used in agriculture. By November 2006, the use of lindane had been banned in 52 countries and restricted in 33 others. Seventeen countries, including the US and Canada, allowed either limited agricultural or pharmaceutical use.

In 2009, an international ban on the use of lindane in agriculture was implemented under the Stockholm Convention on Persistent Organic Pollutants. The production of the lindane isomers α- and β-hexachlorocyclohexane was also banned. Although the US has not ratified the Convention, it has similarly banned agricultural uses while still allowing its use as a second-line lice and scabies treatment. The Persistent Organic Pollutants Review Committee of the Stockholm Convention on Persistent Organic Pollutants considers the use of lindane in agriculture as largely redundant with other, less toxic and less persistent pesticides. In the case of pharmaceutical use, the Committee noted that "alternatives for pharmaceutical uses have often failed for scabies and lice treatment and the number of available alternative products for this use is scarce. For this particular case, a reasonable alternative would be to use lindane as a second-line treatment.

Lindane is also known as gamma-HCH since it is made up of at least 99% of the gamma isomer of hexachlorocyclohexane (HCH). 'Technical HCH' can include varying proportions of alpha, beta, delta and epsilon HCH isomers sold under various trade names (table.28) which have been shown to have serious short and long term health effects and their tendency to bio-accumulate along food chains.

Table 28: Commercial insecticide of HCN ingredient

Name of insecticide	Hexachlorocyclohexane (HCN)
Trade names	Agrocide, Ambrocide, Aparasin, Aphitiria, Benesan, Benexane, benhexachlor, benzene hexachloride, BHC, BoreKil, Borer-Tox, Exagama, Gallogama, Gamaphex, gamma-BHC, Gamma-Col, gamma-HCH, Gammex, Gammexane, Gamasan, Gexane, hexachlorocyclohexane, HCH, Isotox, Jacutin, Kwell, Lindafor, Lindagronox, Lindaterra, Lindatox, Lintox, Lorexane, New Kotol, Noviagam, Quellada, Steward, Streunex, and Tri-6.
Nature	Stomach poison and Fumigant
Active ingredient	γ-HCH
Formulations	Suspension, emulsifiable concentrate, fumigant, seed treatment, wettable and dustable powder, and ultra-low volume (ULV) liquid.

Environmental contamination

Lindane is a persistent organic pollutant. It has relatively long persistence in the environment, and can bioaccumulate in food chains. The production of lindane generates large amounts of waste hexachlorocyclohexane isomers and it is estimated that "every ton of lindane manufactured produces about 9 tons of toxic waste. The production and agricultural use of lindane are the primary causes of environmental contamination but now the levels of lindane in the environment have been decreasing with decreasing agricultural usage patterns.

Lindane as a global issue

The raw active ingredient is formulated into hundreds of commercial products around the world. Although it has been banned for all uses in 7 countries, severely restricted in another four. Under the Prior Informed Consent (PIC) procedures, a further 23 countries have prohibited its import.

Lindane is a Persistent Organic Pollutant (POP) candidate. The POP Convention seek to phase out the production and use of a range of chemicals which are classified as Persistent Organic Pollutants. Concern over these derives from the fact that many are used in tropical zones, but percolate through the global environment to end up in cold regions such as Scandinavia, Canada and Antarctica. POP pesticides include many which have been banned in several countries and some whose production is believed to have ceased world-wide.

The following countries have banned lindane for all uses:

Country	Year	Reasons for actions taken
Finland	1998	High risk to human health and the environment
Indonesia	1985	May induce adverse effects to environment and human health
Korea	1986	Harmful to human health
Netherlands	1979	Persistence of impurities and its high bioconcentration in the food chain
New Zealand	1990	Environmental reasons
Saint Lucia	1987	High residual effects in soil water
Sweden	1988	Suspected carcinogenic properties and persistence

The following countries have severely restricted the use of lindane

Country	Year	Reasons for actions taken
Australia		Environmentally unacceptable due to persistence
Austria	1992	High persistence in the environment, its bio-accumulation in the food chain and in human tissues. Lindane is a suspected carcinogen. There is evidence that HCH isomers act as a tumour promoter previously initiated by other chemicals
Cyprus	1987	Risk associated with human health and contamination of the environment due to its persistence and accumulation of residues in human tissue
Norway		All products withdrawn by importer. No import since 1991.
Sri Lanka	1986	

It is a banned pesticide in India.

21.3.2.2. Organophosphates

Those insecticides having phosphorus atoms come under the category of organophosphates. These insecticides can also be named as phosphate insecticides or organic phosphate etc. These insecticides have two most important properties such as higher potency and low residual life. These properties could bring them more popular day by day.

The organophosphates (OPs) are derivatives of phosphoric or phosphonic acid. OPs vary greatly in toxicity, residue levels, and excretion. These have been developed for plant and animal protection, and in general offer a distinct advantage by producing little tissue and environmental residue. Some of the OPs developed initially as pesticides are also used as anthelmintics. Five such compounds include dichlorvos, trichlorfon, haloxon, naphthalophos, and crufomate. The first two are primarily used against parasitic infestations in horses, dogs, and pigs; the latter three are used against parasites in ruminants.

Many of the OPs now used as pesticides (eg, chlorpyrifos, diazinon, fenitrothion, malathion, parathion, etc) are not potent inhibitors of cholinesterase until activated in the liver by microsomal oxidation enzymes; they are generally less toxic, and intoxication occurs more slowly. Certain OPs preparations are microencapsulated, and the active compound is released slowly; this increases the duration of activity and reduces toxicity, but the toxic properties are still present.

The Organophosphates (OPs) inhibit the cholinesterase (ChE) enzyme of the nervous system of both vertebrates and insects.

The organophosphates can be divided into three classes such as:

a. Aliphatic derivatives

b. Phenyl derivatives

c. Heterocyclic derivatives

a. Aliphatic derivatives

The literary meaning of aliphatic is carbon chain and the linear arrangement of carbon atoms. These are the simplest structure of OP molecules. e. g. malathion, trichloroform, dimethoate, disulfoton, demeton, dichlorvos, acephate, phorate, mevinphos, monocrotophos, etc. (The underline pesticides in this group are banned chemicals for crop use).

i. Dichlorvos ($C_4H_7Cl_2O_4P$):

Dichlorvos (table.29) is an aliphatic organophosphate. The IUPAC name is 2, 2-dichorovinyl dimethyl phosphate. Its insecticidal property was first described in 1951.

Table 29: Trade names of Dichlorvos and pests controlled, but banned now

Trade names	Vapona, Nuvan, Nogos
Nature	Contact insecticide

Pest controlled by DICHLORVOS 76% EC		mL/ha
Paddy	BPH Cut worm/ Army worm, Leaf roller/ folder	375 500 500
Wheat	Caterpillar	500
Soybean	Leaf eating caterpillar	225-300
Castor	Hairy caterpillar	625
Groundnut	Red hairy caterpillar	375-750
Mustard	Painted bug	500
Sunflower	Caterpillar Cabbage looper Semi looper	500 500 500
Cucurbit	Red pumpkin beetle	500
Cashew	Apple borer	0.05%

ii. Phosphamidon ($C_{10}H_{19}ClNO_5P$):

The IUPAC name is 2-chloro-2-diethyl-carbamoyl-1-methylvinyl dimethyl phosphate. It is contact insecticide, which is banned now.

In 1956 it was introduced by Ciba. The trade name is Dimecron.

iii. Mevinphos ($C_7H_{13}O_6P$)

The IUPAC name is 2-methoxy-carbonyl-1-methylvinyl dimethyl phosphate. It is also known as methyl 3-(dimethoxy-phosphinyloxy-but-2-enoate. It is contact insecticide

Its insecticidal property was found by Corey and was introduced by Shell chemical in 1953.

b. Phenyl derivatives

The insecticides belonging to this class contain a benzene ring with one of the ring hydrogens displaced by attachment to the phosphorus moiety and other frequently displaced by S, Cl, CH_3, CN, NO_2, etc. As compare to aliphatic derivations these are more stable and their dissipation rate is slower.

e. g. parathion (both ethyl and methyl), runnel, stirofos(Gardona), sulprofos (Bolstar), etc.

i. Parathion ($C_{10}H_{14}NO_5PS$)

The IUPAC name is O, O-diethyl O-4-nitrophenyl phosphorothioate (I). The other name is O, O-diethyl O-p-nitrophenyl phosphorothioate.

It was introduced by the American Cynamid Co. Under the trade name of Thiophos in1947. In 1948, Folidol trade name was given by Bayer.The active ingredient is parathion (Table.30).

Table 30: Trade names of parathion insecticide and pest controlled

Name of insecticide	Parathion	
Trade names	Thiophos, Alkron, Alleron, Aphamite, Corothion, E-605, ENT 15108, Ethyl parathion, Etilon, Fosferno 50, Niran, Orthophos, Panthion, Paramar, Paraphos, Parathene, Parawet, Phoskil, Rhodiatox, Soprathion, Stathion and Thiophos Folidol	
Nature	contact, stomach and fumigant actions	
Active ingredient	Parathion	
Formulations	dust, emulsion concentrate, granular, ULV liquid, and wettable powder formulations	
Pest controlled		
Parathion, 2% DP		mL/ha
Paddy	Ear Head Caterpillar, Leaf roller, Ear head bug	500
Cotton	Aphid Leaf hopper, Thrips	300 500
Black gram	Pod borer	500
Green gram	Pod borer	500
Mustard	Sawfly, Aphids	300
Parathion 50% EC		
Paddy	Gall midge, Green leaf hopper, Hispa, Leaf roller, Stem borer & Whorl maggot	750 500 250 400 500
Wheat	Cutworm	300
Cotton	Aphid Leaf hopper Thrips	500 250 500

c. Heterocyclic derivatives

The meaning of heterocyclic is those cyclic molecules which are made up of different or unlike atoms. In these compounds one or more carbon atoms is displaced by nitrogen, oxygen, sulphur and the ring may have 3, 5 or 6 atoms. e. g. chlorpyriphos, diazinon, azinphosmethyl, methidathion, etc.

i. Chlorpyriphos ($C_9H_{11}C_{13}NO_3PS$):

Its IUPAC name is: O, O-diethyl O-3, 5, 6-trichloropyridin-2-yl phosphorothioate. It is a crystalline organophosphate insecticide.

It was introduced in 1965 by Dow Chemical Company and is known by many trade names, including Dursban and Lorsban (table.31). It acts on the nervous system of insects by inhibiting acetylcholin esterase.

Table 31: Trade names of chlorpyriphos and pest controlled

Name of insecticide	Chlorpyriphos
Trade names	Dursban Lorsban
Nature	Contact insecticide
Active ingredient	Chlorpyriphos
Formulations	EC, DP,G
Pest controlled with different formulations	

Chlorpyrifos 1.5% DP		mL/ha
Paddy	Stem borer Green leaf hopper Brown plant hopper Leaf folder, Gall midge Grass hopper	375
Bengal gram	*Helicoverpa armigera*	375
Paddy	Hispa Leaf roller Gall midge, Stem borer Whorl maggot	250 375 250 250
Beans	Pod borer, Black bug	600
Gram	Cut worm	500
Sugarcane	Black bug Early shoot & stalk borer Pyrilla	150 250-300 300
Cotton	Aphid, Bollworm , White fly, Cut worm	250 750
Ground nut	Aphid Root grub	200 225
Mustard	Aphid	100

Brinjal	Shoot & fruit borer	200
Cabbage	Diamond back moth	400
Onion	Root grub	1000
Apple	Aphid	0.05%
Ber	Leaf hopper	0.03%
Citrus	Black citrus, Aphid	0.02%
Tobacco	Ground beetle	350

ii. Diazinon ($C_{12}H_{21}N_2O_3PS$):

IUPAC name: O,O-Diethyl O-[4-methyl-6-(propan-2-yl)pyrimidin-2-yl] phosphorothioate, INN - Dimpylate).

It is a colorless to dark brown liquid, a thiophosphoric acid ester (table.32) developed in 1952 by Ciba-Geigy, a Swiss chemical company (later Novartis and then Syngenta).

Table 32: Trade names of Diazinon pesticide, pest controlled, but banned now

Name of insecticide	Diazinon
Trade names	Basudin, Dazzel, Gardentox, Kayazol, Knox Out, Nucidol, and Spectracide
Nature	Contact insecticide
Active ingredient	Diazinon
Formulations	Dust, granules, seed dressings, wettable powder, and emulsifiable solution
Pest controlled	Cockroaches, silverfish, ants, and fleas sucking and leaf eating insects

Diazinon 0.5% m/m+ Pyrethrum 0.1% m/m	
Insects	Dosage ml/ sq. m
Cockroach, Housefly	25-50

iii. Demeton

Demeton is used as a systemic insecticide against sucking insects and mites. Demeton is used mainly as a foliage spray and has a relatively long residual life. It is a mixture of demeton-O and demeton-S and is highly toxic to mammals. Demeton poisoning were reported in several hundred cattle grazing near cotton treated with this insecticide. The corresponding individual analogs of demeton i.e. demeton-O-methyl and demeton-S-methyl are also used for similar purposes but are less toxic than demeton (Table.33).

Table 33: Trade names of Demeton and pest controlled.

Name of insecticide		*Demeton*
Trade names		Basudin, Dazzel, Gardentox, Kayazol, Knox Out, Nucidol, and Spectracide
Nature		Contact insecticide
Active ingredient		Diazinon
Formulations		dust, granules, seed dressings, wettable powder, and emulsifiable solution
Pest controlled		cockroaches, silverfish, ants, and fleas sucking and leaf eating insects
Demeton – Methyl 25% EC		mL/ha
Paddy	Blue leaf hopper White leaf hopper	125 250
Maize	Shoot fly	250
Sorghum	Shoot fly	250
Cotton	Aphid, Jassid/ leaf hopper	300
Ground nut	Aphid/ Leaf minor	250
Mustard	Aphid	250
Sesamum	Leaf hopper	300
Bhindi	White fly Jassid/ Leaf beetle	250 400
Chilli	Aphid Mites Thrips	400 500 250
Onion	Thrips	300
Tomato	White fly	250
Potato	Aphids	250
Apple	Sanjose scale Wooly Aphid	0.07% 0.025%
Banana	Tingyi bug Aphids	0.025% 0.05%
Mango	Hoppers	0.025%
Peaches	Leaf curl aphids	0.025%
Coffee	Green bug Leaf minor	625 1000
Tobacco	White fly/Aphids	250

iv. Dimethoate

Dimethoate (Table. 34) is used extensively in horticulture as a systemic insecticide, but it also kills insects by contact.

Table 34: Trade names of Dimethoate and pest controlled

Name of insecticide		***Dimethoate***
Trade names		Cekuthoate, Chimigor 40, Cygon 400, Daphene, De-Fend, Demos NF, Devi
Nature		Contact insecticide
Active ingredient		Diamethoate
Formulations		dust, granules, seed dressings, wettable powder, and emulsifiable solution
Pest controlled		mL/ha
Pearlmillet	Milky weed bug	180-200
Maize	Stem borer Shoot fly	200 350
Sorghum	Midge	500
Castor	Jassids, Mites Semi looper	250 350
Mustard	Leaf minor, Aphid, Sawfly	200
Safflower	Aphid	200
Okra	Aphid Leaf hopper, Jassid	700 600
Brinjal	Shoot borer	200
Cabbage & Cauliflower	Aphid, Painted bug Mustard aphid	200
Chillies	Mite	300
Onion	Thrips	200
Potato	Thrips	200
Tomato	Aphids White fly	200 300
Apple	Stem borer	0.03%
Apricot	Aphid	0.03%
Banana	Aphid, Lace wing bug	0.03%
Citrus	Black citrus aphid	0.03%
Fig	Fig jassid Mealy bug	0.03% 0.03%
Mango	Hopper	0.05%
Rose	Scale Thrips	750 400

v. Malathion

Malathion (Table.35) is one of the safest organophosphates because of its selective toxicity. It is highly toxic to insects but much less toxic to mammalian species. Malathion is excreted in cow's milk.

Table 35: Trade names of Malathion and pest controlled

Name of insecticide		*Malathion*
Trade names		Celthion, Cythion, Dielathion, Karbofos, Maltox, El 4049, Emmaton, Fyfanon and Exathion
Nature		Contact insecticide
Active ingredient		Malathion
Formulations		dust, granules, seed dressings, wettable powder, and emulsifiable solution
Pest controlled		-
MALATHION 5% DP		mL/ha
Paddy	Rice Hispa	1250
Sorghum	Earhead midge	1000
MALATHION 50% EC		
Paddy	Rice Hispa	575
Sorghum	Earhead midge	500
Pea	Pod borer	750
Soybean	Leaf weevil	750
Castor	Jassids Semi looper	750 1000
Sunflower	White fly	500
Okra	Aphid Jassids Spotted Boll Worm	500 625 750
Brinjal	Mites	750
Cabbage	Mustard aphid	750
Cauliflower	Head borer	750
Radish	Stem borer	750
Turnip	Tobacco caterpillar	600
Tomato	White fly	750
Apple	Sanjose scale, Wooly aphid	0.05%
Mango	Mealy scale, Mango hooper	0.075%
Grape	Beetle	500

vi. Methyl parathion

Methyl parathion (Table.36) is less toxic than parathion (diethyl parathion). Methyl parathion is excreted in cow's milk. Methyl parathion produces reproductive and developmental toxicity.

Table 36: Pest controlled by Methyl parathion, but banned now

METHYL PARATHION 50% EC		mL/ha
Paddy	Gall midge, Green leaf hopper, Hispa, Leaf roller, Stem borer & Whorl maggot	750 500 250 400 500
Wheat	Cutworm	300
Cotton	Aphid Leaf hopper Thrips	500 250 500

vii. Diethyl Parathion

Diethyl Parathion (parathion) is widely used for control of plant pests and is approximately one-half as toxic as tetraethyl pyrophosphate. Most cases of occupational insecticide poisonings in people have been attributed to parathion or its degradation products. Parathion is used extensively to control mosquitoes and insects in orchards and on market garden crops besides field crops (Table.37). Normally, because so little is used per acre, it presents no hazard to livestock. However, because of the potency of parathion, care should be taken to prevent accidental exposure. Parathion does not produce significant residues in animal tissues.

Table 37: Pest controlled by Diethyl Parathion

Diethyl Parathion 2% DP		mL/ha
Paddy	Ear Head Caterpillar, Leaf roller,	500-25000
	Ear head bug	300-500
Cotton	Aphid Leaf hopper, Thrips	15000-25000
Black gram	Pod borer	500-25000
Green gram	Pod borer	500-25000
Mustard	Sawfly, Aphids	300-15000

viii. Phorate

Phorate (Table.38) is closely related to demeton. It is a systemic insecticide and miticide.

Table 38: Trade names of Phorate and pest controlled, but banned now

Active ingredient		***Phorate***
Trade names		AC 8911, Agromet, Geomet, Granutox, Phorate 10G, Rampart, Thimenox, Thimet, Vegfru Foratox, Timet and Vegfru
Nature		contact, and fumigant action
Formulations		granules
Pest controlled	As below	
Phorate 10% G		g/ha
Pearl millet	Shoot fly	3000
	White grub	2500
Barley	Aphid	1000
Maize	Shoot fly	3000
	Stem borer	1000
Paddy	Gall fly, Hispa, Leaf hopper, Plant hopper, Stem borer	1000
	Root weevil	750
Sorghum	Shoot fly, Aphids White grub	1875 2500
Wheat	Shoot fly	1875
Black gram	Stem fly, White fly	1000
Green gram	Stem fly Jassids	1000 1500
Pigeon pea	Jassids Stem fly	1500 1000
Soybean	Stem fly	1500
Sugarcane	Top borer White grub	3000 2500
Cotton	Aphid, Jassids, Thrips, White fly	1000
Groundnut	Aphid, Leaf minor White grub	1500 2500
Mustard	Mustard aphid, Painted bug	1000 1500
Seasamum	Jassids, White fly	1000
Brinjal	Aphid, Jassids, Lace wing bug, Red spider mite. Thrips	1500 1000
Cauliflower	Aphid	2000

Chillies	Aphid , Mite, Thrips	1000
Potato	Aphid	1000
Tomato	White fly	1500
Apple	Woolly aphid	10- 15/plant
Banana	Aphid	2.5 -1.25/ plant
Citrus	Leaf minor	1500

21.3.2.3. Carbamates

i. Aldicarb

Aldicarb is a carbamate insecticide and the active substance in the pesticide. It is available in the market as **Temik.** It is effective against thrips, aphids, spider mites, fleahoppers, and leafminers, but is primarily used as a nematicide. Aldicarb is a cholinesterase inhibitor which prevents the breakdown of acetylcholine in the synapse. In case of severe poisoning, the victim dies of respiratory failure.

Aldicarb was one of the most widely used pesticides internationally, and was also one of the most environmentally toxic. Aldicarb poisoning from agricultural water runoff has led to the destruction of healthy ecosystems and the irreversible poisoning of fertile agricultural land. Poisoning from this pesticide is also believed to be linked to high cancer rates in communities located around the Aral Sea.

Aldicarb was effective where resistance to organophosphate insecticides has developed, and was extremely important in potato production, where it was used for the control of soil-borne nematodes and some foliar pests. Its high level of solubility restricts its use in certain areas where the water table was close to the surface.

Aldicarb was one of the "dirty dozen" pesticides that the environmental group Pesticide Action Network North America targeted in 1985. EPA put a ban on the main aldicarb pesticide, Temik 15G (Produced by Bayer CropScience), in 2010, requiring an end to distribution by 2017. Use on citrus and potatoes was banned beginning in 2012, with a complete phase out of the product by 2018. Despite this, a new aldicarb pesticide named Meymik 15G, produced by AgLogic Chemical, was approved by the EPA in December 2011. It is registered for use on cotton, dry beans, peanuts, soybeans, sugar beets, and sweet potatoes.

"Tres Pasitos", a mouse, rat, and roach killer that contains high concentrations of aldicarb, has been illegally imported into the United States from Mexico and other Latin American countries. The product is highly toxic to animals and people, and according to the EPA "should never be used in [the] home."

Aldicarb is manufactured by Bayer CropScience, but was formerly owned and produced by Union Carbide. Union Carbide's agricultural chemicals division was sold to Rhône-Poulenc. Later, Aventis Cropscience was formed from Hoechst AG and Rhone-Poulenc Agrochemical, which lasted until Bayer acquired it in 2002.

In November 2009, corn treated with Temik was placed in and around peanut fields in Eastland County, Texas, near the town of Cisco. The corn was eaten by feral hogs, deer, and other animals, prompting the Texas Parks and Wildlife Department to issue a hunting ban.

ii. Carbaryl

Carbaryl (1-naphthyl methylcarbamate) is a chemical in the carbamate family and is used chiefly as an insecticide. It is a white crystalline solid commonly sold under the brand name **Sevin**, (table.39) a trademark of the Bayer Company. Union Carbide discovered carbaryl and introduced it commercially in 1958. It remains the third-most-used insecticide in the United States for home gardens, commercial agriculture, and forestry and rangeland protection. About 11 million kilograms were applied to U.S. farm crops in 1976. As a veterinary drug, it is known as carbaril.

Table 39: Pest controlled by insecticide Carbaryl, but banned now

Name of insecticide	Carbaryl
Trade names	Sevin
Nature	Cholinesterase inhibitor
Active ingredient	Carbaryl
Formulations	DP,G, WP
Pest controlled	thrips, aphids, spider,mites, lygus, fleahoppers, and leafminers,

Carbaryl 5% DP		mL/hact
Paddy	Leaf roller/folder Brown plant hopper,	1250 1000
Cotton	Spotted bollworm American bollworm Pink Bollworm Jassids Aphids	1000
Sorghum	Earhead midge	1000
Okra	Jassid	1000
Cabbage	Cabbage borer	600
Cauliflower	Cabbage borer	600
Carbaryl 10% D.P.		
Paddy	Blue Jassid, Case worm	2500

Sorghum	Aphid, Earhead caterpillar	2500 2000
Tur	Plume moth, Pod fly	2000
Cotton	Aphid, American bollworm, Stem weevil Thrips	2500
Sesamum	Til leaf roller	2500
Okra	Fruit borer, Jassids	2500
Cabbage	Diamond back moth Army worm	2500
Carbaryl 4% G.R.		
Maize	Stem borer	250
Carbaryl 50% WP		
Maize	Shoot fly	750
	Stem Borer	700
Paddy	Brown Plant Hopper, Stem Borer	1000
	Green Leaf Hopper	750
Cotton	Aphids, Jassids, Thrips, Leaf Roller, Spotted Bollworm, Pink Bollworm, American Bollworm	1000
Jute	Semi Looper	1000
Sorghum	Hoppers, Aphids, Stem Borer	1000
	Shoot fly	750
Tomato	Fruit Borer	1000
Chillies	Thrips	1000
Brinjal	Fruit Borer, Jassids	1000
Okra	Fruit Borer	1000
Cauliflower	Cabbage borer	800
Cabbage	Cabbage Borer	800
Wheat	Army worm	1000

The development of the carbamate insecticides has been called a major breakthrough in pesticides. The carbamates do not have the persistence of chlorinated pesticides. Although toxic to insects, carbaryl is detoxified and eliminated rapidly in vertebrates. It is neither concentrated in fat nor secreted in milk, so is favored for food crops, at least in the US. It is the active ingredient in Carylderm shampoo used to combat head lice until infestation is eliminated.

Risk

Carbaryl kills both targeted (e.g. malaria-carrying mosquitos) and beneficial insects (e.g. honeybees), as well as crustaceans. Although approved for more than 100 crops in the US, carbaryl is illegal in several countries, including the United Kingdom, Austria, Denmark, Sweden, Iran, Germany, and Angola. Carbaryl is often produced using methyl isocyanate (MIC) as an intermediary. A leak of MIC used in the production of carbaryl caused the Bhopal disaster, the largest industrial accident in history.

Carbaryl is a cholinesterase inhibitor and is toxic to humans. It is classified as a likely human carcinogen by the United States Environmental Protection Agency (EPA).

iii. Carbofuran

Carbofuran is one of the most toxic carbamate pesticides. It is marketed under the trade names Furadan, by FMC Corporation and Curater, among several others. It is used to control insects in a wide variety of field crops, including potatoes, corn and soybeans. It is a systemic insecticide, but also has contact activity against pests.

The technical or chemical name of carbofuran is 2, 3-dihydro-2, 2-dimethyl-7-benzofuranyl methylcarbamate and its CAS number is 1563-66-2. It is manufactured by the reaction of methyl isocyanate with 2, 3-dihydro-2, 2-dimethyl-7-hydroxybenzofuran.

Carbofuran usage (table.40) has increased in recent years because it is one of the few insecticides effective on soybean aphids, which have expanded their range since 2002 to include most soybean-growing regions of the U.S. The main global producer is the FMC Corporation.

Table 40: Pest controlled by insecticide Carbofuran, but banned now.

Carbofuran 3% CG		g/ha
Barley	Aphid Jassid Cyst nematode	1000 1250 1000
Bajra	Shoot fly	1500
Sorghum	Shoot fly, Stem borer	1000 250
Jute	Nematodes	1000
Groundnut	Pod borer White grub	1500 1000
French bean	White grub	700
Potato	Aphid, Jassids	500 1000
Tomato	White fly	1200
Apple	Woolly aphid	5/tree
Citrus	Nematode Leaf miner	360 1500

Maize	Stem borer Shoot fly Thrips	1000 1000 1000
Paddy	Brown plant hopper, Gall midge, Stem borer, GLH, Hispa, Nematodes	750 750 750 1500
Mustard	Mustard leaf miner White fly	2000 1000
Soybean	Root knot nematode	1500
Sugarcane	Top borer	2000
Bhindi	Jassids	1000
Chillies	Aphid , Thrips	1000
Cabbage	Nematode	1000
Wheat	Ear cockle nematode Cereal cyst nematode	3000 2000
Brinjal	Root knot nematode Reniform nematode	2000 2000
Banana	Rhizome weevil Aphid Nematode	1 g/ suckers 50g/suckers 1.5g/suckers
Peach	Leaf curl aphid	1000
Mandarins	Soft greens scale	0.4g/plant
French bean	White grubs Grey & Stem weevil	750 1000
Pea	Shoot fly & Aphid	1000
Tea	Cock chafer grub	0.3g/plant

Bans

Carbofuran is banned in Canada and the European Union. In 2008, the United States Environmental Protection Agency (EPA) announced that it intends to ban carbofuran. In December of that year, FMC Corp., the sole US manufacturer of carbofuran, announced that it had voluntarily requested that the United States Environmental Protection Agency cancel all but six of the previously allowed uses of that chemical as a pesticide. With this change, carbofuran usage in the US would be allowed only on maize, potatoes, pumpkins, sunflowers, pine seedlings and spinach grown for seed. However, in May 2009 EPA cancelled all food tolerances, an action which amounts to a de facto ban on its use on all crops grown for human consumption. Kenya is considering banning carbofuran, but it is legal to buy over-the-counter in Kenya.

iv. Carbosulfan

Carbosulfan is an organic compound belonging to the carbamate class. At normal conditions, it is brown viscose liquid. It is not very stable and decomposes slowly at room temperature. Its solubility in water is low but it is miscible with xylene, hexane, chloroform, dichloromethane, methanol and acetone. Carbosulfan (Table.41) is used as an insecticide. The European Union banned use of carbosulfan in 2007.

Table 41: Pest controlled by Carbosulfan

Carbosulfan 6% G		g/ha	
Rice	Stem borer Gall midge Green leaf hopper Leaf folder	1000	
	Stem borer	1000	
Carbosulfan 25% EC			
Rice	Green leaf hopper White plant hopper Brown plant hopper Gall midge Stem borer	200-250	
Chilli	Leaf folder White aphid	200-250 200-250	
Carbosulfan 25% DS			
Cotton	Jassid, Aphids and Thrips	15 gm/kg Seed.	60gm/kg Seed

v. Fenoxycarb

Fenoxycarb is a carbamate which works as insect growth regulator. It has a low toxicity for bees, birds, and humans, but is toxic to fish. Fenoxycarb is non-neurotoxic and does not have the same mode of action as other carbamate insecticides. Instead, it prevents immature insects from reaching maturity by mimicking juvenile hormone.

vi. Methiocarb

Methiocarb is a carbamate pesticide which is used as a bird repellent, insecticide, acaricide and molluscicide since the 1960s. Methiocarb has contact and stomach action on mites and neurotoxic effects on molluscis. Seeds treated with methiocarb also affects birds.

Like the other carbamates, methiocarb inhibits reversibly acetylcholinesterase activity resulting in a cholinergic stimulation making methiocarb a potent neurotoxin. It can even cause acute toxicity in humans if exposed for long periods of time or a sufficient dose. It is not listed as a carcinogen. Methiocarb is also a known poison to water organisms.

Methiocarb is used as toxin for different purposes. It ranges from snails, insects, rodent and even as a bird repellent. As an insecticide it is effective for thrips and has a low dose that is lethal for these animals.

As snail bait methiocarb has the same effectiveness as methomyl for 1% (mass percent) and 2%, but the LC_{50} of methiocarb is higher than the one of methomyl 0.93:0.31. These both reached an average mortality of 85%, by the use of 2% methiocarb/methomyl bait.

Environmental toxicity of Methiocarb

Because methiocarb is widely used as an insecticide on crops, environmental risks were also studied to establish safety risks for human health. In plants the major metabolites were methiocarb sulfoxide and methiocarb sulfoxide phenol. Environmental fate in water and soil has been determined from the metabolites formed by (an)aerobic degradation, photolysis, adsorption and leaching of methiocarb. In soil the half life of methiocarb sulfone phenol is 20 days, methiocarb sulfoxide phenol is 2 days, methiocarb 1.5 days and methiocarb sulfoxide 6 days.

Methiocarb is mainly metabolized to methiocarb phenol and minor to methiocarb sulfoxide and methiocarb sulfoxide phenol. Also after 217 days no methiocarb or metabolites are present anymore in the soil. This is because a lot of gets metabolized to CO_2. In water no methiocarb was present already after 32 days. The half life of methiocarb in water is strongly pH dependent but at pH 7 the half life is about 28 days.

21.3.2.4. Neonicotinoid

Neonicotinoids are a class of neuro-active insecticides chemically similar to nicotine. In the 1980s Shell and in the 1990s Bayer started work on their development.

The neonicotinoid family includes acetamiprid, clothianidin, imidacloprid, nitenpyram, nithiazine, thiacloprid and thiamethoxam. Imidacloprid is the most widely used insecticide in the world. Compared to organophosphate and carbamate insecticides neonicotinoids cause less toxicity in birds and mammals than insects. Some breakdown products are toxic.

In the late 1990s neonicotinoids came under increasing scrutiny over their environmental impacts. Neonicotinoid use was linked in a range of studies to adverse ecological effects, including honey-bee colony collapse disorder (CCD) and loss of birds due to a reduction in insect populations. In 2013, the European Union and a few non EU countries restricted the use of certain neonicotinoids.

As of 2011, seven neonicotinoids from different companies are on the market (table.42).

Table 42: Marketed products of Neonicotinoids pesticide

Name	Company	Products
Imidacloprid	Bayer CropScience	Confidor, Admire, Gaucho, Advocate
Thiamethoxam	Syngenta	Actara, Platinum, Cruiser
Clothianidin	Sumitomo Chemical/Bayer CropScience	Poncho, Dantosu, Dantop
Acetamiprid	Nippon Soda	Mospilan, Assail, ChipcoTristar
Thiacloprid	Bayer CropScience	Calypso
Dinotefuran	Mitsui Chemicals	Starkle, Safari, Venom
Nitenpyram	Sumitomo Chemical	Capstar, Guardian

21.4. Cyclodiene Insecticides

21.4.1. Chlordane

This pesticide first registered in the United States in 1948. In 1988, all chlordane uses, except its use for fire ant control in power transformers, were voluntarily canceled in the United States. Chlordane was used against insects on food and non-food agricultural crops, residential lawns and gardens, and in buildings. **It was particularly used against termites in a variety of buildings, including homes.**

Chlordane is a mixture of over 50 closely related chemicals sold under various trade names (table 43). Variations in the makeup of chlordane can result in differences in toxicity. It is a thick clear to amber liquid. It may be odorless or exhibit a mildly irritating odor. Chlordane still can be legally manufactured in the United States, but it can only be sold to or used by foreign countries.

Table 43: Trade names of Chlordane, pest controlled, but banned now.

Name of insecticide	Chlordane
Trade names	Belt, Corodane, Chlortox, Niran, Octachlor, Octa-Klor, Sym-klor, Toxichlor
Nature	Contact insecticide
Formulations	wettable and dustable powders, emusifiable concentrates, granules
Pest controlled	leaf hoppers, cutworms, thrips, Japanese beetles

Mode of action

Chlordane affects the nervous system. It does this by blocking important chemical signals and enzymes resulting in overstimulation. It is toxic to Animals. Chlordane is moderately toxic if ingested. Chlordane is moderately toxic if the skin is exposed. Chlordane is irritating to the skin and eyes. Chlordane ranges from very low to moderately toxic if inhale.

In humans, exposure to chlordane may cause headaches, nausea, vomiting, poor balance, tremors, and mental confusion. Exposure to high chlordane levels can result in the sudden onset of convulsions. Convulsions may also occur up to 48 hours following a high level exposure. Convulsions may or may not be preceded by other symptoms. Convulsions can recur periodically several days after exposure.

Following exposure, chlordane levels are initially highest in the kidneys and liver. Chlordane and its breakdown products then move to fat. Chlordane does accumulate in the body. Chlordane cause reproductive or birth defects. Animals Scientists fed young male and female rats a diet containing chlordane after weaning found that the rats had difficulty mating and producing healthy litters.

Chlordane breaks down slowly in the environment and can accumulate in living organism. It does not readily breakdown by water or light. The soil half-life for chlordane is estimated at 350 days but can range from 37 days to 3500 days. Studies show that chlordane has a low potential to move in soil. Despite its low potential to move in soil, researchers have detected chlordane in groundwater.

Why were uses of chlordane cancelled?

Chlordane uses were canceled based on concerns regarding its potential to cause cancer and its slow break down in the environment.

21.4.2. Heptachlor

Heptachlor is a non-sytemic stomach and contact organochlorine insecticide first synthesized in 1946 and banned in 1988. It acts as a nonsystemic stomach and contact insecticide.

Trade names

Biarbinex, Cupincida, Drinox, E 3314, Fennotox, Velsicol 104, Aahepta, Agroceres, Hepta, Heptachlordane, Heptagran, Heptamul, Heptox, Gold Crest H-60, Rhodiachlor, Velsicol104, Basaklor, Soleptax, Termid

Uses

Heptachlor as an insecticide was used heavily in the 1960s and 1970s in household and agricultural settings. It was used extensively to control termites and pests on corn crops. It was also used against spittlebugs, weevils, wireworms, cornborers, root worms, boll weevils, leaf hoppers, cutworms, thrips, Japanese beetles, grasshoppers, mosquitoes, and plum curculio.

Regulatory Status

Heptachlor is a moderately toxic compound in EPA toxicity class II. Phase out of heptachlor use in U.S.began in 1978. In 1988, the EPA canceled all uses of heptachlor in the U.S.The only commercial use still permitted is for fire ant control in power transformers. Heptachlor is still available outside the U.S.

Heptachlor is an organochlorine cyclodiene insecticide, first isolated from technical chlordane in 1946. During the 1960s and 1970s, it was used primarily by farmers to kill termites, ants, and soil insects in seed grains and on crops, as well as by exterminators and home owners to kill termites. Before heptachlor was banned, formulations available included dusts, wettable powders, emulsifiable concentrates, and oil solutions.

An important metabolite of heptachlor is heptachlor epoxide, which is an oxidation product formed from heptachlor by many plant and animal species. Both technical-grade and pure heptachlor smell generally of mothballs or camphor. Heptachlor is not highly flammable, combustible, or soluble.

Studies link heptachlor to cancer, endocrine disruption, and developmental toxicity. Heptachlor is extremely persistent and is found throughout the United States nearly twenty years after its cancellation. Heptachlor is readily converted to more potent heptachlor epoxide once it enters the environment or the body.

21.4.3. Endosulfan

Endosulfan is a chlorinated hydrocarbon insecticide and acaricide of the cyclodiene subgroup which acts as a poison to a wide variety of insects and mites on contact. Although it may also be used as a wood preservative, it is used primarily on a wide variety of food crops including tea, coffee, fruits, and vegetables, as well as on rice, cereals, maize, sorghum, or other grains.

Trade Names

Afidan, Beosit, Cyclodan, Devisulfan, Endocel, Endocide, Endosol, FMC 5462, Hexasulfan, Hildan, Hoe 2671, Insectophene, Malix, Phaser, Thiodan, Thimul, Thifor, and Thionex.

Regulatory Status

Endosulfan is a highly toxic pesticide in EPA toxicity class I. It is a Restricted Use Pesticide (RUP). Labels for products containing endosulfan must bear the Signal Words DANGER - POISON, depending on formulation

Formulations of endosulfan include emsulsifiable concentrate, wettable powder, ultra-low volume (ULV) liquid, and smoke tablets. It is compatible with many other pesticides and may be found in formulations with dimethoate, malathion, methomyl, monocrotophos, pirimicarb, triazophos, fenoprop, parathion, petroleum oils, and oxine-copper. It is not compatible with alkaline materials. Technical endosulfan is made up of a mixture of two molecular forms (isomers) of endosulfan, the alpha- and beta-isomers.

22

Botanical Insecticides

22.1. Pyrethrum and Related Compounds

Pyrethrum is the most widely used botanical insecticide. The active ingredient, pyrethrin, is extracted from a chrysanthemum plant, grown primarily in Kenya, Rwanda, Tanzania and Ecuador. Pyrethrum is the powdered, dried flower head of the pyrethrum daisy, *Chrysanthemum cinerariaefolium*

The insecticidal properties of pyrethrum flowers (genus chrysanthemum) have been recognized since the middle of 1st century, when commercial sale of "insect powder" from Dalmatian pyrethrum flower heads began (McLaughlin, 1973). In addition to their insect killing activity, their lack of persistence in the environment and rapid "knock down" activity whereby flying insects become uncoordinated and unable to fly makes it very useful. Pyrethrum used to be a major insecticide for both domestic and agricultural use despite its poor light stability. Its usefulness was extended by introduction of piperonyl butoxide and other compounds as synergists, which greatly reduced the unit cost of crop treatment.

The term "pyrethrum" is the name for the crude flower dust itself, while the term "pyrethrins" refers to the six related insecticidal compounds that occur naturally in the crude material, the pyrethrum flowers.

Pyrethrum and Pyrethrins

	Structure.
Cinerin.I	$C_{20}H_{28}O_3$
cinerin.II	$C_{21}H_{28}O_5$
cinerin.III	$C_{21}H_{30}O_3$
jamolin.I	$C_{21}H_{30}O_3$
jamolin.II	$C_{22}H_{30}O_5$
pyerethrin.I	$C_{21}H_{28}O_3$
pyrethrin. II	$C_{22}H_{28}O_5$

These pyrethrins are extracted from crude pyrethrum dust as a resin that is used in the manufacture of various insecticidal products.

The insecticide pyrethrum is marketed with various trade names (table.44).

Table 44: Insecticide pyrethrum and pyrethrin and the pest controlled

Name of insecticide	Pyrethrum
Trade names	Buhach, Chrysanthemum Cinerariaefolium, Ofirmotox, Insect Powder, Dalmation Insect Flowers, Firmotox, Parexan and NA 9184. Pyrenon,RedArrow (pyrethrins/rotenone) and pyrellin
Nature	"knockdown" paralysis
Active ingredient	Pyrethrin
Formulations	EC, Solution
Pest controlled	As below

Diazinon 0.5% m/m+ Pyrethrum 0.1% m/m	
Insects	Dosage ml/ sq. m
Cockroach, Housefly	25-50

Active ingredient	Pyrethrin
Trade names	Talstar, Bifen IT, Permethrin SFR, Tengar SFR, Dragnet SFR, Steri-Fab Demon WP, Cyper WP, Cyper TC, Cynoff WP, CB Air Devil, Cy-kick CS, Cy-kick aerosol, Tempo SC,
Nature	"knockdown" paralysis
Formulations	WP
Pest controlled	Ants, roaches, spiders, stinging insects, pantry pests, bed bugs, as well as insects that we consider to be beneficial such as lady bugs and honey bees.

Mode of action

Pyrethrins exert their toxic effects by disrupting the sodium and potassium ion exchange process in insect nerve fibers and thereby interrupting the normal transmission of nerve impulses. Pyrethrins are extremely fast acting and cause an immediate "knockdown" paralysis in insects, however many insects are able to metabolize (break down) pyrethrins quickly. After a brief period of paralysis, these insects may recover rather than die. To prevent insects from metabolizing pyrethrins and recovering from poisoning, most products containing pyrethrins also contain the synergist, piperonyl butoxide (PBO). Without PBO the effectiveness of pyrethrins is greatly reduced. The insect mortality is increased by mixing pyrethrins with a synergist, piperonyl butoxide (PBO) to create pyrethrum PBO; however, is not certified to be used on organic crops.

Most insects are highly susceptible to pyrethrin at very low concentrations. The compound acts rapidly on insects, causing immediate knock down. Flying insects drop almost immediately after exposure. Although it causes fast knock down but insect death does not occur; and many insects recover after the initial knockdown phase. Pyrethrins are highly irritating to insects; as a result, they may be used as "flushing agents," causing insects to come out of hiding, a desirable circumstance when you need to identify an insect that is hiding in the turf grass such as grubs or sod web worm.

Pyrethrum is non-toxic to most mammals, making it among the safest insecticides in use. The Environmental Protection Agency has approved it for more uses than any other insecticide.

It's sold under trade names Pyrenon, Red Arrow (pyrethrins/rotenone) and Pyrellin (Pyrethrins/rotenone). Pyrethrum has been assigned a CAUTION rating.

Pyrethroids

Pyrethroids are synthetic materials designed to imitate natural pyrethrum. They can be used on houseplants, but because they are synthetic, they are not registered to use on organic crops.

Development of synthetic pyrethroid with increased light stability and insecticidal activity allows it to be used as foliar insecticide while the natural pyrethrins are now used mainly as domestic insecticides.

Pyrethroids are synthetic insecticides that share some similarities with natural pyrethrins. There are more than 1,000 different pyrethroids in use today, though less than a dozen are available in the United States. Pyrethroids can be found in many types of products such as pet sprays, pet shampoos, human head lice treatments, topical mosquito repellents and of course insecticide sprays for homes, farms, and businesses.

How do synthetic pyrethroids work?

Pyrethroids work by disrupting an insect's nervous system causing a weakened state followed by death. Pyrethroids are broad spectrum insecticides and most insects that come in contact with either the wet spray or the dry deposit of the active ingredient left by the spray will be eliminated. This includes ants, roaches, spiders, stinging insects, pantry pests, bed bugs, as well as insects that we consider to be beneficial such as lady bugs and honey bees.

Difference between pyrethroids and pyrethrins?

Pyrethrins, also called pyrethrum, are produced from the oil of the chrysanthemum flower from Kenya, Africa. Due to frequent shortages of the essential oil needed to create the natural organic pyrethrum, scientists produced synthetic pyrethroids. Originally pyrethroids were made to synthetically mimic pyrethrins. Over the years and as industry needs changed scientists were able to produce the pyrethroid with more complicated molecules, making the synthetic insecticide more potent. Today's pyrethroids have very low or no odor at all, they last longer, can be used in lower concentrations, they are manufactured without petroleum by-products, these have a low order of toxicity in mammals and are available in varying formulations. Pyrethrins, while they have pretty much stayed the same over the years, still have their place in pest control. Pyrethrins can be used in environments where a long lasting residual is not desired or where natural, organic pesticides or insecticides are needed.

Pyrethroids are available in many formulations. They are sold as wettable powders, emulsifiable concentrates, granules, dusts, aerosols, microencapsulates, flowables and tablets. Some formulations only last for 1-2 days in the environment while others last for as long as 60-90 days. Some formulations can be used as an effective "flushing" agent, driving roaches or other insects from hiding spots or harborages so pest control technicians know where to treat. Choosing the correct formulation depends on the environment that is being treated as well as the insects that needs to be eliminated.

Pyrethroids can be divided into two groups

1. First Generation Pyrethroids

Active ingredient: bifenthrin, permethrin, d-phenothrin

Product Trade Names: Talstar, Bifen IT, Permethrin SFR, Tengard SFR, DragnetSFR, Steri-Fab. Some insects such as German roaches or bed bugs may in some instances be able to recover from the effects of the first generation pyrethroids after knockdown but before death.

2. Second Generation Pyrethroids

Active ingredient: cypermethrin, cyfluthrin, deltamethrin, fenvalerate, lambda-cyhalothrin

Product Trade Names: Demon WP, Cyper WP, Cyper TC, Cynoff WP, CB Air Devil, Cy-kick CS, Cy-kick aerosol, Tempo SC, Tempo WP, Suspend SC, Deltagard G, D-Force HPX aerosol, Evercide MC, Conquer, Demand CS, Demand G ,Cyonara 9.7, SurrenderPestabs, Deltadust, Tempodust.

The second generation pyrethroids contains a cyano group that enhances insecticidal activity and molecular stability ensuring effectiveness.

Other known insecticidal pyrethroids (table.45) under organophosphates are Allethrin, Bifenthrin, Permethrin, Cyhalothrin, Lambda-cyhalothrin, Cypermethrin, Cyfluthrin, Deltamethrin, Ftofenprox, Fenvalerate, Phenothrin, Prallethrin, Resmethrin, Tetramethrin, Transfluthrin.

Table 45: Pest controlled by Pyrethroid Cypermethrin

Cypermethrin 10% EC		**mL/acre**
Cotton	Spotted bollworm American bollworm Pink bollworm	50-70 50-70 50-70
Cabbage	Diamond backmoth	60-70
Okra	Fruit borer	50-70
Brinjal	Fruit & shoot borer	50-70
Wheat	Shoot fly	50
Sunflower	Bihar hairy caterpillar	60-70

Cypermethrin 25% EC		
Cotton	Bollworms, Jassids, Thrips	40-70 20-30
Bhindi	Shoot & fruit borer Jassids	37-50 37-50
Brinjal	Shoot & fruit borer Jassids, Epilachna grub	37-50

22.1.1. Rotenone

Structure.$C_{23}H_{228}O_6$

Source: Rotenone is insecticidal compound that occurs in the roots of *Lonchocarpus* species in South America, Derris species in Asia, and several other related tropical legumes. Commercial rotenone was at one time produced from Malaysian Derris. Currently the main commercial source of rotenone is Peruvian *Lonchocarpus*, which often is referred to as cube root. Rotenone is extracted from cube roots in acetone or ether. Extraction produces a 2-40% rotenone resin which contains several related but less insecticidal compounds known as rotenoids. The resin is used to make liquid concentrates or to impregnate inert dusts or other carriers. Most rotenone products (Table.46) are made from the complex resin rather than from purified rotenone itself. Alternatively, cube roots may be dried, powdered and mixed directly with an inert carrier to form an insecticidal dust.

Table 46: Trade names of Rotenone pesticide and pest controlled.

Active ingredient	Rotenone
Trade names	Chem-Fish, Cuberol, Fish Tox, Noxfire, Rotacide, Sinid and Tox-R. It is also marketed as Curex Flea Duster, Derrin, Cenol Garden Dust, Chem-Mite, Cibe Extract and Green Cross Warble Powder
Nature	Powerful inhibitor of cellular respiration
Formulations	Crystalline preparations (approximately 95% pure), emulsified solutions (approximately 50% pure), and dusts (approximately 0.75 - 5% pure).
Pest controlled	Used in home gardens for insect control, for lice and tick control on pets, and for fish eradications as part of water body management.

Mode of action

Rotenone is a powerful inhibitor of cellular respiration. In insects rotenone exerts its toxic effects primarily on nerve and muscle cells, causing rapid cessation of feeding. Death occurs several hours to a few days after exposure. Rotenone is extremely toxic to fish, and is often used as a fish poison (piscicide) in water management programs. It is effectively synergized by PBO or MGK 264. Formulations include crystalline preparations (approximately 95% pure), emulsified solutions (approximately 50% pure), and dusts (approximately 0.75 - 5% pure).

Mammalian toxicity

Although rotenone is a potent cell toxin, mammals detoxify ingested rotenone efficiently via liver enzymes. As with pyrethrins, rotenone is more toxic by inhalation than by ingestion. Exposure to high doses may cause nausea, vomiting, muscle tremor, and rapid breathing. Very high doses may cause convulsions followed by death from respiratory paralysis and circulatory collapse. Direct contact with rotenone may be irritating to skin and mucous membranes. Treatment of poisoning is symptomatic. Chronic exposure to rotenone may lead to liver and kidney damage. Although some rodent testing has shown that chronic dietary exposure to rotenone may induce tumor formation, the most recent US EPA registration standard considers rotenone to be non-carcinogenic.

22.2. Sabadilla (veratrine alkaloids)

Structure

Cevadine:$C_{32}H_{49}NO_9$

Veratridine: $C_{36}H_{51}NO_{11}$

Source: Sabadilla is derived from the ripe seeds of *Schoenocaulon officinale*, a tropical lily plant which grows in Central and South America. Sabadilla is also sometimes known as cevadilla or caustic barley. When sabadilla seeds are aged, heated, or treated with alkali, several insecticidal alkaloids are formed or activated. Alkaloids are physiologically active compounds that occur naturally in many plants. In chemical terms they are a heterogeneous class of cyclic compounds that contain nitrogen in their ring structures. Caffeine, nicotine, cocaine, quinine, and strychnine are some of the more familiar alkaloids. The alkaloids in sabadilla are known collectively as veratrine or as the veratridine alkaloids. They constitute 3-6% of aged, ripe sabadilla seeds. Of these alkaloids, cevadine and veratridine are the most active insecticidally.

European white hellebore (*Veratrum album*) also contains veratridine in its roots. Hellebore was once commonly used in Europe and the U.S. for insect control, but is now unavailable commercially and is not registered by the US EPA.

Mode of action

The toxic alkaloids present in sabadilla affect nerve cell membrane of the insect, causing loss of action and function of nerve cell membrane leading to insect paralysis and death. Sabadilla (table.47) kills insects of some species immediately, while others may survive in a state of paralysis for several days before dying. Sabadilla is effectively synergized by PBO or MGK 264.

Table 47: Pest controlled by Sabadilla

Active ingredient	Sabadilla
Trade names	Red Devil Natural Guard
Nature	Affect nerve cell function
Formulations	emulsified solutions, dusts
Pest controlled	Caterpillars, leaf hoppers, thrips, stink bugs and squash bugs.

Mammalian toxicity

Sabadilla, in the form of dusts made from ground seeds, is the least toxic of the registered botanicals. Purified veratrine alkaloids are quite toxic, however, are considered on par with the most toxic synthetic insecticides. Sabadilla can be severely irritating to skin and mucous membranes, and has a powerful sneeze-inducing effect when inhaled. Ingestion of small amounts may cause headaches, severe nausea, vomiting, diarrhea, cramps and reduced circulation. Ingestion of very high doses may cause convulsions, cardiac paralysis, and respiratory failure. Sabadilla alkaloids can be absorbed through the skin or mucous membranes. Systemic poisoning by sabadilla preparations used as insecticides has been very rare or non-existant.

22.3. Ryania

Ryania comes from the woody stems of *Ryania speciosa*, a South American shrub. Powdered Ryania stem wood is combined with carriers to produce a dust or is extracted to produce a liquid concentrate. The most active compound in ryania is the alkaloid ryanodine, which constitutes approximately 0.2% of the dry weight of stem wood.

Mode of action

Ryania is a slow-acting stomach poison. Although it does not produce rapid knockdown paralysis, it does cause insects to stop feeding soon after ingesting it. Little has been published concerning its exact mode of action in insect systems. Ryania is effectively synergized by PBO and is reported to be most effective in hot weather.

22.4. Nicotine

Structure. $C_{10}H_{16}N_2O_4S$

Source: Nicotine is a simple alkaloid derived from tobacco, *Nicotiana tabacum*, and other Nicotiana species. Nicotine contitutes 2-8% of dried tobacco leaves. Insecticidal formulations generally contain nicotine (table.48) in the form of 40% nicotine sulfate and are currently imported in small quantities from India by other nations.

Table 48: Pest controlled by Nicotine pesticide

Active ingredient	Nicotine
Trade names	Black leaf 40
Nature	Nerve toxin
Formulations	emulsified solutions, dusts
Pest controlled	aphids, thrips, spider mites and other sucking insecticides

Mode of action

In insects and mammals nicotine is an extremely fast-acting nerve toxin. It competes with acetylcholine, the major neurotransmitter, by bonding to acetylcholine receptors at nerve synapses and causing uncontrolled nerve firing. This disruption of normal nerve impulse activity results in rapid failure of those body systems that depend on nervous input for proper functioning. In insects, the action of nicotine is fairly selective, and only certain types of insects are affected.

Mammalian toxicity

Despite the fact that smokers regularly inhale small quantities of nicotine in tobacco smoke, nicotine in pure form is extremely toxic to mammals and is considered a Class I (most dangerous) poison. Nicotine is particularly hazardous because it penetrates skin, eyes, and mucous membranes readily and both inhalation and dermal contact may result in death. Ingestion is slightly less hazardous due to the effective detoxifying action of the liver.

Symptoms of nicotine poisoning are extreme nausea, vomiting, excess salivation, evacuation of bowels and bladder, mental confusion, tremors, convulsions, and finally death by respiratory failure and circulatory collapse. Poisoning occurs very rapidly and is often fatal. Treatment for nicotine poisoning is symptomatic, and only immediate treatment, including prolonged artificial respiration, may save a victim of nicotine poisoning. Nicotine has been responsible for numerous serious poisonings and accidental deaths because of its rapid penetration of skin and mucous membranes because of the concentrated form in which it is used.

a. Nicotine Sulphate (40 EC) as Botanical Pesticide

Nicotine sulphate is an insecticide of plant origin. It is very effective against a wide range of insect-pests affecting the crops of economic importance as well as on ectoparasites affecting live stock.

Chemical Formula: $(C_{10}H_{14}N_2)2H_2SO_4$

Chemical Composition: Nicotine sulphate 40% contains Nicotine Alkaloid 400gm/1000gm i.e. 40%(w/w); Vegetable Oils, Tar, etc. 20gm /1000gm i.e. 2% (w/w); Water and other ingredients 580gm/1000 gm i.e. 58% (w/w)

Mode of Action: Nicotine sulphate acts upon the central nervous system of the insects. It has both contact as well as fumigation action.

Uses: Nicotine sulphate 40% is used to kill aphids, bugs, worms, leafhoppers and similar sucking insects which attack and destroy fruit, vegetables, crops and even flowers. It is also effective against Lice, Mite and Ticks which are a menace to livestock.

Advantages: While synthetic pesticides leave toxic remnants on fruits and vegetables, Nicotine Sulphate being highly bio-degradable does not leave any objectionable residue on market produce. Hence, it is an extremely safe pesticide. Nicotine Sulphate being an insecticide of plant origin does not pollute the environment like other chemical insecticides. No immunity has ever been developed by any pest against this product during its ninety years of continuous usage all over the world.

22.5. Citrus Oil Extracts

Limonene and Linalool

Crude citrus oils and the refined compounds d-limonene (hereafter referred to simply as limonene) and linalool are extracted from orange and other citrus fruit peels. Limonene, a terpene, constitutes about 90% of crude citrus oil, and is purified from the oil by steam distillation. Linalool, a terpene alcohol, is found in small quantities in citrus peel and in over 200 other herbs, flowers, fruits, and woods.

Terpenes and terpene alcohols are among the major components of many plant volatiles or essential oils. Other components of essential oils are ketones, aldehydes, esters, and various alcohols. Essential oils are the volatile compounds responsible for most of the tastes and scents of plants. Many of the essential oils also have some physiological activity.

Mode of action

Limonene is thought to cause an increase in the spontaneous activity of sensory nerves. This increase activity of sensory nerves sends spurious information to motor nerves which results in twitching, lack of coordination, and convulsions. Massive over stimulation of motor nerves leads to rapid knockdown paralysis. Adult fleas and other insects may recover from knockdown, unless limonene is synergized by PBO. Linalool is also synergized by PBO

22.6. Other Essential Plant Oils

Herbal Repellents and Insecticides

Essential oils are volatile, odorous oils derived from plant sources. Although they are used mainly as flavorings and fragrances in foods, cosmetics, soaps, and perfumes, some of them also have insect repellent or insecticidal properties. Many essential oils have GRAS (Generally Regarded As Safe) status; however, when applied topically at high concentrations they tend to be irritating to skin and mucous membranes. They are sometimes used as topical counter irritant to relieve or mask pain. Many of the essential oils that have low dermal toxicity may be toxic by ingestion.

The most common essential oils used as repellents are the oils of **cedar, lavender, eucalytus, pennyroyal, and citronella**. They are used mostly on pets and humans to repel fleas and mosquitoes. With the exception of pennyroyal, these essential oils are thought to pose little risk to people or pets, though they should not be used above recommended rates. Some herbal paste products that contain essential oils are recommended for daily use or "as often as needed." These products should be used moderately and with careful observation to spot early signs of skin irritation or possible toxic effects.

Citronella is sold mainly in the form of candles to be burned outdoors to repel mosquitoes from back yards or other small areas. It is also contained in some "natural" mosquito repellent lotions. Before the development of synthetic repellents, citronella was the most effective mosquito repellent available. Despite its wide usage, there is little scientific information available regarding its efficacy or mammalian toxicity.

22.7. Neem

Neem is a botanical pesticide derived from the neem tree (*Azadirachta indica*) which is native to Southeast Asia and India. This tree supplies at least two compounds, azadirachtin and salannin, that have insecticidal activity and other unknown compounds with fungicidal activity. The use of this compound is new in the United States, but neem has been used for more than 4,000 years for medicinal and pest control purposes in India and Africa. It is not highly toxic to mammals.

Properties

Substances with pesticidal properties are found in all parts of the neem tree. However, the greatest concentrations of these substances are found in the seed. Azadirachtin, the active ingredient in many pesticides currently available, is extracted from the seed kernels. Azadirachtin consists of more than 25 different but closely related compounds. In another extraction process, neem oil is extracted from the seed kernel. In agriculture, neem oil, fruit and the different byproducts such as seed cake are used as biopesticides, fungicides and organic manures.

Mechanisms of Action

Azadirachtin act as a feeding deterrent against a number of insect pests including beetles. It reduces the level of the insect hormone Ecdysome by disrupting the insect's molting process so that the immature larvae cannot develop into adults. After treatment with neem-based pesticides, you may see insects with crippled, distorted wings or the immature larvae and nymphs remain in an immature stage and then die. Some soft-skinned insect larvae may be killed by direct contact with the spray. Adults are not killed by the growth regulating properties of azadirachtin but mating and sexual communication may be disrupted which results in reduced fecundity.

Neem oil works in a number of different ways. The oil forms a coating on the insect's body, blocking the breathing openings and suffocating the insect. It also has a repellent effect on certain insects and mites. Neem oil prevents the germination and penetration of some fungal spores. One percent neem oil treatment was effective in managing powdery mildew on hydrangeas, lilacs and phlox.

Neem as Bio-pesticide

The importance of neem as bio-pesticide was realized by the modern scientific community, as early as 1959, when a German scientist in Sudan found that neem was the only tree that remained green during a desert locust plague. The pesticidal characteristics of neem is largely attributable to Azadirachtin found in the neem extracts which is a growth regulator as well as a powerful feeding and ovipositional deterrent. Azadirachtin is non-volatile and an insect cannot prevent it by smell but has to taste it, in order to respond to it. A taste of azadirachtin stimulates at least one 'deterrent neuron' in insects which show an anti-feedant response. The strength of 'deterrent neuron' responses has been correlated with the strength of anti-feedant responses. Neem oil can also suffocate mites, whiteflies, aphids and other types of soft bodied insects on contact. So it is clear that neem does not kill the insect on contact, but inhibits feeding and reproduction of the pests. These multiple modes of action make it unlikely that insects and plant pathogens can develop resistance to neem. Also certain pest such as **floral thrips, diamond back moth and several leaf miners** which develop resistance to the inorganic pesticides or that are inherently difficult to control with conventional pesticides are effectively controlled or managed with neem.

More than 60 insect pests may be affected by azadirachtin including **aphids, beetles, caterpillars, lacebugs, leafhoppers**, leafminers, **mealybugs, psyllids, thrips** and **whiteflies**. Due to its insect growth regulating properties, it is most effective against the immature stages of insects. For example, the immature larvae of many species in the lepidoptera family (moths and butterflies) are particularly sensitive to azadirachtin. Neem products may be registered for use on certain fruits, herbs and vegetables in addition to ornamentals. For edible crops, some neem-based products may be used up to the day of harvest.

Uses

A neem-based pesticide, **Margosan-O,** has been shown to control gypsy moths, leaf miners, sweet potato whiteflies, western flower thrips, loopers, caterpillars and mealybugs. This product is labeled for use on ornamentals, foliage plants, trees, shrubs and other non-food crops.

Margosan-O works as an insect growth regulator. The treated insect usually can't molt to its next lifestage and dies. It also may deter egg laying. Do not expect a quick kill.

Neem also can be purchased under the trade name **Bioneem** (by Safer Products).

The Emulsifiable Concentrate (EC) of neem

The emulsifiable concentrate (EC) of Neem prepared mostly from the Neem oil is used as the pesticide of choice in organic agriculture and considered as medium to broad spectrum pesticides. Among the other known botanical pesticides such as Rotenone and Pyrethrins, neem is found to be superior due to a number of reasons. It is also compatible with a list of other synthetic pesticides, which enables its usage as a component of integrated pest management (IPM).

Neem as a pesticide is the best and non-replacable component of IPM due to its following characteristic.

- Neem pesticide is a natural product, absolutely non toxic, 100% biodegradable and eco friendly.
- It is suited for mixing with other synthetic pesticide and in fact enhances their action.
- None or lesser quantity of synthetic pesticides needs to be used, thereby reducing the environmental load.
- Several synthetic pesticides being single chemical compounds cause easy development of resistant in species of pests. Neem consists of several compounds hence development of resistance is impossible.
- Neem does not destroy natural predators and parasites of pests thereby allowing these natural enemies to keep a check on the pest population.
- Neem also has systemic action and seedlings can absorb and accumulate the neem compounds to make the whole plant pest resistant.
- Neem has a broad spectrum of action active on more than 200 species of pests.
- Neem is harmless to non target and beneficial organisms like pollinators, honey bees, mammals and other vertebrates.

How Neem works as a pesticide

a. As an insect growth regulator

It is a very interesting property of neem products and unique in nature, since it works on juvenile hormone. The insect larva feeds when it grows and it sheds the old skin and again starts growing. This particular shedding of old skin is the phenomenon of ecdysis or moulting is governed by an enzyme ecdysone. When the neem components, especially Azadirachtin enter into the body of larvae, the activity of ecdysone is suppressed and the larva fails to moult, remains in the larval stage and ultimately dies. If the concentration of Azadirachtin is not sufficient, the larva manages to enter the pupal stage but dies at this stage and if the concentration is still less the adult emerging from the pupa is 100 % malformed, absolutely sterile without any capacity for reproduction.

b. As a feeding deterrent

The most important property of neem is feeding deterrence. When an insect larva sits on the leaf and it is hungry, it wants to feed on the leaf. This particular trigger of feeding is given through the maxillary glands give a trigger, peristalsis in the alimentary canal is speeded up, the larva feels hungry and its starts feeding on the surface of the leaf. When the leaf is treated with neem product, because of the presence of Azadirachtin, Salanin and Melandriol there is an anti–peristaltic wave in the alimentary canal and this produces something similar to vomiting sensation in the insect. Because of this sensation the insect does not feed on the neem product treated surface. Its ability to swallow is also blocked.

c. As an ovi-positional Deterrent

Another way in which neem reduces pests is not by allowing the female to deposits eggs. This property is known as Ovipositional deterrence, and comes in very handy when the seeds in storage are coated with neem Kernel powder and neem oil. The seeds or grains obtained from the market are already infested with some insects. Even these grains could be treated with neem seed kernel extract or neem oil; after this treatment the insects will not feed on them. There will be no further damage to the already damaged grains and at the same time when the female comes to the egg laying period of its life cycle, egg laying is prevented.

d. Other pesticidal activity includes

(1) The formation of chitin (exoskeleton) is also inhibited.

(2) Mating as well as sexual communication is disrupted.

(3) Larvae and adults of insects are repelled.

(4) Adults are sterilized.

(5) Larvae and adults are poisoned.

Mode of applications of Neem pesticide

The mode of applications, the dosages and the points to be noted during application are explained on the product label. The EC formulations are mostly applied as aerial sprays. It is also applied as soil drench to some plant species which translocates it through the plant tissue.

Neem can be effectively used in Hydroponics, Aeroponics, greenhouses, shade houses, ornamentals, interior-scapes, horticulture, landscapes, nurseries, turf farms, shrubs, flowering shrubs, fruits and nut trees.

Use of Neem as a bio-fungicide

As a fungicide, neem oil is mainly used as a preventative and when disease incidence is just starting. It coats the leaf surface which in turn prevents the germination of the fungal spores. Neem oil is effective against rots, mildews, rusts, scab, leaf spot and blights.

Use of Neem in storage pest control

Throughout the tropics much of the food harvested is lost during storage. More affluent farmers can and do spray their stored food crops with chemical pesticides to prevent worms, beetles and other infestations. Neem offers the impoverished farmers and even affluent farmers wanting to replace pesticides with a natural and inexpensive alternative. A light coating of neem oil or kernel powder protects stored food crops for up to twenty months from all types of infestations with no deterioration or loss of palatability. Even today in rural India, people store their grains mixed with dried neem. Neem not only protects but also prevents the further proliferation of storage pest if it has already infested the grains. The insects stop feeding on them due to the anti-feeding property of neem. On the other hand, the ovipositional deterrent of neem prevents the female insects from laying the egg during its egg laying period of its life cycle.

Neem and environment

The natural insecticides, fungicides and bio-pesticides made out of neem have many advantages. Research studies indicate that they are not harmful to humans or animals. The pests will not develop resistance over generations while the beneficial insects like butterflies, ladybugs, etc are spared. The soil is enriched, and neem extracts leave no residue in the environment.

Neem's "soft" pesticide character improves the lives of impoverished farmers throughout the tropical range of the neem tree. According to the World Health Organization, misuse or overuse of chemical pesticides result in poisoning of about 500,000 people resulting about a million illnesses and about 20,000 deaths in the Third World alone. Substituting crude neem extracts for expensive chemical controls saves both money and lives. In developing countries, most of the earnings of farmers are spent on crop protection chemicals. These results in a vicious cycle of continued debt and thereby the poverty, ill health and environmental degradation. To break this cycle and to improve the farmer's lives, agencies such as CARE, AID and AFGRO are actively promoting the introduction and use of neem in Southeast Asia, Africa, the Caribbean and both South and Central America. The farmers are educated about the means and methods for easily making a cheap, safe and effective product that protects their crops from over 200 different insect pests.

Farmers in Mexico and Haiti and shepherds in Australia have begun switching to simple neem-based sprays from the usual synthetic chemical pest controls. This has allowed the farmers to export mangos and other fruit to the United States without the chemical residues that often stopped their shipments at inspection stations. Neem-based sprays have similarly allowed the shepherds in Australia to produce pesticide-free wool that is being sold to European buyers for a considerable premium over the standard wool impregnated with chemical pesticides. Neem extracts have been approved by the U.S. Environmental Protection Agency for use on food crops and today more and more of the governmental agencies approve the usage of neem on food crops.

Neem Cake has multiple effects on the soil in controlling soil borne fungi, nematodes besides providing nutrition. The effects also last for the subsequent year. Hence, farmers may be encouraged to use Neem cake as much as possible.

Various neem base formulations (table.49) are available in the market as bio-pesticide for control of crop specific pests (table.50).

Table 49: Neem based formulations as Bio-pesticide

Trade Name	Formulation
ACHOOK	0.5% Azadiradione, Nimbocinol and Epinimbocinol
APHIDIN	Soapy humus extract
AZA	0.15% Azadirachtin
AZADIT	0.3% Azadirachtin
BIOSOL	Oil
BIONEEM	0.03% Azadirachtin
BIOPEST	0.15% Azadirachtin
ECONEEM	0.3% Azadirachtin
FIELD MARSHAL	0.3% EC Azadirachtin
FORTUNA AZA	0.15% Azadirachtin
JAI NEEM	0.03% Azadirachtin
JAWAN	0.03% Azadirachtin
JEEVAN CROP PROTECTOR	0.05% Azadirachtin
JUERKEN	0.03% Azadirachtin
LIMONOOL	75% EC (0.03%) Azadirachtin
MARGOCIDECK	20% EC(Azadirachtin 0.15% w/w)
Margosan-o	80% EC Azadirachtin
NEEMARK	0.03% Azadirachtin
NEEMAZAL-F	5% Azadirachtin
NEEMGOLD	0.15% Azadirachtin
NEEM GUARD	0.03% Azadirachtin
NEEM RICH	0.03% Azadirachtin
NEEMOLIN	0.15% Azadirachtin
NEEMPOURN	0.15% Azadirachtin
NEEMOSAN	0.03% Azadirachtin
NEEMOX	80% EC Azadirachtin
NEEMTA	0.15% Azadirachtin
NEETHRIN	Oil 0.03% Azadirachtin
NIMBA	Wettable Powder 0.15% Azadirachtin
NIMBECIDINE	0.03% Azadirachtin
NIMBOSOL	0.15% Azadirachtin
NIMBASAL	0.15% Azadirachtin
NIMLIN	0.03% Azadirachtin
NEEM BASE	20%EC 20 WDP Azadirachtin
NEEM HIT	0.15% Azadirachtin
NEEM OIL	0.03% EC Azadirachtin
NEEM PLUS	0.15% Azadirachtin
NEEM TOP	0.03% Azadirachtin
NEEMOL	0.03% Azadirachtin
NIM-76	0.05% Azadirachtin
NEEM OIL	2% Azadirachtin

Trade Name	Formulation
NEEMAX	Seed Powder 300 ppm (Azadirachtin)
NEEMOCIDE	0.03% Azadirachtin
NEEMACTINE	0.15% Azadirachtin
NEEMBO BAS	0.03% Azadirachtin
NEEMBO BAS	0.15% Azadirachtin
NIMBITOR (ZA-199)	0.03% Azadirachtin
PHYTOWIN	0.03% Azadirachtin
RAKSHAK	0.15% Azadirachtin
REPLIN 555	93% EC Azadirachtin
RD-9 REPELIN	100% EC Azadirachtin, Pongamia, Anona
SHAKTIMAN	0.03% Azadirachtin, Nimbim
SUKRINA	0.03% Azadirachtin, Meliontrial
SWATICURE	0.03% Azadirachtin
SUNEEM	0.03% Azadirachtin
TRIC	Oil, 0.03% Azadirachtin
VAPACIDE	0.03% Azadirachtin
WELLGRO	0.03% Azadirachtin

Table 50: Important Target Pests for Neem Products

Crop	Pest
Cotton	White fly, Aphids, American boll warm, Tobacco caterpillar
Okra, Brinjal	Green leaf hopper (Jassids)
Tomato,Cabbage, Cauliflower	Common Fruit borer, whitefly, dimond backmoth, head borer
Paddy	leaf hopper, brown plant hopper
Ground Nut	Leaf miner, tobacco caterpillar
Lemon (Acid lime)	Leaf miner
Mango	Mango hopper
Castor	Red Hairy Caterpiller

Doses and Mode of application:

Neem Oil

Dosage: 450 ml per acre
Dilution: 5 ml/ ltr water
Frequency: twice during crop period

Benefits of Neem based pesticides

Neem based pesticides does not persist in the environment and is degraded by ultraviolet light and rainfall. Many neem-products tend to have low mammalian toxicity. Because many neem products degrade quickly, they may have less of an effect on non-target beneficial organisms compared to some of the more traditional pesticides. Researchers found few effects on many insect predators including spiders, earwigs, and ants but flower fly larvae are very sensitive to neem sprays.

Neem-based materials are compatible with integrated pest management (IPM). Regular scouting helps to insure the early detection of immature stages of many insects. Azadirachtin is compatible with insecticidal soap, superior horticultural oil and *Bacillus thuringiensis* (*Bt*).

As mentioned previously, azadirachtin has a number of different modes of action. It is less likely that insects or pathogens will develop resistance to neem products compared to materials with a single mode of action.

Limitations of Neem based pesticides

Botanical pesticides, such as neem, have limited persistence in the environment. Temperature, ultraviolet light, rainfall and other environmental factors can degrade neem. Repeated applications may be needed to achieve the desired result. Because it is an insect growth regulator, it is only effective against the immature stages of insects. Rescue treatments will not be effective. You will not see an immediate knockdown effect and insects may continue to feed. However, due to its repellant effects, insect feeding will be reduced. Phytotoxicity (damage to plants) may be of concern for certain formulations of neem-based products with flowers being particularly sensitive. Newly transplanted plants with limited root development or plants that are wilted or under stress should not be treated.

23

Invertebrate Animal Based Insectcide

Nereistoxin (NTX) was originated from a marine annelid worm ***Lumbriconereis heteropoda*** and its analogue pesticides (table 51) including cartap, bensultap, thiocyclam and thiobensultap have been commonly used in agriculture, because of their low toxicity and high insecticidal activity.

Table 51: Nereistoxin pesticide and pest controlled

Active ingredient	Nereistoxin (NTX)
Trade names	Cartap Bensultap Thiocyclam Thiobensultap
Nature	Block nicotinic acetylcholine receptor
Formulations	WSP Dust Granules
Pest controlled	aphids, thrips, spider mites sucking insecticides

However, NTX has been reported about its inhibitory neuro toxicity in human and animal body, by blocking nicotinic acetylcholine receptor and it cause significant neuromuscular toxicity, resulting in respiratory failure.

24

Pesticides for Control of Slug and Snails

Slugs and snails occasionally cause serious damage to seedlings, tender, low-growing leafy vegetables; or ripening fruit that are on the ground. Feeding damage, hollowed-out areas, can be found anywhere on fruit, but is usually concentrated near the stem. Slugs leave a tell tale slime trail (silvery trail) behind on the surface of fruit or leaves.

Slugs and snails are active at night or cloudy days. Slugs and snails are favor by continuously moist soil and organic mulch. Their eggs are laid in groups in moist soil, and they use the organic mulch to overwinter. Slugs can complete their entire life cycle in a field.

If slugs are a problem, their hiding places, i.e., boards, stones, weedy areas, should be eliminated. Heavy mulching will also create a favorable habitat for slugs and should be thinned so the soil can become warm and dry. This will reduce the number of slugs.

Raised beds that can dry out more readily than flat beds reduce slug problems. Use of black plastic mulch will discourage slug build-up because it causes the soil to heat up and dry out.

Metaldyhyde bait (Clean Crop, G30-40 lb./A or 7.5 G 15-20 lb./A) can be used and is usually very effective. Follow label instructions carefully for application methods for each particular vegetable crop.Apply bait in evening after a rain or irrigation.

For an organic alternative, diatomaceous earth can be spread around plants (a 1 inch high x 3 inch wide band). Control of slugs with diatomaceous earth has been poor to fair.

Use of tobacco dust for the snail management is also effective. Apply the tobacco dust directly on the snails feeding on/ in crop (Mhaske et al. 2005).

25

Ready-Reckoner for Insect Pest Specific Insecticide

The application of insecticide to control a particular crop pest should have options of the pesticides, as some may be less effective than other due to various reasons. The development of pesticide resistant in a pest against a particular pesticide, make it ineffective in the control of pest, and therefore the pesticide from other group to control the same pest is necessary to be known. Table 52, indicates the Insect pest specific insecticide for control of various insect pests.

Table 52: Insect Pest Specific Insecticides

Sr.No	Name of Insect Pest	Name of Insecticides
1.	Leaf eating caterpillar	Azadirachtin, Chlorantraniprole, Emamectin benzoate, Spinosad, Lambda-cyhalothrin, Chlorpyriphos.
2	Stem and Shoot borer	Azadirachtin, Chloropyriphos, Cypermethrin, Dimethoate, Lamda-cyhalothrin, Thiodicarb, Deltamethrin, Fenvalerate, Phosalone, Pyridalyl, Carbosulfan, Methoxychlor,
3	Heliothis	Flubendiamide, acephate, dimethoate, chlorantraniliprole, fipronil, Emamectin benzoate,
4	Semilooper	Malathion, Dimethoate, Chlorpyriphos, Quinalphos. Profenophos, Triazophos, Spinosad, Fipronil.
5	Cutworm	Chlorpyrifos,
6	Bollworm	Azadirachtin, Cypermethrin, Malathion, Chlorpyrifos,
7	Fruit borer	Fenvalerate, Azadirachtin, Indoxacarb, lambda-Cyhalothrin, Novaluron, Quinolphos, Phosalone, Trichloforon, Deltamethrin, Emamectin benzoate, Fipronil, Flubendamide, Indoxacarb, Novaluron, Spinosad, Thiodicarb, Cypermethrin,
8	Fruit flies	Methoxychlor, Spinosad+ molasses/fruit juice as attractant, Maldison with protein bait, Trichlorton, Fipronil, Abamectin,Spinetoram,Thiamethoxam+lambda cyhalothrin

Sr.No	Name of Insect Pest	Name of Insecticides
9.	Fruit Moth	Spinosyn, esfenvalerate, malathion. Chlorantraniliprole, acetamiprid, spinetoram, spinosad, pyriproxyfen, cyantraniliprole, phosmet, methoxyfenozide,
10	Bark eating caterpillar	Quinalphos, Fenvalerate,
11	Army worm	Chlorantraniliprole, emamectin benzoate, spinetoram, Bifenthrin, Deltamethrin, gamma-cyhalothrin, imidacloprid+cyfluthrin.
12	Hairy caterpillar	Cypermethrin, deltamethrin, fluvalinat, Azadirachtin, Phosalone, Fenitrothion, Chlorpyrifos.
13	Pod borer	Lambda cyhalothrin, Parathion, Malathion, Chlorpyrifos,
14	Bugs	Dimethoate, Demeton methyl, Chlorpyrifos,
15	Beetle	Malathion, Demeton methyl, Chlorpyrifos, Methoxychlor,
16	Stem weevil	Malathion, Methoxychlor,
17	Shoot fly	Demeton methyl,
18	Gall midge	Carbosulfan,
19	Tobacco caterpillar	Novaluron, Malathion,
20	Diamond black moth	Azadirachtin, Chlorantranilprole, Chlorfenpyr, Chlorpyrifos, Cypermethrin, Difenthiuron, Fenvalerate, Fipronil, Flufenoxuron, Indoxacarb, Lufenuron, Metaflumizone, Novaluron, Permethrin, Pyridalyl, Spinosad,
21	Aphid	Acetamiprid, Azadirachtin, Dimethoate, Carbosulfan, Fipronil, Imidacloprid, Oxydemeton methyl, phosalone, Quinalphos, Permethrin, Malathion, Nicotin pesticide, Parathion, Demeton methyl, Chlorpyrifos, Methoxychlor.
22	Jassid	Dimethoate, Imidacloprid, Cypermethrin, Deltamethrin, Permethrin, Quinalphos. Azadirachtin, Carbosulfan, Malathion, Demeton methyl,
23	Wooly aphid	Malathion, Demeton methyl,
24	Thrips	Acetamiprid, Dimethoate, Emamectin benzoate, Ethion, Fenpropathrin, Fipronil, Imidacloprid, Oxydemeton methyl, Thiacloprid, Nicotin pesticide, Cypermethrin, Carbosulfan, Parathion, Demeton methyl,
25	Mites	Dicofol, Fenzaquin, Flufenzine, Malathion, Difenthiuron, Dimethoate, Emamectin benzoate, Ethion, Fenpropathrin, Milebemectin, Oxydemeton methyl, Quinalphos, Nicotin pesticide, Demeton methyl,
26	Mealybug	Malathion, Dimethoate, Demeton methyl
27	White flies	Azadirachtin, Imidacloprid, Malathion, Thiamethoxam, Difenthiuron, Carbosulfan, Fenpropathrin, Demeton methyl, Chlorpyrifos,
28	Hopper	Quinalphos, Azadirachtin, Carbosulfan, , Parathion, Malathion, Dimethoate, Demeton methyl, Chlorpyrofos, Methoxychlor,

Sr.No	Name of Insect Pest	Name of Insecticides
29	Red spider mite	Spiromesifen, Dicofol
30	Yellow mite	Buprofezin, Fenazaquin, Fenpyroximate, Hexythiazox, Spiromesifen
31	Leaf roller/leaf folder	Carbosulfan, Parathion, , Chlorpyrifos, ,
32	Web forming caterpillar	Permethrin, Lambda-cyhalothrin, Cyfluthrin, bifenthrin, Fluvalinate, Esfenvalerate, Acephate, Malathion, Spinosad.
33	Leaf minor	Azadirachtin, Dimethoate, Demeton methyl,
34	Root aphid	Parathion,
35	Root Mealybug	Chlorpyrifos, Deltamethrin, Organophosphate, Bendiocarb, bifenthrin, cyfluthrin, fenpropathrin, kinoprene
36	Root Weevils	Fipronil, Thiametoxam, Thiacloprid, Acetamiprid, Chlorpyrifos, Fenitrothion, Guthion, Ficam, Orthene, Cypermethrin,
37	Root Maggot	Chlorpyrifos, Pyrethrins, Cyantranilprole, Ecotrol G
38	Fungus Gnat larvae	Imidacloprid, Pyrethrins, Permethrin,
39	Root borer	Chlorpyrifos, Malathion, Ethylene dichloride,
40	Termite	Chlorpyrifos, Permethrin, Cypermethrin, Bifenthrin, Fenvalerate, Imidacloprid, Fipronil Baits containing hexaflumuron, sulfluramid, or diflubenzuron
41	Grub	Cypermethrin, Chlorpyrifos,
42	Gaint Snail and slugs	Bait containing Metaldehyde, Methiocarb , Iron phosphate
43	Locust	Fenitrothion, Chlorpyrifos, Lambda cyhalothrin+deltamethrin, Malathion, Deltamethrin.

26

Insect Attractants for Insect Pest Management

Many chemical and visual lures attract insects and therefore can be used as attractants to monitor or directly catch and reduce insect populations. These attractants are used in ways that do not injure other animals or humans or result in residues on foods or feeds, and can be used in an environmentally sound manner in pest management programs. The effective use of attractants and traps requires knowledge of basic biological principles and the pest or crop-specific details involved in individual applications. These includes chemical attractants, visual lures (such as light), and attractant-baited and unbaited traps.

26.1. Herbal Attractants

The essential oil and a number of extracts of *Rosmarinus officinalis* L. in solvents of increasing polarity have been isolated, their components identified and tested in pest control as herbal attractants. Ethanol and acetone extracts of these attract grape berry moth, *Lobesia botrana*. However, none of the extracts had a significant effect on western flower thrips, *Frankliniella occidentalis*, which is attracted by 1, 8-cineole, a major essential oil component (Katerinopoulos *et al*., 2005). Lemon essential oil is distilled from the peels of *Citrus limonum*. It has a light yellow color and a characteristic lemon aroma. Lemon essential oil contains several terpenes and geraniol, which have all been shown to attract thrips, mealybugs, scale, and Japanese beetles. Adding this oil to the insect-a-peel, thrips/leafminer blue trap, the yellow aphid/whitefly sticky trap will attract these unwanted pests and capture them on the trap.

Geraniol and eugenol are effective attractants and are used as lures in traps for the Japanese beetle, *Popillia japonica while* methyl-eugenol has been used to trap oriental fruit fly, *Dacus dorsalis* (Vargas *et al*., 2000). Cinnamyl alcohol, 4- methoxy-cinnamaldehyde, cinnamaldehyde, geranylacetone and α-terpeol are also attractive to adult corn rootworm beetles, *Diabrotica* sp. (Hammack, 1996; Petroski and Hammack, 1998). Compositions of cis-jasmone were found to effectively attract adult Lepidoptera. The cis-jasmone may be used alone or in combination with one or more other volatiles of the Japanese honeysuckle flower, particularly linalool and/or phenylacetaldehyde. By attracting the adult Lepidoptera to attracticidal baits and/or field traps, the attractants are useful for the control and monitoring of these agricultural pests (Pair and Horvat, 1997).

Similarly, natural essential oils have shown a high attractiveness for greenhouse whitefly, *Trialeurodes vaporariorum.* Greenhouse whitefly reacted particularly intensively to sandalwood oil, basil oil, and grapefruit oil. After the application of aromatic substances on yellow sticky traps, the number of insects caught increased significantly amounting to 487.64, 483.20, and 333.09%, respectively (Górski, 2004). Thus, natural essential oils or their constituents could be useful in the monitoring of pests, at least greenhouse whitefly in the present case.

26.2. Chemical Attractants

Insects use many different semeiochemicals, chemicals that convey messages between organisms (The Greek word "semeio" means sign.). Although semeiochemicals may seem analogous to tastes or smells perceived by humans, the use of such compounds by insects is characterized by a high degree of sensitivity and specificity. Receptor systems that ignore or screen out countless irrelevant chemical messages are nonetheless able to detect messenger compounds at extremely low concentrations. Detection of a chemical message triggers very specific unlearned behaviors or developmental processes. Chemicals that act as attractants or carry other messages across distances are volatile (quick to evaporate) compounds. When released into the air, they can be detected by certain insects (those receptive to a specific compound) a few inches to hundreds of yards away. Chemicals that carry messages over considerable distances are most often used in pest management. First, semeiochemicals may carry messages either within or between species.

Pheromones are semiochemicals that are produced and received by members of the same species. A range of behaviors and biological processes are influenced by pheromones, but pest management programs most often use compounds that attract a mate (sex pheromones) or call others to a suitable food or nesting site (aggregation pheromones). Other pheromones regulate caste or reproductive development in social insects (honey bees and termites for example), signal alarm (in honey bees, ants, and aphids), mark trails (ants), and serve other functions.

Allelochemicals are semiochemicals that affect one or more species other than the producer. Of known allelochemicals, volatile compounds similar to those given off by food sources (plants or animals) are important in pest managements. Feeding attractants are examples of kairomones, allelochemicals produced by one species but used to advantage by another species. For example, carbon dioxide given off by humans and other animals is used as a kairomone by female mosquitoes seeking a blood meal. In contrast, allomones are allelochemicals that favor the producer. For example, secretions that deter predators are allomones.

26.3. Attractant-Baited Traps to Monitor Pest Populations

The most common use of chemical attractants is in traps to monitor insect populations. Although not all of the compounds used in this manner are pheromones, many publications refer to all attractant-baited traps as pheromone traps. For use in monitoring, chemical attractants usually are impregnated or encased in a rubber or plastic lure that slowly releases the active component(s) over a period of several days or weeks. Traps containing these lures are constructed of paper, plastic, or other material. Most traps use an adhesive-coated surface or a funnel-shaped entrance to capture the target insect. Traps for some pests (such as the apple maggot) are coated with an adhesive that also contains the chemical attractant. Attractant-baited traps are used in monitoring programs for at least three purposes to detect the presence of an exotic pest (an immigrant pest not previously known to inhabit a state or region); to estimate the relative density of a pest population at a given site; and to indicate the first emergence or peak flight activity of a pest species in a given area, often to time an insecticide application or to signal the need for additional scouting.

The use of traps to detect exotic pests has been demonstrated in widely publicized efforts to detect and eradicate pests such as the gypsy moth and the Mediterranean fruit fly whenever infestations are detected in new areas. Attractant-baited traps can be used to signal the need for additional sampling efforts or to time insecticide applications and eliminate unnecessary spraying. One example of the use of pheromone traps to trigger further sampling involves the black cutworm, a common but sporadic pest of seedling corn in the Midwest. Pheromone traps baited with a specific sex attractant are used in a statewide sampling program to monitor the annual spring migration of black cutworm moths from southern states into Illinois. In area where counts of male moths in traps indicate the potential for damaging infestations of cutworm larvae, producers are urged to check for cutworm density and crop damage in fields of seedling corn.

26.4. Attractant-Baited Traps to "Trap out" Pest Populations

Pheromone traps are so effective for catching certain insects, numerous traps placed throughout a pest's environment can sometimes remove enough insects to substantially reduce the local population and limit the damage it causes. Efforts to "trap out" insect pests (a process also termed removal trapping or mass trapping) have utilized species-specific aggregation pheromones that attract both male and female beetles or species-specific sex pheromones that attract male moths. When aggregation pheromones are used to attract adult beetles of both sexes, traps may reduce the feeding damage caused by the adult insects and reduce reproduction by capturing adults before they lay eggs. When sex pheromones are used to capture moths, success depends upon capturing males before mating occurs. Although mass trapping programs using chemical attractants have targeted such important pests as bark beetles, codling moth, apple maggot, Japanese beetle, and Indian meal moth, field-scale successes have been limited. For mass trapping to adequately reduce pest populations, a large number of very efficient traps are usually needed.

26.5. Attractants to Disrupt Insect Mating

To disrupt insect mating, a species-specific sex attractant is broadcasted throughout the area. In an environment permeated with artificially applied sex pheromone molecules, male insects that rely on pheromones to locate females are unable to do so. They either follow an artificial signal to a frustrating destination or their sensory receptors become overloaded by constant exposure to pheromone molecules, leaving the insect temporarily unable to detect additional pheromone messages. The way in which artificial attractants might "out-compete" female moths and prevent their success in attracting a mate. In field applications of mating disruption techniques attractants have been applied to fields or forests in hollow plastic fibers, capsule like pellets, and attractant-impregnated plastic strings or ties. Although mating disruption programs are not widely used, trials have been successful against the oriental fruit moth, pink bollworm in cotton, grape berry worm, tomato pinworm, and several pests of forest conifers. The trial use of pheromones to disrupt mating for codling moth control in apples has produced mixed results.

Mating disruption programs are most successful where large areas are treated, where the treated area is isolated from sources of pests that might immigrate, and where the pest population is low. When pest densities are low, artificial attractants are more likely to out compete a high percentage of female insects in attracting males. For insect attractants to be broadcast into the environment for direct control, the attractants (regulated as pesticides) must be evaluated and approved by the US Environmental Protection Agency. The sex attractant of the oriental fruit moth has been approved for such use in plastic "ropes" to be tied onto the limbs of fruit trees.

26.6. Attractants in Poison Baits

Combining insect attractants with poisons (insecticides) is a practice that has been used in pest management for many years. In the early 1900s, for example, poisoned bran baits were used for grasshopper control; hoppers that were attracted to the treated bran and fed on it were killed by an insecticide that could not be applied safely, economically, or effectively in any other manner. Because pests are lured to toxic compounds that are combined with attractants, poisoned baits can sometimes be used effectively at low rates and often in a manner that does not leave residues on plants or animals. Insecticidal baits are used currently in the control of several pests including the house fly, slugs, certain ants, cockroaches, and yellow jackets.

26.7. Lights to Attract Insects

A great number of insect species are attracted to light of various wavelength. Although different species respond uniquely to specific portions of the visible and nonvisible spectrum (as perceived by humans), most traps or other devices that rely on light to attract insects use fluorescent bulbs or bulbs that emit ultraviolet wavelengths (black lights). Hundreds of species of moths, beetles, flies, and other insects, most of which are not pests, are attracted to artificial light. They may fly to lights throughout the night or only during certain hours. Key pests that are attracted to light include the European corn borer, codling moth, cabbage looper, many cutworms and armyworms, diamondback moth, sod webworm moths, peach twig borer, several leaf roller moths, potato leafhopper, bark beetles, carpet beetles, adults of annual grubs (*Cyclocephala*), house fly, stable fly, and several mosquitos. The mosquitoes *Ochlerotatus* (formerly *Aedes*) *triseriatus, Ochlerotatu* (also formerly *Aedes*) *hendersoni*, and *Aedes albopictus* are not attracted to light, however lights and light traps are used with varying degrees of success in monitoring populations and in mass trapping.

26.8. Colored Objects to Attract Insects

Specific colors are attractive to some day-flying insects. For example, yellow objects attract many insects and are often used in traps designed to capture winged aphids and adult whiteflies. Red spheres and yellow cards attract apple maggot flies. Like other attractants, colored objects can be used in traps for monitoring or mass trapping. Yellow plastic tubs filled with water, for example, are used to monitor the flights of aphids in crops where these insects are important vectors of plant viruses. Aphids attracted to the yellow tub land on the water and are unable to escape. Yellow, sticky-coated cards or plastic cups are widely used in mass trapping programs to help control whiteflies in greenhouses. Although recommended trap densities in greenhouses are based on studies involving only a few crops, recommendations of 1 trap per 5 square yards or 1 trap every 3 to 4 feet along benches are common. Yellow sticky traps capture adult whiteflies, not wingless nymphs. Research indicates that red sphere traps and chemical attractants can be used to "trap out" apple maggot flies and limit damage to fruit. For mass trapping programs to work, traps must be in place before flies begin to move into orchards (in early June), and a great number of traps must be used (one every 15 feet in the trees at the perimeter of the orchard). Mass trapping for apple maggot control is still an experimental approach, and commercial producers should not adopt a mass trapping program if complete control of apple maggot damage is necessary.

26.9. Other Traps

Several unique types of traps are used for the control of various species of flies. House fly traps containing foods or chemical attractants lure house flies to a reservoir from which they cannot escape. These traps capture thousands of house flies around livestock facilities, but the overall population in such areas is usually not reduced by a meaningful level unless a great number of traps are used. The effectiveness of such traps must be judged not by the number of flies in the traps but by the number of flies still present in the area. (These traps do not capture stable flies, the biting flies that are most annoying to livestock). Another type of trap useful to gardeners and farmers is the pitfall trap. Perhaps its best known use is in slug control. Bowl, cups, or other containers are set into the soil surface. Beer or a fermented mixture of flour, sugar, yeast, and water is added to the container to attract slugs; slugs that enter the container are unable to escape and "drown" in the liquid. Similar pitfall traps containing a preservative (not an attractant) are sometimes used to sample populations of insects active at the soil surface. Relatively new pitfall traps are now available for detecting beetle infestations in stored grains. These traps can be used with or without an attractant to provide a very sensitive measure of insect presence in warm grain.

Insect attractants and traps (table.53) are useful tools for monitoring insect populations to determine the need for control or the timing of control practices. In some instances, attractants and traps also can be used to control insect populaions directly by mass trapping or mating disruption. Using attractants and traps to monitor and control insect populations can improve the effectiveness of insecticide applications and sometimes reduce the use of broad-spectrum, more toxic compounds.

Table 53: List Of Marketed Traps, Lure and Attractants

Product Name	Manufacturer	Target Pest
Water Pheromone Trap	Harmony Ecotech Pvt. Ltd., Cherapalli, Hyderabad.	Tomato leaf miner
Cucurbit Pheromone Lure	Agri Phero Solutions, Fateh Nagar, Hyderabad.	Melon Fruit fly
Roller Sticky Trap	Ponalab Biogrowth Private Limited, Rajaji Nagar, Bengaluru	Aphids,whitefly, thrips, Fruit fly, Leaf miner
Biolure – LO	Biotech International Limited, Delhi	*Lucinodes orbonalis* on Brinjal
Pheromone lures and Traps	Pest control India Pvt. Ltd., Ashwin Nagar, Nasik	Thrips, whitefly, fruit fly
Biolure – SI	Biotech International Limited, Delhi	*Scirpophaga incertulus* on Paddy
Bee Pheromone Lure	Green Life Pheromone System, Rangareddy.	Bees and wasps
Yellow Sticky Sheet	Agri Phero Solutions, Fateh Nagar, Hyderabad.	Aphids, whitefly, thrips, Fruit fly, Leaf miner
Pink Bollworm Pheromone Lure	Agri Phero Solutions, Fateh Nagar, Hyderabad.	Pink bollworm
Funnel Trap	Agri Phero Soluntions, Fateh Nagar, Hyderabad.	Thrips, whitefly, fruit fly

27

Insect Repellent for Insect Pest Management

A number of plant substances have been considered for use as insect anti-feedants or repellents, but apart from some natural mosquito repellents, little commercial success has ensued for plant substances that modify arthropod behavior. Several factors appear to limit the success of botanicals, most notably regulatory barriers and the availability of competing products that are cost-effective and relatively safe compared with their predecessors.

27.1. Botanicals as Repellent

At present there are four major types of botanical products used as insect repellent viz. 1. Pyrethrum 2. Rotenone 3. Neem and 4. Essential oils, along with three others in limited use viz. Ryania, nicotine, and sabadilla. These are dealt in details in Botanical insecticide chapter.

Different plant Essential oils are also used as insect repellent.

27.1.1. Plant Essential Oils as a Repellent

A number of plants containing essential oils have been traditionally used for protection of stored commodities, especially in the Mediterranean region and in southern Asia, but interest in their essential oils was renewed with emerging demonstration of their fumigant and contact insecticidal activities to a wide range of pests in the 1990s. Plant essential oils are produced commercially from several botanical sources, many of which are members of the mint family (Lamiaceae). The oils are generally composed of complex mixtures of monoterpenes, biogenetically related phenols, and sesquiterpenes. Examples include 1, 8-cineole, the major constituent of oils from rosemary (*Rosmarinus officinale*) and eucalyptus (*Eucalyptus globus*); eugenol from clove oil (*Syzygium aromaticum*); thymol from garden thyme (*Thymus vulgaris*); and menthol from various species of mint (*Mentha* species).

The rapid action of some essential oils against some pests is indicative of a neurotoxic mode of action, and there is evidence for interference with the neuromodulator octopamine by some oils and with GABA-gated chloride channels by others.

Some of the purified terpenoid constituents of essential oils are moderately toxic to mammals, but, with few exceptions. The oils themselves or products based on oils are mostly nontoxic to mammals, birds, and fish. However, both pollinators and natural enemies are vulnerable to poisoning by products based on essential oils. Owing to their volatility, essential oils have limited persistence under field conditions; therefore, although natural enemies are susceptible via direct contact, predators and parasitoids reinvading a treated crop one or more days after treatment are unlikely to be poisoned by residue contact as often occurs with conventional insecticides. In the United States, commercial development of insecticides based on plant essential oils has been greatly facilitated by exemption from registration for certain oils commonly used in processed foods and beverages. This opportunity has spurred the development of essential oil-based insecticides, fungicides, and herbicides for agricultural and industrial applications and for the consumer market, using rosemary oil, clove oil, and thyme oil as active ingredients. Interest in these products has been considerable, particularly for control of greenhouse pests and diseases and for control of domestic and veterinary pests, with several private companies (e.g., EcoSMART Technologies, Inc., United States) moving toward or into the marketplace. Another factor favoring development of botanical insecticides based on plant essential oils is the relatively low cost of the active ingredients, a result of their extensive worldwide use as fragrances and flavorings.

27.1.1.1. Citronella (Cymbopogon nardus)

The essential oil has been used for over fifty years both as an insect repellent and an animal repellent. Combining few drops each of citronella, lemon (*Citrus limon*), rose (*Rosa damascena*), lavender and basil essential oils with one litre of distilled water is effective to ward off indoor insect pests. The larvicidal activity of citronella oil has been mainly attributed to its major monoterpenic constituent citronellal (Zaridah *et al.*, 2003).

27.1.1.2. Catnip (Nepeta cateria)

The essential oil is highly effective for repelling mosquitoes, bees and other flying insects. The most active constituent in catnip has been identified as nepetalactone. It repels mosquitoes ten times more than DEET. It is particularly effective against *Aedes aegypti* mosquito, a vector for yellow fever virus. Oil of *Trachyspermum* sp. is also larvicidal against *A. aegypti* and southern house mosquito, *Culex quinque fasciatus* (at LC50 = 93.19–150.0 ppm) (Vrushali *et al.*, 2001).

27.1.1.3. Ocimum Sanctum

It cause 20% mortality to 3rd instar of *S. litura* larvae (Sharma *et al.*, 2001). At a topical dose of 100 μg/ larvae, > 90% larval mortality has been reported when essential oil of *Satoreja hortensis*, *Thymus serpyllum* and *Origanum creticum* (LD50 = 48.4–53.4) were applied to 3rd instars *S. litura* (Isman and Machial, 2006). Similar studies were reported by Sharda *et al.* (2000) where essential oil of *Ageratum conyzoides* caused 43.0–68.75% mortality at 0.025–0.25 μl concentration. Tripathi *et al.*, (2003) has reported toxicity of essential oil of *Aegle marmelos* by topical application to *S. litura* larvae with LD50 = 116.3 μg/ larvae. Essential oil of *Lippia alba* induces growth inhibition (GI50 = 6.9–11.0 mg/g diet), where both relative growth and feeding consumption rates of *S. litura* were conspicuously reduced (Tripathi *et al.*, 2003).

27.1.1.4. Dill Oil

It is obtained from dill plant (*Anethum sowa*) as an industrial by-product of dill which is rich in carvone. The other major constituent of *A. sowa* is dillapiole which is well known for its insecticide synergistic properties. It also occurs to the extent of about 40–60% in *Anethum graveolens* seed oil and more than 51% in spearmint oil (*Mentha spicata*).

The turmeric (*Curcuma longa*) leaves, the unutilized part of turmeric plant, on hydrodistillation yields oil rich in α-phellandrene (70%). This oil induces growth inhibition and larval mortality against *Spilosoma obliqua* (Agarwal *et al.*, 1999). The leaf oil is also ovicidal and nymphicidal against *Dysdercus koenigii* and induces moderate knockdown effect against *T. castaneum*. Curcumene and ginger oil at 0.2% concentration induces 86% inhibition of the mycelial growth of the test fungus *Rhizoctonia solani.*

Residual LC50 values for nootkatone did not differ significantly at 4 week post-treatment from the observations made at the initial 24 hour treatment. The ability of these natural products to kill arthropods at relatively low concentrations also represents an alternative to the use of synthetic pesticides for control of disease vectors (Panella *et al.*, 2005; Dietrich *et al.*, 2006).

Repellency of oils of lemon, eucalyptus, geranium, and lavender have also been recorded against *Ixodes ricinus* (Acari: Ixodidae) in the laboratory and field (Jaenson *et al.*, 2006). The oils of *Cymbopogon nardus* (citronella), *Pogostemon cablin* (patchuli), *Syzygium aromaticum* (clove) and *Zanthoxylum limonella* were the most effective and provided 2 h of complete repellency. Among three essential oil constituents viz. eugenol, cineole and citronellal, the later was found to be most effective against *A. aegypti* mosquito (Coats *et al.,* 1991). Lemon grass oil ointment containing 15% v/w citral exhibited

50% repellency which lasted for 2–3 h (Oyedela *et al.,* 2002). It has now been reported that a component of the essential oil of the catnip plant (*Nepeta cateria*), the nepetalactone repels mosquitoes 10 times more effectively than DEET as it takes about one-tenth as much nepetalactone as DEET to have the same effect. *Tagetes erecta* is a potential plant whose essential oil from flowers has been effective repellent against insects (Ray *et al*., 2000). Accordingly ocimene from *T. minuta* has also repellent properties which need to be exploited in detail. Turmerone and *ar*-turmerone (dehydroturmerone), the major constituents of turmeric rhizome powder oil are strong repellents to stored grain pests. The turmeric oil has been reported to provide protection to wheat grains against red flour beetle, *T. castaneum* (Herbst) (Chahal *et a*l., 2005). The fruit oil of *Piper retrofractum* has also shown high repellency (52–90%) against *T. castaneum* at 0.5–2% concentration.

Most of the repellent available in the market (table.54) are used for animals to save the crop from their menace.

Table 54: List of Marketed Repellents

Product	Manufacturer/distributor	Mode of action	Active Ingradient	Application Method	Restriction	Registered use
Shot –Gun Deer and Rabbit Repellent	Bonide Products Inc. 2 Wurz Ave. Yorkville, NY 13495 (325) 736-8231	Taste	Thiram 11.0%	Spray/ Brush	For fruit trees,Limit application on normal bearing stages	Ornamental shrubs, trees, nursery stock and fruit trees
Bulb saver	Bonide Products Inc. 2 Wurz Ave. Yorkville, NY 13495 (325) 736-8231		Thiram 11.0%	Dip	Do not use on edible tubers and bulbs	Flower bulbs and tubers
Plant Pro-Tech (Garlic) Unit	Plant Pro-Tech Inc. P.O. Box 902 Palo Cedro, CA 96073 (800) 572-0055	Odor and taste if Repellent unit is eaten	Garlic and Chilli Powder	Repellent unit can be attached to plant or placed in ground	None	
Tree Guard	Norteck	Taste	Bitrex 0.2%	Spray		Non food crops
Deerbuster's Deer Repellent Spray	Trident Enterprises 9735A, Bethel Road Frederick, MD 21702	Taste	Putrescent egg solid 25% White Pepper 1.9% Garlic 1.9%	Spray		Non food crops
Hot Sauce Animal Repellent	Miller Chemical and fertilizer Corp. 333 Hanover, PA 17331 (717) 632 921	Taste	Capsacin 2.5%	Spray	Do not use on food crops after fruit appearance	Ornamental trees and shrubs,fruit and nut trees,nursery stock, deer,rabbit, pine mice
Deer-A-Way Big Game Repellent	Intagra Inc. 8906 Wentworth Ave.S. Minneapolis, MN 55420 (612) 881 5535	Taste/ odor	Putrscent whole egg solid 37%	Spray ready-to-apply powder	Apply to conifers and ornamental trees only	Conifer seedlings (spray only), dormant ornamental flowers (powder only)

Product	Manufacturer/distributor	Mode of action	Active Ingradient	Application Method	Restriction	Registered use
Hinder Deer and Rabbit Repellent	Pace International 1011 Western Ave. Suite 505 Seattle, WA 98104 (800) 247-8711	Taste/ Odor	Ammonia 1.5% Mixed rosin and fatty acid 13.0%			
Deer Off	Deer Off 1492 High Ridge Road Stamford, CT 06903 (203) 968-8485	Taste/ Odor	Putrscent whole egg 6.25% Capsacin and related product 0.0045% Garlic 0.005%			Edible parts, flowers, grasses, bulbs, shrubs, seedlings, trees
Gustafson 42-S Repellent	Gustafson LLC 1400 Preston Road Suite 400 Plano, TX 75093	Taste	Thiram 42%	Spray,Brush or Dip	Do not apply on fruit trees or other edible fruit that will bear within one year of application	Fruit trees, shrubs, ornamentals,nursery stock
Magic circle Rabbit Repellent	J.C. Erlich Chemical Corp. 500 Spring Ridge road Reading, PA 19612 (610) 372-9700	Taste	Thiram 20%	Spray	Do not use on plant parts that are to be used for food	For use on lawns,trees, flowers and shrubs
Repel Rabbit/ Deer Repellent	Farnam Companies Inc. 301 West Osborn P.O.Box 34820 Phoenix, AZ 85067-4820 (602) 285-1660	Taste	Thiram	Spray	Do not use on edible crops	Ornamentls, flowers. Dormant fruit trees
Dr. T's Deer Blocker	Nature Products P.O. Box 682 Pethan, GA 31799 (800) 299-8485	Taste/ Odor	Putrscent whole egg 6.25% Capsacin 0.0045% Garlic 0.005%			Food crops,bulbs, and vegetable garden

28

Chemical Molecules as Weedicides/ Herbicides for Weed Management

Weeds can be defined as plants which are undesirable, persistent, damaging and interfere with growth of other crop plants thus affecting agriculture, natural resources and economy of the country. These plants influence the produce of farmers in several ways. The agriculture in tropics is maximally affected by weeds as compared to other pest and disease damaged. Out of the total annual loss of agricultural produce in India, weeds share almost 37 % followed by insects pest (29 %), diseases (22%) and other pests (12%). At global level the losses in agricultural production by weeds are reported to the tune of 34% followed by Pests (18%) and diseases (16%).

Bridges (1994) reported that impact of weeds on the US economy exceeds $20 billion annually. Annual loss of about Rs. 1,980 crores to Indian agriculture is caused by weeds, which is higher than cumulative losses caused by insects, pests, and diseases (Yadav and Malik 2005). Soil deterioration in terms of nutrient depletion is another problem. It has been reported that weeds deplete nutrients from soil to the level of 11.0, 3.0 and 10.0 kg of Nitrogen, phosphorus and potash/ha respectively (Gautham and Mishra, 1995). Berca stated that 'weeds eat the food of about 1 billion inhabitants (Pacanoski, 2007).

Therefore, weed management is necessary to realise the maximum agricultural production from per unit of cultivated crops.

The global herbicide market was $ 35.72 billion in 2021 at a compound annual growth rate (CAGR) of 6.2% and the market is expected to reach $ 47.09 billion in 2025 at a CAGR of 7 percent.

Herbicides are classified/grouped in various ways e.g. according to the chemical family, activity, method of application, site of action or timing of application.

28.1. Classification Based on Translocation

Systemic/Translocated: These herbicides are extensively translocated in the plant through its vascular system along with water, nutrients and other materials from site of absorption to sites of action. Systemic herbicides are more effective on perennial weeds than contact herbicides. Unlike contact herbicides which are fast acting, systemic herbicides require longer time period (days or weeks) to kill weeds. **Glyphosate and glufosinate are nonselective systemic herbicides. 2, 4-D and dicamba are examples of selective systemic herbicides.**

Non-systemic/Contact: These herbicides kill only the portion of plant tissue that is in contact. These are not translocated through the plant. Uniform spray coverage and particle size are essential for adequate application. They are less effective on perennial plants, which are able to regrow from rhizomes, roots or tubers. Repeated application of contact herbicide is needed to kill regrowth of underground plant parts. **These are comparatively fast acting herbicides e.g. bromoxynil and bentazon are contact herbicides.**

28.2. Classification Based on Time of Application

Pre-plant herbicide: Pre-plant herbicides are those herbicides which are applied to soil before planting and gets mechanically incorporated into the soil. These are non-selective herbicides. The objective for incorporation is to prevent dissipation through photodecomposition and/or volatility. The herbicides kill weeds as they grow through the herbicide treated zone. Volatile herbicides have to be incorporated into the soil before planting the pasture.Agricultural crops grown in soil treated with a pre-plant herbicide include tomatoes, corn, soybeans and strawberries. **Soil fumigants like metam-sodium and dazomet are in use as pre-plant herbicides.**

Pre-emergence herbicide: Pre-emergence herbicides are those which are applied before the weed seedlings emerge through the soil surface. Herbicides do not prevent weeds from germinating but they kill weeds as they grow through the herbicide treated zone by affecting the cell division in the emerging seedling. **Dithopyr and Pendimethalin are Pre-emergence herbicides.** Weeds that have already emerged before application or activation are not affected by pre-emergence herbicides as their primary growing point escapes the treatment.

Post-emergence herbicide: Post-emergence herbicides are those which are applied after weed seedlings have emerged through the soil surface and generally require multiple applications for adequate control. They can be foliar or root absorbed, selective or nonselective, contact or systemic. Liquid formulations of herbicides are more effective than granular formulations. Application of these herbicides is avoided during rain because the problem of being washed off to the soil makes it ineffective. **2,4-D is a selective, systemic, foliar absorbed Post-emergence herbicide.**

28.3. Classification Based on Method of Application

Soil Applied: Herbicides applied to the soil are usually taken up by the root or shoot of the emerging seedlings and are used as pre-plant or pre-emergence treatment. Several factors influence the effectiveness of soil-applied herbicides. Weeds absorb herbicides by both passive and active mechanism. Herbicide adsorption to soil colloids or organic matter often reduces its amount available for weed absorption. Positioning of herbicide in correct layer of soil is very important, which can be achieved mechanically and by rainfall. Herbicides on the soil surface are subjected to several processes that reduce their availability. Volatility and photolysis are two common processes that reduce the availability of herbicides. Many soil applied herbicides are absorbed through plant shoots while they are still underground leading to their death or injury. **Thiocarbamates (e.g. EPTC) and dinitroanilines (e.g. trifl uralin) are soil applied herbicides.**

Foliar Applied: These are applied to portion of the plant above the ground and are absorbed by exposed tissues. These are generally post-emergence herbicides and can either be translocated (systemic) throughout the plant or remain at specific site (contact). External barriers of plants like cuticle, waxes, cell wall etc. affect herbicide absorption and action. **Glyphosate, 2, 4-D and dicamba are foliar applied herbicide.**

28.4. Classification Based on Specificity

Selective Herbicides: They control or suppress certain plants without affecting the growth of other plants species. Selectivity may be due to translocation, differential absorption, physical (morphological) or physiological differences between plant species. **2, 4-D, mecoprop, dicamba control many broadleaf weeds but remains ineffective against turfgrasses.**

Non-selective Herbicides: These herbicides are not specific in acting against certain plant species and kill all plant material with which they come into contact. They are used to clear industrial sites, waste ground, railways and railway embankments. **Paraquat, glufosinate, glyphosate are non-selective herbicides.**

28.5. Classification Based on Site of Action

Herbicides are also classified according to their site of action, and this classification is comparatively better to handle the herbicide resistance management program. Retzinger and Mallory-Smith (1997) proposed herbicide classification according to site of action with a view that it would help in dealing with herbicide resistance management. In order to differentiate herbicides with the same site of action each class was given a group number. The International Herbicide Resistance Action Committee (HRAC) also published a classification system based on letters for each group (Schmidt 1998). Mallory-Smith and Retzinger (2003) updated the classification and included some herbicides listed in the Weed Science Society of America 2002 Herbicide Handbook

The choice of herbicide and adjuvant to be used will depend on the target weed, site and environmental conditions, cost of chemicals, and in some cases, on state regulations. Following group of weedicides are generally used in the weed management (table.55).

Table 55: EPA register herbicide available for weed control

Common name	Partial list of trade name	Target weed species(general)
2,4-D	Hi-Dep, Weeder-64, Weed RHAP, Amine-4, Aquakleen (Amines)	Foliage applied, Selective, Some broadleaves, woody and aquatic plants susceptible, Thistles, Sulfur cinquefoil, dyers wood, knapweed, purple loosestrife, tall buttercup, whitetop.
Aminopyrid	Milestone	Foliage applied, selective, many broadleaf weeds, tolerated by most grasses, perennial and biennial thistles, knapweeds, sulfur cinquefoil.
Chlorsulfuron	Telar	Foilage applied, selective, some broad leaf plants and grasses susceptible. Dyer's wood, thistles, common tansy, houndstongue, whitetop, tall buttercup, toadflax.
Clopyralid	Stringer, Curtail, Transline, Reedem	Foliage applied, Selective, many broadleaf and woody species susceptible. Thistles, yellow starthistle, hawkweeds, knapweeds, rush skeleton weed, oxeye daisy.
Dicamba	Banvel, Clarity, others	Foliage applied, Selective, some broadleaf plants, bush and vines susceptible. Houndtongue, yellow starthistle, common crupina, hawkweed, oxeye daisy, tall buttercup, blueweed, leafy spurge, tansy ragwort, knapweed.

Common name	Partial list of trade name	Target weed species(general)
Diuron	Diuron 4L	Applied pre or post-emergence, broad spectrum, most annual and perennial broadleaf plants, grasses and some woody vegetation. Annual weeds and broadleaves for infrastructure maintenance.
Glyphosate	Roundup, Rodeo, Accord, Glyphomate	Foliage applied, Non selective, Most plants are susceptible, Broad spectrum for broadleaf plants and grasses. Purple loosestrife, fieldbind weed, yellow starthistle, thistles, cheatgrass, common crupina, toadflax.
Hexazinone	Velpar, Pronone 10G	Broad spectrum control with some selectivity for conifers. Cheatgrass, oxeye daisy, yellow starthistle, thistles.
Imazapic	Plateau	Foliage applied, selective, some broadleaf plants and grasses susceptible. Cheatgrass, leafy spurge, toadflax.
Imazapyr	Arsenal, Chopper	Applied pre or post emergence, broad spectrum, most annual and perennial broadleaf plants, grasses and woody vegetation. Dyers wood, fieldbind weed.
Methsulfuron methyl	Escort, Ally	Applied pre or post emergence, selective, some broadleaf weeds and annual grasses. Houndstongue, thistle, sulfur cinquefoil, common crupina, dyers wood, purple loosestrife, common tansy, whitetop.
Picloram	Tordon, Grazon, pathway	Foliage applied, selective, most annual and perennial broadleaf and woody plants are susceptible. Grasses are tolerant. Thistles, yellow starthistle, common crupina, hawkweed, knapweeds, rush skeleton weed, common tansy, toadflax, leafy spurge.
Sulfomefuron methyl	Oust	Applied pre or post emergence, broad spectrum, many annual and perennial grasses and broad leaf plants, woody vegetation tolerant Cheatgrass, whitetop, oxeye daisy, tansy rugwort, muskthistle.
Triclopyr	Garlon, Reedem,Remedy	Foliage applied, selective, woody plants, some broad leaf plants are root-sprouting species are susceptible. Grasses are tolerant. Hawkweed, sulfur cinquefoll, purple loose strife, knapweed, oxeye daisy, thistle.

28.6. Most Commonly used and Recommended Herbicides

28.6.1. Glyphosate

a. RoundUp Pro®

Andropogon virginicus (broomsedge), *Paspalum conjugatum* (buffalograss), *Melinis minutiflora* (molasses grass) and *Setaria palmifolia* (palmgrass) is controlled by using a 2% solution of RoundUp Pro® with water soluble packets of blue Turfmark® dye by foliar applications in Hawaii. A surfactant is already included in the RoundUp Pro® formulation so there is no need to add any other adjuvants. Similarly *Panicum repens* (torpedo grass) and *Urochloa distichya* (Tropical signalgrass) is controlled by using a 2% solution of RoundUp Pro® with SunEnergy® surfactant (applied at 1oz/gallon) by foliar applications.

b. Rodeo®

Phragmites australis (common reed) and *Rosa multiflora* (multiflora rose) have a 90% kill rate using a 2% solution of Rodeo® with 0.5% TL-90® non-ionic surfactant, applied with a backpack or ATV-mounted sprayer.

Similarly *Mimosa pigra* (catclaw mimosa), *Lygodium japonicum* (Japanese climbing fern), *Panicum repens* (torpedo grass), *Paederia foetida* (skunkvine), *Lantana camara* (lantana), *Solanum viarum* (tropical soda apple) and *Imperata cylindrica* (cogon grass) have excellent control (>95% kill) with a 4% solution of Rodeo® plus a 0.3% solution of either Silken® or Kinetic® organosilicone surfactant.

Phalaris arundinacea (reed canarygrass) have good control by first mowing in late spring-early summer at the onset of flowering, followed by applying a foliar spray of Rodeo® at 2% solution with either 0.5% Bio-88® or R-11® nonionic surfactant in fall, before the first frost. The formulation can be applied with a backpack sprayer or an ATV with a boom attachment.

Typha spp. (cattails) have nearly 100% kill in West Virginia by combining 2.5 gallons Rodeo®, 1 quart Surf-Ac 820® nonionic surfactant plus Blazon® blue turf dye and 7.25 gallons of water to make 10 gallons of tank mix.

c. Accord®

Hypericum perforatum (St. Johnswort), *Lythrum salicaria* (purple loosestrife), and *Phalaris arundinacea* (reed canarygrass) is controlled by Accord® herbicide at 2.5% a.i. with Hi-Light Dye® tablets (1 tablet per gallon mix). The formulation is applied to St. Johnswort foliage by either wicking using a modified exterior sponge PVC adapted to a Solo® backpack sprayer, or by using a backpack sprayer. For purple loosestrife and reed canarygrass, first cuts the stems and then applies Accord® at 5% a.i. solution with the Hi-Light Dye®, and apply the mix using either a backpack sprayer or a sponge wicking applicator to the stem and cut surface.

Rhamnus frangula (glossy buckthorn) is controlled using a cut-stump herbicide treatment. First cuts each stem 6 inches above the ground surface, and within 5 minutes, apply Accord® at 14% a.i. mix directly to the cut surface using a sponge-tipped applicator. Accord® can also be sprayed onto foliage using a 2% a.i. mix.

28.6.2. Triclopyr

a. Garlon 3A®

Polygonum cuspidatum (Japanese knotweed) had 100% kill by using a 3 to 5% solution of Garlon 3A® with 1 oz/gallon Hasten® ethylated seed oil. For treatments near water, a 3-5% solution of Garlon 3A® with 1 oz/gallon of R-11® nonionic surfactant is used. First cut the stems in spring, then foliar spray the regrowth with a backpack sprayer in fall. 95% kill of fennel in California is achieved by using 1 lb a.i./acre of Garlon 3A® with a 0.25% solution of Pro-Spreader® activator nonionic surfactant.

Dioscorea bulbifera (air potato) had good control with a 2.5% solution of Garlon 3A® plus a 0.3% solution of either Kinetic® or Silken® surfactant, applied as a foliar spray onto leaves. *Rosa multiflora* (multiflora rose), *Elaeagnus umbellata* (autumn olive) and *Ailanthus altissima* (tree of heaven) can be controlled by application of undiluted Garlon 3A® with no additional adjuvant. The use of undiluted Garlon 3A® with no adjuvant on tree of heaven, using a girdle and squirt (cut into bark with a girdling knife, squirt in herbicide using a spray bottle) technique causes about 95% mortality.

Wedelia trilobata (trailing daisy) had moderate control using repeated treatments of a 2% solution of Garlon 3A® with1 oz/gallon CideKick II® surfactant. They also add TurfMark® dye (1 to 2 oz/gallon) for these foliar treatments.

Tibouchina herbacea (glorybush) and *Ulex europaea* (gorse) two invasive species in Hawaii is controlled using a 2% solution of Garlon 3A® combined with a 0.2% solution of Breakthru® organosilicone surfactactant as a foliar spray.

Garlon 3A® or Garlon 4® also control *Senna pendula* (climbing cassia), *Colubrina asiatica* (Asiatic colubrina), *Schinus terebinthifolius* (Brazilian peppertree), *Casuarina equisetifolia* (Australian pine), and *Cupaniopsis anacardioides* (Carrotwood) woody invaders by using either a cut-stump treatment with a 50% solution of Garlon 3A® (in water), or a basal bark treatment with 10% Garlon 4® mixed with 90% JLB® oil solution. For both types of treatments, no other surfactants are used, but Turfmark® dye is added at a rate of 1 to 2 oz/gallon tank mix.

b. Garlon 4®

Rhamnus cathartica (common buckthorn) can be controlled with a solution of 20% Garlon 4® and 80% mineral oil using the basal bark application technique. Adds Basal Red® dye at 3 oz/15 gallons to the tank mix. Similarly a cut-stump treatment using a solution of 25% Garlon 4® with 75% Diluent Blue® *can control Tamarisk* spp (salt cedar, tamarisk).

28.7. Commonly Used and available Herbicides in India

1) 2, 4-D (2,4, Dichlorophenoxy Acetic acid)

It is selective translocated herbicide and most widely used to control dicot weeds (Broad Leaf Weeds) in cereals and sugarcane as post emergence application.

It is cheap, easily available, easy to apply and non-poisonous to human beings and animals. It is available in three formulations as Sodium Salt (**Fernoxone 80 WP or Agrosodium**), Dimethylamine (**Agrodor 96 or Weedar 96**) and Ethylester (**Knockweed 36 EC or Knock weed 4G**). Sodium salt is white powder with characteristics odour, low volatile and is soluble in water. The dimethylamine is light brown liquid forms emulsion in water and is non-volatile. Ethylester is light yellow liquid may be low or high volatile and forms milky emulsion with water and it is absorbed by weeds at faster rate than other forms so more effective for control of weeds. It is applied @ 1 to 1.5 kg a.e / ha in 500 to 600 litres as post emergence spray.

2) 2, 4, 5-T (2, 4, 5 – Trichlorophenoxy Acetic Acid) and 2,4,5-TP

(2, 4, 5-Trichloropropionic Acid)

Both the herbicides are similar to 2, 4-D in properties and mode of action and useful for controlling bushes and woody weeds.

3) Dicamba (3, 6 –Dichloro-2-Methoxybenzoic Acid)

It is selective translocated herbicides and available by trade name **Banvel (50 EC)**. Useful to control broad leaf weeds and used @ 0.5 to 3 kg a.i/ha as post-emergence spray.

4) Simazine (2, Chloro-4,6 –bi(Ethylamino) s-triazine)

Atrazine (2-Chloro -4-(ethylamino -6- isoprophylamino-5-triazine) and Simazine herbicides are selective, translocated and used as pre-emergence spray for killing broad leaf weeds and grasses in sugarcane, maize, jowar, potato, grapes etc.

The effective rates are 0.5 to 2.5 kg a.i /ha in 500 to 600 litres of water. Simazine is available in the market by trade name. **Tafazine 50 W.P and Atrazine as Atrataf or Artex 50 WP.** Atrazine is more soluble in water than simazine. Therefore, atrazine is more useful in dry farming areas.

5) Mertibuzine (4-Amino -6-tert-butyl- 3-(Methylthio) as triazine – 5(4H)-one)

It is broad spectrum selective herbicides with both, soil and shoot activity and found useful in sugarcane, potato, tomato, soybean, etc.

The effective rates are 0.2 to 1 kg a.i /ha. It is available in the market by trade name **sencor as 70 WP**.

6) Paraquat (1,1- diethyl-4-bipyridinium ion) and Diquat (6,7-dihydrodipyrido[1,2-a:2,1-C] pyrazinediium ion)

Both the herbicides are contact, non-selective with zero persistence in the soil. Useful in sugarcane before germination and as direct spray after germiantion in sugarcane and orchards. Also useful as fallow application before sowing and also in non-cropped areas. 1 to 2 kg a.i /ha as post-emergence spray.

Paraquat is available as trade name gramoxone 24 SL and diquat as reglone 20 EC.

7) Benthiocarb or Thiobencarb (S-(4-Chlorobenzyl) N,N – Diethyl – Thicarbamate)

It is effective for control of Echinochloa spp. in rice. It is used as pre-emergence spray @ 1 to 2 kg a.i/ha. **It is available in the market by trade name satrun 50 EC or Saturn 10 G or Bolero 10.**

8) Alachlor (2-Chloro – 2, 6- Diethyl- N-(Methoxymethyl) Acetanilide)

It is very useful for controlling annual grasses and certain broadleaf weeds in groundnut, soyean, cotton, potatoes, maize, sugarcane and vegetables. It is applied @ 1 to 2 kg a.i /ha. **It is available by trade name alachlor** as 50 % EC. or 10% granules.

9) Butachlor N (Butoxymethy 2,6 –diethylacetanilide)

It is selective pre-emergence herbicides used in seedbed and transplanted rice for control of Echinochloa spp. and most other annual weeds. The common rates are 1 to 2 kg a.i /ha. Upto 3 to 5 days after sowing in seeded rice and upto 10 days after transplanting in transplanted rice. **It is available by trade name Machete as 50 EC or 5% granules.**

10) Fluchloralin (N-(2-Chloroethyl)-2, 6-dinitro-N-prophyl) – 4 (tri-fluromethyl) amine)

It is selective pre-emergence herbicide effective against large number of annual weeds in cotton, soybean, jute, chickpea, sunflower, onion and certain solanaceous vegetables.

Pre-planting soil incorporation of the herbicide give better results as it is volatile. It is applied @ 1 to 1.5 kg a.i /ha and **available by trade name basalin as 45% EC.**

11) Pendimethalin (N-(1-ethylprophyl)- 3, 4- dimethyl -2,6 -dinitro-benzenamine)

It is primarily selective pre-plant, soil incorporated herbicide found effective against number of annual weeds particularly grasses in cotton, soybean, groundnut and pea. It is also used as pre-emrgence herbicide in maize, rice, small grains, onion and potato. The effective rates are 1 to 1.5 kg a.i /ha. **It is available by trade name stomp 30 EC.**

12) Glyphosate (N-(Phosphomethyl) (glycine)

It is post-emergence, non-selective, translocated herbicide. It is highly shoot mobile and weak residue herbicide and can be used against perennial weeds few days before sowing or as post-emergence spray in non-cropped areas.

Very effective to control perennial weeds like nutgrass, hariali, etc. Recommended dose is 1 to3 kg a.i /ha as post-emergence spray when weeds are in active stage of growth. **It is available by trade name Roundup or**

Glycol 41 as 41 S.L. It is useful for killing perennial weeds effectively in fruit crops and sugarcane as directed spray.

28.8. Spreader and Stickers for Herbicide

Spreaders are compounds that causes the surface tension of the pesticide to be reduced in such a way that it easily spreads into a very thin film over a surface. Like surfactants, spreaders and stickers increase the efficiency of the pesticide dramatically. They may contain fatty acids, latex, aliphatic alcohols, crop oils such as cottonseed, or inorganic oils. Each formulation is very different.

Stickers are very much like thickening agents or oils in that they cause the pesticide solution to adhere to the leaf surface, resisting rain, evaporation and runoff. Some products use emulsified polyethylenes, others use polymerized resins, fatty acids or petroleum distillates. Stickers are commonly used in field crops (like corn and soybeans) where residue on leaves is not a problem. However, in greenhouses, they can cause a mess, especially on leaves with indentations such as Pileas, or hairy plants such as Dusty Miller. Some spreader/sticker cause phytoxicity in tender annuals and herbs.

28.9. Adjuvants, Surfactant, Wetting & Spreading Agents to Enhance Efficiency of Herbicide

An adjuvant is any compound that is added to the herbicide formulation or tank mix to facilitate the mixing, application, or effectiveness of that herbicide. Adjuvants are already included in the formulations of some herbicides available for sale (e.g. RoundUp®), or they may be purchased separately and added into a tank mix prior to use.

Adjuvants are chemically and biologically active compounds, and they may improve the effectiveness of the herbicide they are added to, either increasing its desired impact and/or decreasing the total amount of formulation needed to achieve the desired impact. Some herbicides require the addition of an adjuvant to be effective. Some adjuvants enhance the penetration of herbicide into plants by ensuring adequate spray coverage and keeping the herbicide in contact with plant tissues, or by increasing rates of foliar and/or stomatal penetration.

The U.S. Environmental Protection Agency (EPA) regulates the inclusion of certain ingredients in adjuvant formulations, but it does not stringently test and regulate the manufacture and use of adjuvant products (as they do for herbicides and other pesticides). As such, there is little information on the effects of these different adjuvants, other than that provided by the manufacturer. A herbicide label may specify what types of adjuvant are appropriate or advisable to use with that herbicide, but it will not suggest specific brands. Therefore, there is no good single resource or system to help you determine which specific adjuvant product (if any) to use for each application situation. However, it is worth checking the label of any adjuvant you are considering to see if it is registered in certain states. These states regulate adjuvants and require the disclosure of their ingredients, results from efficacy trials, and data from environmental and toxicological studies.

28.9.1. Types of Adjuvants

There are many ways to classify adjuvants, and there is currently no standard system used by all adjuvant or herbicide manufacturers. A good review of different adjuvant terms and definitions can be found in Hazen (2000) or in Van Valkenburg (1982). In this chapter, we divide adjuvants into two primary types based on their functions i.e.

28.9.1.1. Activator Adjuvants

a. **Surfactant -**
    - Nonionic
    - Ionic
    - Atmospheric

b. **Oil adjuvants**
    - Petrolium oil concentrates
    - Vegetable oil

28.9.1.2. Utility Adjuvants (including Spray Modifiers)

- Wetting agents (spreaders)
- Dyes
- Drift control & foaming agents
- Thickening agent
- Deposition agents (stickers)
- Water conditioner
- Compatibility agents
- pH buffers
- Humectants

- Defoaming & antifoam agents
- UV absorbants

28.9.1.1. Activator Adjuvants

Activator adjuvants are compounds that when added to the spray tank, enhance herbicide activity (Penner 2000a). Activator adjuvants include surfactants, oil carriers such as phytobland (not harmful to plants) oils, crop oils, crop oil concentrates (COCs), vegetable oils, methylated seed oils (MSOs), petroleum oils, and silicone derivatives, as well as nitrogen fertilizers. Some brands of herbicide formulations already include activator adjuvants (e.g. RoundUp Ultra® contains the herbicide glyphosate and a surfactant, and Pathfinder II® which contains the herbicide triclopyr, an oil carrier which is an activator, and a dye which is a utility adjuvant).

Activator adjuvants work to enhance the activity of the herbicide, often by increasing rates of absorption of the herbicide into the target plant(s). Utility adjuvants, sometimes called spray modifiers, work by altering the physical or chemical characteristics of the spray mixture to improve its ease of application, its ability to remain on the plant surface rather than rolling off, or its persistence in the environment (McWhorter 1982).

Activator adjuvants enhance the activity of the herbicide, often by increasing rates of absorption of the herbicide into the target plant(s).

28.9.1.1.1. Oil Adjuvants

Oil adjuvants can increase the penetration of oil-soluble herbicides into plants, and are commonly used when conditions are hot and dry, and/or when leaf cuticles are thick. They are derived from either refined petroleum (mineral) oils or from vegetable oils (including seed oils), and do not readily mix with water. Therefore, when an oil adjuvant is combined with water in a spray tank, a surfactant emulsifier must also be added, which distributes the oil droplets (micelles) uniformly throughout the mix. These "emulsifiable oil" adjuvant combinations typically contain both a non-phytotoxic oil (typically ranging 80 to 99%) and a surfactant (1 to 20%), and are added to the spray tank usually as just 1% of the total spray volume (Hess 1999). Emulsifiable oil adjuvant blends can enhance the absorption of an oil-soluble herbicide into the plant more than an oil adjuvant by itself. Adding a surfactant to the mixture not only emulsifies the oil in the water-based spray solution, but also lowers the surface tension of the spray solution. These adjuvants can also increase herbicide absorption through the plant cuticle, increase spray retention on leaf surfaces, and reduce the time needed for the herbicide formulation to become rainfast

(Pringnitz 1998; Miller & Westra 1996). The exact mode of action of these oils is unknown, but they enhance the spread of droplets on plant surfaces (Gauvit and Cabanne 1993, Green 2001). They may also split open the cuticle and increase both the fluidity of cuticular components and herbicide diffusion rates (Santier & Chamel 1996, Green 2001).

Two types of emulsifiable oil adjuvants are "crop oils" and "crop oil concentrates" (COC). Crop oils contain up to 5% surfactant and COCs may contain up to 20% surfactant (Hess 1999). COCs enhance spreading and penetration and are used primarily with grass-specific herbicides (Miller & Westra 1996). Crop oils and COCs do not necessarily contain oil derived from crop plants (although some do), but are so named because they are intended for application to crops.

28.9.1.1.2. Petroleum Oils

Petroleum oils or petroleum oil concentrates are highly refined oils, which are often used as carriers of oil-soluble herbicides. They are typically used in low quantities (generally 0.25 to 1 gallon/acre), and when used as carriers, can reduce surface tension, increase wetting and spreading, give quicker absorption, improve rainfastness, and reduce loss of carrier during and after application (Bohannan & Jordan 1995, Green 2001). Petroleum oil concentrates may include paraffinic and napthalenic oils. Paraffinic oil can smooth epicuticular wax, or cause cracks in the cuticle, allowing increased herbicide penetration (Foy & Smith 1969, Green 2001). Paraffinic oils are sometimes referred to as dissolving waxes, but in fact, paraffinic oils are poor solvents and only soften wax.

28.9.1.1.3. Vegetable Oils

Vegetable-derived oils (from soybeans, cottonseeds, etc.) also decrease surface tension, but they are not as effective as other surfactants at increasing spreading, sticking, or penetration (Miller & Westra 1996). Vegetable oils are generally of two types: triglycerides or methylated oils. Triglycerides are essentially oil-surfactant hybrids, and are generally called "seed oils." These seed oils are extracted from plants by pressing or solvent extraction, and tend to have higher viscosities than methylated oils. Triglyceride oils usually contain only 5 to 7% surfactant emulsifier, while methylated seed oils contain 10 to 20% surfactant. Methylated seed oils (MSO) are better solvents than petroleum-based oils, but their role as a solvent of cuticular waxes is controversial. The composition of these oils varies depending on the seed source and can influence efficacy (Nalewaja 1994). Esterified seed oils are vegetable seed oils with a surfactant or an emulsifier already added. They have good spreading and penetration properties, but tend to be more expensive than other oil adjuvants.

28.9.1.2. Utility Adjuvants

Utility adjuvants, which are sometimes called spray modifiers, alter the physical or chemical characteristics of the spray mixture making it easier to apply, increasing its adherence to plant surface so that it is less likely to roll off, or increasing its persistence in the environment. When deciding which type of adjuvant to use, remember to always read and follow the directions on the herbicide label.

Oils are sometimes used alone as contact herbicides and in other situations as adjuvant carriers for synthetic herbicides. Salts may also be used as activator adjuvants, often to fertilize and enhance the growth of the target plant in the short-term, which can increase the uptake and effect of the herbicide in the slightly longer term. Salt adjuvants of this type are used extensively in crop agriculture and in some rangelands, but are rarely appropriate in wildlands.

Utility adjuvants are added to improve the application of the formulation to the target plants. By themselves, they do not directly enhance herbicidal activity (McMullan 2000). Instead, they change the physical or chemical properties of the tank mix in ways that make it easier to apply to the target plant(s), minimize unwanted effects, and broaden the range of conditions under which a given herbicide formulation can be effective.

Most utility adjuvants are typically not used in wildland situations, since herbicides applied in wildlands are generally not applied aerially, with large booms, or in tank mixtures with several herbicides and other additives.

28.9.1.2.1. Use of Ammonium (Nitrogen) Fertilizers as utility adjuvant

Ammonium or nitrogen fertilizers are often added to herbicide mixes in range and row-crop agriculture situations, where the addition of fertilizer works to both enhance herbicidal effects as well as to stimulate the growth of desirable crop or forage plants. Ammonium fertilizers can function as utility adjuvants, because they help prevent the formation of precipitates in the tank mix or on the leaf surface. They also decrease surface tension, increase spreading of the herbicide on the leaf surface, neutralize ionic charges, and increase herbicide penetration into the leaf (Nalewaja & Matysiak 2000). Ammonium fertilizers are used primarily with broad leaf specific herbicides (Miller & Westra 1996; Wanamarta et al. 1993). Ammonium fertilizers used as adjuvants include urea-ammonium nitrates (UAN), ammonium sulfates, ammonium nitrates and ammonium polyphosphates. Although their exact mode of action in herbicide control is unknown, they are often used to enhance the post emergence activity of weakly acidic herbicides, primarily by increasing herbicide absorption. The activity of ammonium fertilizers is strongly herbicide and species-specific, and is probably dependent on several mechanisms. Ammonium sulfates are also used to reduce antagonism by hard water ions in spray solutions. Iron, zinc, magnesium, sodium, potassium and calcium ions can react with certain herbicides (such as 2, 4-D and glyphosate) to form precipitates or herbicide salts, decreasing the efficacy of those herbicides (Nalewaja and Matysiak 1993). Ammonium sulfate prevents the formation of the calcium salt of glyphosate (Thelen et al. 1995) and is recommended in most areas with hard water (Hartzler 2001).

28.9.2. Surfactants

Surfactants are adjuvants that reduce surface tension within the external surface layers of water. Surfactants are the most widely used and probably the most important of all adjuvants (Miller & Westra 1998). The name is derived from surface active agents and these compounds facilitate or enhance the emulsifying, dispersing, spreading, sticking or wetting properties of the herbicide tank mix (includes spray modifiers). Surfactants reduce surface tension in the spray droplet, which ensures that the formulation spreads out and covers plants with a thin film rather than beading up. This facilitates herbicide absorption into the plant. Surfactants can also directly influence the absorption of herbicides by changing the viscosity and crystalline structure of waxes on leaf and stem surfaces, so that they are more easily penetrated by the herbicide (Kirkwood 1999; Coret et al. 1993). Some herbicide formulations come with a surfactant already added, but most require the addition of a surfactant for good control results. Surfactants are generally not added to pre-emergent herbicides that are applied directly to soil (Miller & Westra 1998).

28.9.2.1. Types of Surfactants

28.9.2.1.1. Non-ionic Surfactants

Non-ionic surfactants are the most commonly recommended and used adjuvants. Labels for most post-emergent herbicides used in wildlands that do not already contain a non-ionic surfactant often recommend the addition of one. Non-ionic surfactants have no ionic charge and are hydrophilic (water-loving). They are generally biodegradable and are compatible with many fertilizer solutions. Some non-ionic surfactants are waxy solids and require the addition of a cosolvent (such as alcohol or glycol) to solubilize into liquids. Glycol cosolvents are generally preferred over alcohols, as the latter are flammable, evaporate quickly, and may increase the number of fine spray droplets (making the formulation likely to drift when sprayed). The adjuvant label or MSDS should specify the active ingredient (alcohol, glycol, ether, etc) of the adjuvant product.

Organosilicone and silicone surfactants are two types of non-ionic surfactants. Organosilicone surfactants drastically reduce surface tension to the point where the herbicide droplets thin and coalesce to form a thin layer on the leaf surface (known as "superspreading"). They can even reduce surface tension to the point that some of the formulation may be able to slide through the microscopic stomatal openings on leaf surfaces. Once through the stomates however, the herbicide formulation must still penetrate the thin cuticle and cell membranes of the cells that line the cavity below the stomates. Silicone surfactants also decrease surface tension and may allow spray solutions to penetrate the stomates. They can also make the formulation nearly impossible to wash off (rainfast) even if it rains shortly after they are applied (Green 2001; Roggenbuck et al. 1993). Silicone surfactants can also influence the amount/ rate of herbicide that is absorbed through the cuticle.

28.9.2.1.2. Ionic Surfactants

Ionic surfactants possess either a positive (cation) or a negative (anion) charge, and can pair readily with oppositely charged herbicides, increasing the solubility of polar herbicides in water. Ionic surfactants may complex with other compounds in the mix (including contaminants in the water) in unexpected ways, and this can interfere with their function. For this reason, non-ionic surfactants are more commonly recommended. Ionic surfactants are not often used in wildland settings, but are frequently used in agriculture. The most common cationic surfactants used in agriculture may be the tallow amine ethoxylates, which are often used with glyphosate. The most common anionic surfactants are sulfates, carboxylates, and phosphates attached to lipophilic hydrocarbons.

28.9.2.1.3. Amphoteric Surfactants

Amphoteric surfactants contain both a positive and negative charge and typically function similarly to non-ionic surfactants. A commonly used amphoteric surfactant is lecithin (phosphatidylcholine), which is derived from soybeans. There is little published research on the use and efficacy of amphoteric surfactants.

28.9.2.1.4. Anionic Surfactants

Anionic surfactant are negatively charged, and enhance foaming and other spreading properties. For example, shampoo for hair contains sodium or ammonium laureth sulfate, which is the preferred anionic surfactant for hair. Using an anionic surfactant in the greenhouse can cause problems with sprayers that have an agitator, or any system where the foam could disrupt water flow or pump suction.

28.9.2.1.5. Cationic Surfactants

Cationic surfactants are positively charged, and are often very toxic to plants as they can disrupt membrane ion balance. Cationic surfactants are not widely used for pest control, but they are more commonly used in heavy-duty cleaning compounds. Don't grab a bottle of engine wash surfactant used to clean the tractor. The results can be devastating to plant material.

28.9.2.1.6. Natural Surfactants

Natural surfactants are biodegradable, wetting agents and oils that are processed differently from "crop oils" and alkylated sugars. Materials such as coconut oils, palm oils, castor oils, lanolins, wheat amino acids, and many others have been used in the past, but there is little research to verify these products are effective when used in combination with pesticides or in a greenhouses environment.

28.9.2.2. New Classes of surfactants molecules

28.9.2.2.1. Alkyl polyglucosides

It simply refers to modified sugars. It has been found that some modified sugar molecules have surfactant-like properties and may be used as spreader/stickers. Although this class of adjuvant has been around for some time, it is only now becoming popular due to its organic origins, biodegradability, and the fact that it is environmentally friendly. There are low foaming and high foaming types, and many are very good wetting agents. Originally developed as environmentally safe cleaning agents (car wash cleaners) in England, they have transitioned to being used in laundry detergents and as agronomic adjuvants. These products have very low potential for phytotoxicity because they are derived from plant sugars. Current work indicates they work well with glyphosate herbicide.

28.9.2.2.2. Organosilicates

These were developed in the 1970s, and have many uses including silicone-based sprays for waterproofing. About 10 years ago, their use as spray adjuvants for crop production was discovered. Organosilicate surfactants are very good at increasing the "rainfastness" of pesticides. They also reduce the surface tension and allow everything from micronutrients to fungicides to enter the leaf stomates. These products have phenomenal wetting abilities. Depending on the pesticide they are used with and the application rate, results may be good or bad, so be careful when using these products. Phytotoxicity can occur if these products are used at too high rate or used when temperatures are above 32.2^{0}C. Some research suggests that the wetting properties of these surfactants are so good that they can also allow bacteria and fungi to more easily invade plants (via stomata). A sanitary facility, careful applications, and rotation with other surfactants is required to realize the benefits with these products. Applications of organosilicone surfactants in above 32.2^{0}C weather can cause plant damage.

28.9.3. Wetting & Spreading Agent

Wetting agents or spreading agents lower surface tension in the spray droplet, and allow the herbicide formulation to form a large, thin layer on the leaves and stems of the target plant. Since these agents are typically non-ionic surfactants diluted with water, alcohol, or glycols (Hazen 2000), they may also function as activator adjuvants (surfactants). However, some wetting or spreading agents affect only the physical properties of the spray droplets, and do not affect the behavior of the formulation once it is in contact with plants.

28.9.3.1. Dyes

Dyes are commonly used for spot or boom spraying. It's use is recommended for most herbicide treatments in wildlands even if applied with small handheld sprayers or wicks because the presence of a dye makes it far easier to see where the herbicide has been applied and where it has dripped, spilled or leaked. Dyes make it easier to detect missed spots and to avoid spraying a plant or area twice.

28.9.3.2. Foaming Agents and Drift control

Drift control agents are designed to reduce spray drift, which most often results when fine (< 150 μm diameter) spray droplets are carried away from the target area by breezes, including those caused by the aircraft or vehicle carrying the spray equipment (Downer et al. 1998). Drift control agents alter the viscoelastic properties of the spray solution, yielding a coarser spray with greater mean droplet sizes and weights, and minimizing the number of small, easily-windborne droplets (Hewitt 1998). These agents are typically composed of large polymers such as polyacrylamides, and polysaccharides, and certain types of gums. Foaming agents also act as drift control agents. When used with specialized nozzles, these agents create foams with different degrees of stability (Witt 2001). These foams can be placed more precisely than standard liquid sprays, and are sometimes used to mark the edge of spray applications. Foams ensure complete coverage without over spraying. Foaming agents are usually added in quantities of 0.1 to 4.0% of the entire spray mixture (McWhorter 1982).

28.9.3.3. Thickening Agents

Thickening agents can modify the viscosity of spray solutions and are used to reduce drift, particularly for aerial applications (Witt 2001). They are used primarily in agriculture. Thickening agents may include water swellable polymers that can produce a "particulated solution," hydroxyethyl celluloses, and/or polysaccharide gums. Viscosity can also be increased by making invert emulsions (follow directions on individual herbicide labels) of the spray solution. The compatibility of the thickening agent with the tank mix can be influenced by the order of mixing, pH, temperature, and/or the salt content of the tank solution. Thickening agents are typically used in areas where sensitive populations or crops are growing close to treated areas (McWhorter 1982).

28.9.3.4. Deposition Agents (Stickers)

Deposition agents, or stickers, are used to reduce losses of formulation from the target plants due to the droplets evaporating from the target surface, or beading-up and falling off. Spray retention on difficult-to-wet leaf surfaces is regulated by the degree of surface tension and energy dissipation during the spray process. Deposition agents such as guar gum can reduce surface tension while increasing the viscoelasticity of the droplets (Bergeron et al. 2000). Stickers keep the herbicide in contact with plant tissues by remaining viscous, and therefore resist being washed-off by rain or knocked off by physical contact. Stickers are generally the most useful with dry wettable powder and granule formulations (Hazen 2000). Film-forming vegetable gels, emulsifiable resins, emulsifiable mineral oils, vegetable oils, waxes, and water-soluble polymers can all be used as stickers (Witt 2001). Fatty acids (technically anionic surfactants) are frequently used as stickers, and although they are "naturally derived" and are typically considered safe, they may have considerable contact activity. Certain oils may also function as stickers, but only if they have a low degree of volatility (Hazen 2000).

28.9.3.5. Water Conditioners

Water conditioners are frequently added when the water used in the formulation is high in salts in order to minimize or prevent reactions between ions in the spray solution and the herbicide, which would result in the formation of precipitates or salts. When there are many cations present, as in hard water, they can react with the herbicide, decreasing the uptake and effect of the herbicide. For instance, high levels of calcium in water (hard water) reduce the control efficacy of glyphosate (Nalewaja & Matysiak 1993). Similarly, sodium bicarbonate reduces the efficacy of sethoxydim (Matysiak & Nalewaja 1999). A water conditioner, such as ammonium sulfate (which also happens to be a nitrogen fertilizer), can negate this effect for both glyphosate and sethoxydim (McMullan 2000).

28.9.3.6. Compatibility Agents

Compatibility agents prevent chemical and/or physical interactions between different herbicides and fertilizers that could lead to non-homogeneous or unsprayable mixtures when these compounds are combined. For instance, if the herbicides bentazon and sethoxydim are mixed, they may react to form precipitates, resulting in reduced rates of sethoxydim penetration (Wanamarta et al. 1993). In most cases, the herbicide label will state which herbicides may or may not be mixed together. Some herbicides are applied with fertilizers or fertilizer solutions, especially in agricultural settings. Compatibility agents are used to keep these herbicides in suspension, and are generally added with a liquid fertilizer (Witt 2001). Most herbicides can be applied in nitrogen solutions without any compatibility problems, but compatibility may be poor when the water contains high levels of various salts (hard water), or when the water is unusually cool. When 2, 4-D is applied with liquid-nitrogen fertilizers the solution may separate even if mixed vigorously unless a compatability agent is added to the mix.

28.9.3.7. pH Buffers

pH plays a large role in herbicide efficacy. The pH of the tank mix affects the half-life solubility and efficacy of the herbicide, and may determine whether or not precipitates form (McMullan 2000). Being able to buffer or otherwise control changes of pH in the tank mix can be important in preventing herbicides from being degraded by acid or base hydrolysis in aqueous solutions. Some herbicides are sold with a pH buffer already included. Adjuvants that adjust or buffer pH can also improve the herbicide's dispersion or solubilization in the mix, control its ionic state, and increase tank-mixture compatibility. pH buffers are most beneficial when used in extremely alkaline or acid water, which could otherwise have detrimental effects on the herbicide's performance (McWhorter 1982).

28.9.3.8. Humectants

Humectants, like stickers, increase the amount of time that the herbicide is on the leaf, in a form available for uptake (Hazen 2000). When water evaporates from the spray droplet and the herbicide becomes a crystalline residue, it is no longer available for uptake into the leaf. Humectants keep the spray deposit moist and in true solution, and therefore extend the time that it is available for absorption (Hess 1999). They are generally water-soluble and increase the water content of spray deposits by slowing the drying time or by drawing moisture from the environment. Commonly used humectants include glycerol, propylene glycol, diethylene glycol, polyethylene glycol, urea, and ammonium sulfate. Even glucose and molasses were used as humectants in the past, but they are not labeled for such use and should not be added to any herbicide formulation.

28.9.3.10. Defoaming and Antifoam Agents

Defoaming and antifoam agents reduce or suppress the formation of foam in spray tanks (Witt 2001). Many spray mixtures have a tendency to foam excessively, especially when mixed with soft water, which can cause problems during mixing (foam overfill) or when rinsing the sprayer (McMullan 2000). Most defoamer agents are dimethopolysiloxane-based, but silica, alcohol, and oils have also been used for this purpose. Defoaming agents can reduce surface tension, physically burst the air bubbles, and/or otherwise weaken the foam structure. In general, it is easier to prevent foam formation than to eliminate foam after it forms (Green 2001). Antifoam agents are usually dispensed from aerosol cans or plastic-squeeze bottles, and are added directly to the mix at the onset of foam formation. The highest concentration needed for eliminating foam is typically about 0.1% of the entire tank. Some applicators in agricultural settings even use kerosene or diesel fuel at about 0.1% for eliminating foam in spray tanks, but this is not recommended in natural areas.

28.9.3.11. UV Absorbents

Natural sunlight, especially ultraviolet light, may degrade some herbicides (Green 2001). A few adjuvants that protect herbicides from the deleterious effect(s) of sunlight are available. They may do this by either physical or chemical processes, such as by increasing the rate of herbicide uptake into the cuticle, or by absorbing the UV-light themselves.

Various adjuvants, spreaders, Buffers, wetters, drift controllers and stickers (table.56) are available in the market for their use in the tank mix to increase the efficacy of the spray chemicals.

Table 56: List of marketed Adjuvants, Spreaders and Stickers

Active ingredient	Formulation	Concentrate	Product name	Company name	Uses
C-65 methylesters/Klearfac AA-270	EC	406/244 g/l	Dash HC	BASF	Spreader
acid buffer	SL	330 g/l	Qwemibuff	Plaaskem (Gouws and Scheepers)	Buffer
acidifier + buffer	SL	480 g/l	Surebuff	Hygrotech Seed	Buffer
acidifier and buffer	SL	600 g/l	Insure Plus	Plaaskem (Gouws and Scheepers)	Buffer
acidifier/wetter	SL	480/60 g/l	Wettbuff	Hygrotech Properties	Buffer Spreader
acidifying system & ammonium sulphate	SL	505 g/l	Promote	Ag-Chem Solutions	Buffer
alkoxylated fatty alkylamine polymer	SL	1000 g/l	Partner A	Universal Crop Protection	Buffer Spreader
alkoxylated fatty alkylamine polymer	SL	1000 g/l	Partner G	Universal Crop Protection	Buffer Spreader
alkoxylated fatty alkylamine polymer/ ethoxylated sorbitane ester	SL	500/650 g/l	Partner 650	Universal Crop Protection	Wetter Buffer Penetrant
alkoxylated fatty alkylamnine polymer/ ethoxylated sorbitane ester	SL	125/137,5 g/l	Partner 90	Universal Crop Protection	Buffer Spreader
alkyl polyglucoside/organic buffer	SL	198/428 g/l	Upgrade 700	Klub M5	Spreader Buffer
alkyl polyoxyethylene ether/ phosphatidylcholine, methylacetic acid/	SL	800 g/l	LI 700	UAP Crop Care	Spreader Penetrant
alkylaryl polyoxyethylene - glycol phosphate ester, organic acid buffer	SC	85, 497 g/l	Hygrobuff 4	Hygrotech Saad	Buffer
alkylaryl polyoxyethylene glycol phosphate ester	SL	850 g/l	Spray-Aide	Hygrotech Saad	Spreader
alkylated phenol-ethylene oxide	SL	945 g/l	Volcano 90	Volcano Agroscience	Spreader Sticker
alkylated phenol-ethylene oxide condensate	SL	940 g/l	Agral 90	Kynoch Agrochemicals	Spreader Sticker
alkylated phenol-ethylene oxide condensate	SL	940 g/l	Sanawett 90 - 940 SL	Dow AgroSciences	Spreader

Active ingredient	Formulation	Concentrate	Product name	Company name	Uses
alkylated phenol-ethylene oxide condensate	SL	940 g/l	Wett-A-Gral	JCJ Agri Chem	Speader Sticker
alkylated phenolethylene oxide	SL	940 g/l	Allgral	Natural Plant Protection	Spreader Sticker
alkylated phenolethylene oxide condensate/organic buffer system	SL	484,7 g/l	Allbuff	Natural Plant Protection	Buffer Spreader
alkylated phenyl-ethylene oxide	SL	940 g/l	Bladwet 9	Plaaskem (Gouws and Scheepers)	Spreader
alkylated phenyl-ethylene oxide	SL	940 g/l	Foliwett 900	Plaaskem (Gouws and Scheepers)	Spreader
alkylated phenyl-ethylene oxide	SL	940 g/l	Qwemiwet	Plaaskem (Gouws and Scheepers)	Spreader
alkylated phenyl-ethylene oxide condensate	SL	940 g/l	Hygrowett	Hygrotech Saad	Spreader Sticker
alkylated phenyl-ethylene oxide condensate	SL	940 g/l	Techniwett	Plaaskem (Gouws and Scheepers)	Spreader
alkylated phenylethylene oxide	SL	940 g/l	Kynogral 90	Kynoch Agrochemicals	Spreader Sticker
ammonium sulphate	SL		Beef-Up AS	GAP Chemicals	Spreader Penetrant
ammonium sulphate	SL	500 g/l	ASC	Plaaskem (Gouws and Scheepers)	Penetrant
ammonium sulphate	SL	500 g/l	Booster AS	RT Chemicals	Penetrant
ammonium sulphate	SL	500 g/l	Hyperboost	Natural Crop Protection	Penetrant
ammonium sulphate	SL	500 g/l	Imiboost	BASF	Penetrant
ammonium sulphate	SL	500 g/l	Speedup	Cash Chemicals	penetrant
ammonium sulphate & salts of polyacryl, hydroxy carboxylic & propionic acids, phosphate ester	SL	552 g/l	Choice	UAP Crop Care	penetrant

Active ingredient	Formulation	Concentrate	Product name	Company name	Uses
ammonium sulphate/acidifying system/ humectant/wetter/spreader	SL	562 g/l	Power-Up	Ag-Chem Africa	Acidifier Spreader
borax/orange oil	SL	10/50 g/l	Wetcit	Citrus Oil Procuts	Spreader
buffering and acidifying systems	SL	500 g/l	Tenderbuff	Ag-Chem Africa	Buffer
buffering system	SL	340 g/l	pH Green	Ag-Chem Africa	Buffer
buffering system/alkyl polyglucoside	SL	381/86 g/l	Aqua Right 7	Ag-Chem Africa	Spreader Buffer
buffering system/nonyl phenol ethoxylate	SL	340/10 g/l	pH Green Plus	Ag-Chem Africa	Buffer
buffering system/nonyl phenol ethoxylate	SL	511/121 g/l	Aqua Right 3	Ag-Chem Africa	Buffer Spreader
buffering system/nonyl phenol ethoxylate	SL	577/114 g/l	Aqua-Right 5	Ag-Chem Africa	Spreader Buffer
di-1-p-menthene	EC	219 g/l	Hygro-Stic	Hygrotech Saad	Wetter
di-1-p-menthene	EC	875 g/l	Nu-Film-17	Hygrotech Saad	Sticker
di-1-p-menthene	SL	250 g/l	Spray-Stay	Fleuron	Sticker
Dimethene	SL	250 g/l	Ludwig's Spray Stay	Ball Straathoff	Sticker
emulsifiers/methylated vegetable oil	EC	300/700 g/l	Ballista	Bayer	Spreader Penetrant
ethoxylated propoxylated fatty amines	SL	1000 g/l	Volcano-Blend	Volcano Agroscience	Spreader Penetrant
Fluxofenim	EC	964 g/l	Concep 960 EC	Syngenta	Antidote
heptamethyltrisiloxane	SL	1000 g/l	Silwet L-77	Plaaskem (Gouws and Scheepers)	Spreader
inorganic acid & pH indicator/wetting agent	SL	190/36 g/l	Kembuff Spreader	Kempton Chemicals	Spreader Buffer
methylated seed oil		850 g/l	Sadol	Soygro	Spreader
methylated vegetable oil	EC	700 g/l	ZAP	Ag-Chem Africa	Spreader Penetrant
mineral oil	EC	820 g/l	Actipron Super	Ecoguard Distributors	Spreader

Active ingredient	Formulation	Concentrate	Product name	Company name	Uses
mineral oil	EC	830 g/l	BP Spray Mist	BP SA	Spreader
mineral oil	EC	835 g/l	Sunspray Oil	Dow AgroSciences	Spreader
mineral oil	EC	950 g/l	Dubbel-Nat	Viking	Spreader
mineral oil	OL	822 g/l	BP Crop Oil	BP SA	Spreader
mineral oil	OL	882 g/l	BP Agripron Super	BP SA	Spreader
mineral oil (medium grade)	EC	363 g/l	AG Penetrex	Safagric	Spreader
mineral oil/nonylphenolethoxylate	EC	450/250 g/l	Complement	Syngenta	Penetrant Sticker
modified phthaltic glycerol alkyd resin	SL	770 g/l	Latron B-1956	Detekto International	Spreader Sticker
nitrogen solution	SL	920 g/l	Trisure 4	Spectrum Research Services	Spreader
nitrogenic solution	SL	929 g/l	Trisure S	Spectrum Research Services	Spreader
nonionic & anionic components	SL	600 g/l	BB5	Plaaskem (Gouws and Scheepers)	Buffer Spreader
nonionic & anionic components	SL	600 g/l	Bladbuff 5	Plaaskem (Gouws and Scheepers)	Buffer Spreader
nonyl phenol ethoxylate	DC	1000 g/l	Cyspread	BASF	Wetter Spreader
nonyl phenol ethoxylate/glycol ether/fatty acids	SL	900 g/l	Ag-Wett	Villa Crop Protection	Spreader Sticker
nonyl phenol ethoxylate/glycol ether/fatty acids	SL	900 g/l	Aqua-Wet	Ag-Chem Solutions	Spreader Sticker
nonyl phenol ethoxylate/glycol ether/fatty acids	SL	900 g/l	Magnet	Klub M5	Spreader Sticker
organic acid & alkali	SL	220 g/l	Sunbuff	Farmkem	Buffer
organic acid & alkali	SL	500 g/l	Curabuff 500 SL	Dow AgroSciences	Buffer Spreader
organic acid & alkali	SL	650 g/l	Realbuff	Nialcor	Buffer
organic acid & alkali	SL	660 g/l	Pazbuff	Universal Crop Protection	Buffer
organic acid & alkali	SL	660 g/l	Reverbuff	Farmkem	Buffer
organic acid & alkali & wetting agents	SL	536 g/l	Buffernat	Farmkem	Buffer Spreader

Active ingredient	Formulation	Concentrate	Product name	Company name	Uses
organic acid and alkali	SL	660 g/l	Aquabuff	Applied Agricultural Products	Buffer
organic acid and alkali	SL	660 g/l	Commodobuff	Villa Crop Protection	Buffer
organic buffer & alkylated phenylethylene oxide	SL	484,7 g/l	Insure	Plaaskem (Gouws and Scheepers)	Buffer Spreader
organic buffer & alkylated phenylethylene oxide condensate	SL	484.7 g/l	Kynobuff	Kynoch Agrochemicals	Buffer Spreader
organic buffer and alkylated phenol-ethylene oxide condensate	SL	484,7 g/l	Technibuff	Technichem	Spreader Buffer
organic buffer system	SL	467 g/l	Beefabuff	GAP Chemicals	Buffer
organic buffer system	SL	484,7	Technibuff	Plaaskem (Gouws and Scheepers)	Buffer Spreader
organic buffer system	SL	548,5 g/l	Quattro-Buff	Natural Crop Protection	Buffer
organic buffer system	SL	564 g/l	Beefabuff Plus	GAP Chemicals	Buffer
organic buffer system	SL	567 g/l	RT Buff	RT Chemicals	Buffer
organic buffer system & alkylated phenol-ethylene oxide condensate	SL	485 g/l	Wett-A-Buff	JCJ Agri Chem	Buffer Spreader
organic buffering system/nonyl phenol ethoxylate	SL	340/10 g/l	Upgrade 500	Klub M5	Buffer Spreader
organosilicone/synthetic latex	EC	86,5/250 g/l	Designer	UAP Crop Care	Spreader Sticker
oxyalkylated alcohol/synthetic latex	SL	100/450 g/l	Bond	UAP Crop Care	
paraffinic complex	EC	725 g/l	Herbiplus	Total South Africa	Spreader Buffer Penetrant
petroleum hydrocarbons,wetters,spreaders	EC	840 g/l	Voltage	Volcano Agroscience	Spreader Penetrant
phosphoric acid	SL	190 g/l	Kembuff	Kempton Chemicals	Buffer
poly-1-p-menthene	EC	875 g/l	Nu-Film-P	Hygrotech Saad	Sticker

Active ingredient	Formulation	Concentrate	Product name	Company name	Uses
polyacrylamide		250 g/kg	Bandrift Plus	Plaaskem (Gouws and Scheepers)	Drift retardant
polyacrylamide polymer		320 g/kg	38-F Drift Retardant	Orsmond Aerial Spray	Drift control
polyacrylamide polymer/ polysaccharide		270/30 g/kg	41-A Drift Retardant	Orsmond Aerial Spray	Drift control

EC – Emulsifiable Concentrate; SC – Suspension Concentrate; SL – Soluble Liquid; CS – Capsule Suspension
WP – Wettable Powder; SP – Soluble Powder
Source: *http://www.nda.agric.za/doaDev/sideMenu/ActNo36_1947/AR/Adjuvants.htm*

Adjvants recommended for different herbicides

Herbicide	Recommended Adjuvant Type
2-4 D (many brands)	Non-ionic surfactants, Nitrigen Fertilizer or crop oil concentrate
Clopyralid (Transelin, stinger)	Non-ionic surfactants
Fluazifop-P-butyl (Fusilade DX)	Non-ionic surfactants and crop oil concentrate
Fosamine ammonium (Krenite S)	Oil-based surfactants
Glyphosate (Round Up Original)	Adjuvants already added, non-ionic surfactants or ammonium sulfate
Glyphosate (Round Up Ultra)	Adjuvants already added, ammonium sulfate
Glyphosate (Rodeo, Aquamaster, Glypro)	Non-ionic surfactants
Hexazinone (Velpar L)	No recommendations on label
Imazapic (Plateau)	Methylated seed oil or crop oil concentrate, non-ionic surfactant, silicon based surfactant, fertilizer surfactant blend
Imazapyr (Arsenol)	Methylated seed oil or crop oil concentrate, non-ionic surfactant, silicon based surfactant, fertilizer surfactant blend
Picloram (Tordon K, Tordon 22K)	Non-ionic surfactants
Sethoxydim (Poast, Poast Plus)	Adjuvants already added, Methylated seed oil or crop oil concentrate, ammonium nitrate
Sethoxydim (Vantage)	Adjuvants already added, ammonium surfactant
Triclopyr (Garlon 3A, Garlon 4)	Non-ionic surfactants

29

Biological Control Agents for Disease Pest Management

29.1. What is Biological Control

Biological control or biocontrol is a method of controlling pests such as insects, mites, weeds and plant diseases using other organisms. It relies on predation, parasitism, herbivory, or other natural mechanisms, but typically also involves an active human management role. It can be an important component of integrated pest management (IPM) programs.

29.2. Important Early Milestones in Biological Control Program

Biological control techniques as we know them today started to emerge in the 1870s. During this decade, in the US, the Missouri State Entomologist C. V. Riley and the Illinois State Entomologist W. LeBaron began within-state redistribution of parasitoids to control crop pests.

The first international shipment of an insect as a biological control agent was made by Charles V. Riley in 1873, shipping to France the predatory mites *Tyroglyphus phylloxera* to help fight the grapevine phylloxera (*Daktulosphaira vitifoliae*) that was destroying grapevines in France.

The United States Department of Agriculture (USDA) initiated research in classical biological control following the establishment of the Division of Entomology in 1881, with C. V. Riley as Chief. The first importation of a parasitoidal wasp into the United States was that of the braconid *Cotesia glomerata* in 1883–1884, imported from Europe to control the invasive cabbage white butterfly, *Pieris rapae*.

In 1888–1889 the vedalia beetle, *Rodolia cardinalis,* a lady beetle, was introduced from Australia to California to control the cottony cushion scale, *Icerya purchasi*. This had become a major problem for the newly developed citrus industry in California, but by the end of 1889, the cottony cushion scale population had already declined. This great success led to further introductions of beneficial insects into the US.

In 1905 the USDA initiated its first large-scale biological control program, sending entomologists to Europe and Japan to look for natural enemies of the gypsy moth, *Lymantria dispar* , and brown-tail moth, *Euproctis chrysorrhoea*, invasive pests of trees and shrubs. As a result, nine parasitoids (solitary wasps) of the gypsy moth, seven of brown-tail moth, and two predators of both moths became established in the US. Although the gypsy moth was not fully controlled by these natural enemies, the frequency, duration, and severity of its outbreaks were reduced and the program was regarded as successful. This program also led to the development of many concepts, principles, and procedures for the implementation of biological control programs.

Prickly pear cacti were introduced into Queensland, Australia as ornamental plants, starting in 1788. They quickly spread to cover over 25 million hectares of Australia by 1920, increasing by 1 million hectares per year. Digging, burning, and crushing all proved ineffective. Two control agents were introduced to help control the spread of the plant, the cactus moth *Cactoblastis cactorum*, and the scale insect *Dactylopius*. Between 1926 and 1931, tens of millions of cactus moth eggs were distributed around Queensland with great success, and by 1932, most areas of prickly pear had been destroyed.

The first reported case of a classical biological control attempt in Canada involves the parasitoidal wasp *Trichogramma minutum*. Individuals were caught in New York State and released in Ontario gardens in 1882 by William Saunders, a trained chemist and first Director of the Dominion Experimental Farms, for controlling the invasive currantworm *Nematus ribesii*. Between 1884 and 1908, the first Dominion Entomologist, James Fletcher, continued introductions of other parasitoids and pathogens for the control of pests in Canada.

29.3. Biological Control Agents

Natural enemies of insect pests, and disease pathogens are known as biological control agents, which include predators, parasitoids, pathogens, and competitors.

Biological control agents of plant diseases are most often referred to as antagonists.

Biological control agents of weeds include seed predators, herbivores, and plant pathogens.

Biological control can have side-effects on biodiversity through attacks on non-target species by any of the above mechanisms, especially when a species is introduced without a thorough understanding of the possible consequences.

Various types of biological control agents employed in the pest management program are given in table 57.

Table. 57: Types of insects and microorganisms used as biocontrol agents in pest control.

Biocontrol agent	Common examples	Biological action
1.Parasitic insects: living organisms remaining in close association with their hosts and gradually derive their food from the host	*Trichogramma chilonis, Epiricania melanoleuca*	They live and feed internally or externally on the host insect.
	Chrysoperla carnea, Cryptolaemus montrouzieri	Insects which kill and devour the prey
	Bacteria, for example, *Bacillus thuringiensis* Fungi, for example, species of *Trichoderma, Nomuraea, Paecilomyces, Verticillium, Metarhizium, and Beauveria* Viruses, for example, nuclear polyhedrosis virus	Cause diseases in pests

29.3.1. Predators (Predatory insects)

Predators are mainly free-living species that directly consume a large number of prey during their whole lifetime. Given that many major crop pests are insects, many of the predators used in biological control are insectivorous species.

Lady beetles, and in particular their larvae which are active between May and July in the northern hemisphere, are voracious predators of aphids, and also consume mites, scale insects and small caterpillars. The spotted lady beetle (*Coleomegilla maculata*) is also able to feed on the eggs and larvae of the Colorado potato beetle (*Leptinotarsa decemlineata*).

The larvae of many hoverfly species principally feed upon aphids, one larva devouring up to 400 in its lifetime. Their effectiveness in commercial crops has not been studied. The running crab spider *Philodromus cespitum* also prey heavily on aphids, and act as a biological control agent in European fruit orchards.

Several species of entomopathogenic nematode are important predators of insect and other invertebrate pests. Entomopathogenic nematodes form a stress–resistant stage known as the infective juvenile. These spread in the soil and infect suitable insect hosts. Upon entering the insect they move to the hemolymph where they recover from their stagnated state of development and release their bacterial symbionts. The bacterial symbionts reproduce and release toxins, which then kill the host insect.

Phasmarhabditis hermaphrodita is a microscopic nematode that kills slugs. Its complex life cycle includes a free-living, infective stage in the soil where it becomes associated with a pathogenic bacteria such as *Moraxella osloensis*. The nematode enters the slug through the posterior mantle region, thereafter feeding and reproducing inside, but it is the bacteria that kill the slug. The nematode is available commercially in Europe and is applied by watering onto moist soil. Entomopathogenic nematodes have a limited shelf life because of their limited resistance to high temperature and dry conditions. The type of soil they are applied to may also limit their effectiveness.

Species used to control spider mites include the predatory mites *Phytoseiulus persimilis*, *Neoseilus californicus,* and *Amblyseius cucumeris*, the predatory midge *Feltiella acarisuga*, and a ladybird *Stethorus punctillum*.

The bug *Orius insidiosus* has been successfully used against the two-spotted spider mite and the western flower thrips (*Frankliniella occidentalis*).

Predators including *Cactoblastis cactorum* (mentioned above) can also be used to destroy invasive plant species. As another example, the poison hemlock moth (*Agonopterix alstroemeriana)* can be used to control poison hemlock (*Conium maculatum*). During its larval stage, the moth strictly consumes its host plant, poison hemlock, and can exist at hundreds of larvae per individual host plant, destroying large swathes of the hemlock.

29.3.2. Parasitoids (Parasitic insects)

Parasitoids lay their eggs on or in the body of an insect host, which is then used as a food for developing larvae. The host is ultimately killed. Most insect parasitoids are wasps or flies, and many have a very narrow host range.

The most important groups are the ichneumonid wasps, which mainly use caterpillars as hosts; braconid wasps, which attack caterpillars and a wide range of other insects including aphids; chalcid wasps, which parasitize eggs and larvae of many insect species; and tachinid flies, which parasitize a wide range of insects including caterpillars, beetle adults and larvae, and true bugs. Parasitoids are most effective at reducing pest populations when their host organisms have limited refuges to hide from them. Parasitoids are among the most widely used biological control agents.

Commercially, there are two types of rearing systems: short-term daily output with high production of parasitoids per day, and long-term, low daily output systems. In most instances, production will need to be matched with the appropriate release dates when susceptible host species at a suitable phase of development will be available. Larger production facilities produce on a year long basis, whereas some facilities produce only seasonally. Rearing facilities are usually a significant distance from where the agents are to be used in the field, and transporting the parasitoids from the point of production to the point of use can pose problems. Shipping conditions can be too hot, and even vibrations from planes or trucks can adversely affect parasitoids.

Encarsia formosa widely used in greenhouse horticulture, was one of the first biological control agents developed. *Encarsia formosa* is a small predatory chalcid wasp which is a parasitoid of whitefly, a sap-feeding insect which can cause wilting and black sooty moulds in glasshouse vegetable and ornamental crops. It is most effective when dealing with low level infestations, giving protection over a long period of time. The wasp lays its eggs in young whitefly 'scales', turning them black as the parasite larvae pupate.

Gonatocerus ashmeadi (Hymenoptera: Mymaridae) has been introduced to control the glassy-winged sharpshooter *Homalodisca vitripennis* (Hemiptera: Cicadellidae) in French Polynesia and has successfully controlled ~95% of the pest density.

The eastern spruce budworm is an example of a destructive insect in fir and spruce forests. Birds are a natural form of biological control, but the *Trichogramma minutum*, a species of parasitic wasp, has been investigated as an alternative to more controversial chemical controls.

29.3.3. Pathogenic Micro-organisms

Pathogenic micro-organisms include bacteria, fungi, and viruses. They kill or debilitate their host and are relatively host-specific. Various microbial insect diseases occur naturally, but the disease causing pathogens may also be used as biological pesticides. When naturally occurring, these outbreaks are density-dependent in that they generally only occur as insect populations become denser.

a). Bacteria

Bacteria used for biological control infect insects via their digestive tracts, so they offer only limited options for controlling insects with sucking mouth parts such as aphids and scale insects.

Bacillus thuringiensis, a soil-dwelling bacterium, is the most widely applied species of bacteria used for biological control, with at least four sub-species used against Lepidopteran (moth, butterfly), Coleopteran (beetle) and Dipteran (true fly) insect pests.

The bacterium is available to organic farmers in sachets of dried spores which are mixed with water and sprayed onto vulnerable plants such as brassicas and fruit trees. Genes from *B. thuringiensis* have also been incorporated into transgenic crops, making the plants express some of the bacterium's toxins, which are proteins. These confer resistance to insect pests and thus reduce the necessity for pesticide use. If pests develop resistance to the toxins in these crops, *B. thuringiensis* will become useless in organic farming also.

The bacterium *Paenibacillus popilliae* which causes milky spore disease has been found useful in the control of Japanese beetle, killing the larvae. It is very specific to its host species and is harmless to vertebrates and other invertebrates.

b). Fungi

Entomopathogenic fungi constitute the largest single group of insect pathogens among microorganisms. Entomogenous fungi are promising myco-biocontrolling agent for a number of crop pests. Several species belonging to order Lepidoptera, Coleoptera, Homoptera, Hymenoptera, and Diptera are susceptible to various fungal infections. Entomopathogenic fungi have a great potential as myco-biocontrol agents, as they constitute a group with over 750 species from around 90 genera that, when dispersed in the environment, provoke fungal infections in insect populations. Entomopathogenic fungi are among the first organisms to be used for the biological control of pests. Most of them are found within the deuteromycetes and entomophthorales.

Some entomopathogenic fungi have restricted host ranges, for example, *Aschersonia aleyrodes* infects only scale insects, and whiteflies, while other fungal species have a wide host range, with individual isolates being more specific to target pests.

Entomopathogens such as *M. anisopliae* and *B. bassiana* are well characterized in respect to pathogenicity to several insects, and they have been used as agents for the biological control of agriculture pests worldwide. About 11 companies offer at least 16 products based on the entomopathogenic fungi *B. bassiana*. These products are not only used in the coffee crop but also in other crops such as bean, cabbage, corn, potato, and tomato. Several entomopathogenic fungal species, as below, when inundatively introduced into a variety of habitats, can provide effective long-term to short-term control.

Entomopathogenic fungi, which cause disease in insects, include at least 14 species that attack aphids.

Beauveria bassiana is mass-produced and used to manage a wide variety of insect pests including whiteflies, thrips, aphids and weevils.

Lecanicillium spp. are deployed against white flies, thrips and aphids.

Metarhizium spp. are used against pests including beetles, locusts and other grasshoppers, Hemiptera, and spider mites.

Paecilomyces fumosoroseus is effective against white flies, thrips and aphids.

Purpureocillium lilacinus is used against root-knot nematodes, and 89 *Trichoderma* species against certain plant pathogens.

The fungi *Cordyceps* and *Metacordyceps* are deployed against a wide spectrum of arthropods. *Entomophaga* is effective against pests such as the green peach aphid.

Several members of Chytridiomycota **and** Blastocladiomycota **have been explored as agents of biological control. From Chytridiomycota,** *Synchytrium solstitiale* **is being considered as a control agent of the** yellow star thistle (*Centaurea solstitialis*) in the United States.

Lagenidium giganteum belonging to Oomycota is a water-borne mold that parasitizes the larval stage of mosquitoes. When applied to water, the motile spores avoid unsuitable host species and search out suitable mosquito larval hosts. This mold has the advantages of a dormant phase, resistant to desiccation, with slow-release characteristics over several years. Unfortunately, it is susceptible to many chemicals used in mosquito abatement programmes.

Entomopathogenic Fungi used in Insect Pest Management

1. *Beauveria* sp

Beauveria bassiana, is a filamentous fungus, belonging to a class of insect pathogenic deuteromycete also known as imperfect fungus. Strains of *Beauveria* are highly adapted to particular host insects. A broad range of *B. bassiana* spp. have been isolated from a variety of insect worldwide which are of medicinal or agricultural importance.

Beauveria bassiana is a fungus that grows naturally in soils throughout the world and acts as a pathogen on various insect species, causing white muscardine disease, therefore belongs to the entomopathogenic fungi [Sandhu et.al, 1993, 2001;Thakur et.al, 2005., Sandhu and Vikrant,2004., Jain et.al, 2008]. An interesting feature of *Beauveria* sp. is the high host specificity of many isolates. Hosts of medicinal importance include vectors for agents of tropical infectious diseases such as tsetse fly *Glossina morsitans*, and sand fly *Phlebotomus* that transmits *Leishmania* and bugs of genera *Triatoma* and *Rhodnius*, the vectors of Chagas disease.

Hosts of agricultural and forest significance include the Colorado potato beetle, the codling moth, and several genera of termites, American bollworm *Helicoverpa armigera* [Thakur and Sandhu, 2010], *Hyblaeapara* and *Eutectona machaeralis*. Furthermore, the high level of persistence in the host population and in the environment provides long-term effects of the entomopathogenic fungi on pest suppression, if an epizootic is caused.

It is being used as a biological insecticide to control a number of pests such as termites, whitefly, and in malaria-transmitting mosquitoes [Hamlen, 1979; McNeil Donald, 2005]. *B. bassiana* is the anamorph (asexually reproducing form) of *Cordyceps bassiana*. The latter teleomorph (the sexually reproducing form) has been collected only in eastern Asia [Li et.al, 2001]. Rehner and Buckley (2005) have shown that *B. bassiana* consists of many distinct lineages that should be recognized as distinct phylogenetic species. This ubiquitous fungus has long been known to be the most common causative agent of disease associated with dead and moribund insects in nature [Mc.Leod, 1954] and has been scrutinized worldwide as a microbial control agent of hypogeous species [Ferron.1981].

Many curculionidae weevils with a sub-terranean larval stage are highly susceptible to this white muscardine disease [Beavers et.al, 1983]. Like many species of entomogenous fungi, *B. bassiana* is composed of many genetically distinct variants associated with geographical location and host which differ substantially in their ability to produce pathogenesis.

As an insecticide, the spores are sprayed on affected crops as an emulsified suspension or wettable powder. *B. bassiana* parasitizes a very wide range of arthropod hosts and therefore is considered as a non-selective biological insecticide. *B. bassiana* is also applied against the European corn borer *Ostrinia Mubilalis*, pine caterpillars *Dendrolimus* spp., and green leafhoppers *Nephotettix* spp.

2. *Verticillium lecanii*

Another entomopathogenic fungus *Verticillium lecanii* is a widely distributed fungus, which can cause large epizootic in tropical and subtropical regions, as well as in warm and humid environments [Nunez et.al, 2008]. It was reported by Kim et al. [2002] that *V. lecanii* was an effective biological control agent against *Trialeurodes vaporariorum* in South Korean greenhouses. This fungus attacks nymphs and adults and stucks to the leaf underside by means of a filamentous mycelium [Nunez et.al, 2008]. In 1970s, *Verticillium lecanii* was developed to control whitefly and several aphid species, including the green peach aphids (*Myzus persicae*) for use in the greenhouse chrysanthemums [Hamlen, 1974].

Verticillium lecanii was considered as a major parasite which caused a massive decline of cereal-cyst nematode populations in monocultures of susceptible crops [Kerry et.al, 1982]. *Verticillium chlamydosporium* has a wide host range amongst cyst and root-knot nematodes but it is very variable and only some isolates may have potential as commercial biological control agents.

3. *Metarhizium* spp

Metarhizium anisopliae is also very potential pathogen on insect pests and is explored for myco-biocontrol of notorious insect pests [Sandhu et.al, 1993; Sandhu and Mishra, 1994]. A complete bioactivity of *M. anisopliae* has been tested on teak skeletonizer *Eutectona machaeralis* and found *M. anisopliae* to be a potential myco-biocontrol agent of teak pest [Sandhu et.al, 2000]. Hasan et al. [2002] have tested spore production of *M. anisopliae* by solid state fermentation.

4. *Nomuraea* sp

Nomuraea rileyi another potential entomopathogenic fungi is a dimorphic hyphomycete that can cause epizootic death in various insects. It has been shown that many insect species belonging to Lepidoptera including *Spodoptera litura* and some belonging to *Coleoptera* are susceptible to *N. rileyi* [Ignoffo, 1981]. The host specificity of *N. rileyi* and its ecofriendly nature encourage its use in insect pest management. Although, its mode of infection and development have been reported for several insect hosts such as *Trichoplusia ni*, *Heliothis zea*, *Plathypena scabra*, *Bombyx mori*, *Pseudoplusia includes,* and *Anticarsia gemmatalis*, another insect *Spilosoma* was found to be severely attacked by *Nomuraea rileyi*, hence studied in detail for its myco-biocontrol [Mathew et.al, 1998]. Similarly an epizootic of *Nomuraea rileyi* was observed on *Junonia orithya* [Rajak et.al, 1991] which was proved to be the best alternative to manage the hedge plant eater *Junonia orithya.*

5. *Paecilomyces* sp

Paecilomyces is a genus of nematophagous fungus which kills harmful nematodes by pathogenesis, causing disease in the nematodes. Thus, the fungus can be used as a bionematicide to control nematodes by applying to soil.

Paecilomyces lilacinus principally infects and assimilates eggs of root-knot and cyst nematodes. The fungus has been the subject of considerable biological control research following its discovery as a biological control agent in 1979. *P.*

lilacinus has been considered to have the greatest potential for application as a biocontrol agent in subtropical and tropical agricultural soils.

Paecilomyces fumosoroseus (Hyphomycetes) is one of the most important natural enemies of whiteflies worldwide, and causes the sickness called "Yellow Muscardine" [Nunez et.al, 2008]. Strong epizootic potential against *Bemisia* and *Trialeurodes* spp. in both greenhouse and open field environments has been reported. The ability of this fungus to grow extensively over the leaf surface under humid conditions is a characteristic that certainly enhances its ability to spread rapidly through whitefly populations [Wraight et.al, 2000]. Natural epizootics of these fungi suppress *Bemisia tabaci* populations. Epizootics caused by *Paecilomyces fumosoroseus* also lead to substantially reductions in *B. tabaci* populations during or immediately following rainy seasons or even prolonged periods of cool, humid conditions in the field or greenhouse [Faria and Wraight, 2001]. However, in general, epizootics of naturally occurring fungi cannot be relied upon for control. Only a few species of fungi have the capacity to cause high level of mortality, and development of natural epizootics which is not only dependent on the environmental conditions, but also influenced by various crop production practices. Also, epizootics often occur after intense injury has already been inflicted by whiteflies [Faria and Wraight, 2001]. Kim et al. [2002] reported that *P. fumosoroseus* is best for controlling the nymphs of whitefly. These fungi cover the whitefly's body with mycelial threads and stick them to the underside of the leaves. The nymphs show a "feathery" aspect and are surrounded by mycelia and conidia [Nunez et.al, 2008]. *P. furiosus* is also used to control mosquito sp. *Culex pipiens* [Sandhu and Mishra, 1994].

Mode of Action of Entomopathogenic Fungus

Fungi have a unique mode of infection; they reach the haemocoel through the cuticle or possibly through the mouth parts. The death of the insect results from a combination of factors: mechanical damage resulting from tissue invasion, depletion of nutrient resources and toxicosis, and production of toxin in the body of insect.

1. Conidial Attachment with the Cuticle

Attachment of a fungal spore to the cuticle surface of a susceptible host represents the initial event in the establishment of mycosis. It was observed that dry spores of *B. bassiana* possess an outer layer composed of interwoven fascicles of hydrophobic rodlets. This rodlet layer appears to be special to the conidial stage and has not been reported on the vegetative cells. The adhesion of dry spores to the cuticle was suggested to be due to non-specific hydrophobic forces imposed by the rodlets [Boucias, 1988]. Some of these moieties like lectins, a kind of carbohydrate binding glycoproteins, have also been detected on the conidial surface of *B. bassiana*. It was also observed that lectins could be involved in binding between conidia and the insect cuticle. When the pathogen reaches and adheres to the host surface, it proceeds with rapid germination and growth which are profoundly influenced by the availability of water, nutrients, oxygen as well as pH, and temperature, and by the effects of toxic host-surface compound. Fungi with a broad host range germinate in culture in response to a wide range of nonspecific carbon and nitrogen sources [Sandhu, 1955]. Entomopathogenic fungi with restricted host range appear to have more specific requirements for germination [Leger et.al, 1989].

2. Formation of an Infection Structure

Entomopathogenic fungi invade their hosts by infection process: penetration of the host cuticle or put pressure on cuticle by making appressorium and then penetrate by penetration peg. The cuticle has two layers: the outer epicuticle and the procuticle. The epicuticle is a very complex thin structure that lacks chitin but contains phenol-stabilized proteins and is covered by a waxy layer containing fatty acids, lipids and sterols [Hackman, 1984]. The procuticle forms the majority of the cuticle and contains chitin fibrils embedded into a protein matrix together with lipids and quinones [Neville, 1984]. Protein may account for up to 70% of the cuticle. In many areas of the cuticle, the chitin is organized helically giving rise to a laminate structure. Entomopathogenic fungi, *B. bassiana* conidia germinate on the host surface and differentiate an infection structure termed appressorium. The appressorium represents an adaptation for concentrating physical and chemical energy over a very small area so that access may be achieved efficiently. Thus, formation of the appressorium plays a pivotal role in establishing a pathogenic interaction with the host. Appressorium formation may be influenced by host surface topography, and biochemical investigations indicate the involvement of the intracellular second messengers Ca^{2+} and cyclic AMP (cAMP) in appressorium formation [Leger et.al,1991] or in general when the cuticle in hard [Sandhu, 1955].

3. Penetration of the Cuticle

Entomopathogenic fungi need to penetrate through the cuticle into the insect body to obtain nutrients for their growth and reproduction. Entry into the host involves both enzymic degradation and mechanical pressure as evidenced by the physical separation of lamellae by penetrated hyphae. A range of extracellular enzymes that can degrade the major components of insect cuticle, including chitinases, lipases, esterases and at least four different classes of proteases, have been suggested to function during the fungal pathogenesis. Although the complex structure of the insect cuticle suggests that penetration would require the synergistic action of several different enzymes, much of the attention has focused on the cuticle active endoprotease as a key factor in the process.

The production of cuticle-degrading enzymes by *M. anisopliae* during infection structure formation on *Calliphora vomitoria* and *Manduca sexta* has been investigated by biochemical and histochemical analyses both *in vivo* and *in vitro*. Among the first enzymes produced on the cuticle are endoproteases (termed PR1 and PR2) and aminopeptidases, coincident with the formation of appressoria. N-Acetylglucosaminidase is produced at a slow rate as compared to the proteolytic enzymes [Leger et.al, 1989].

These fungi begin their infective process when spores are retained on the integument surface, where the formation of the germinative tube initiates, the fungi starts excreting enzymes such as proteases, chitinases, quitobiases, lipases, and lipoxygenases. *V. lecanii* is capable of penetrating the insect cuticle only with its germ tube while *M. anisopliae* and *B. bassiana* produce specific infection hyphae originating at appressoria. After the successful penetration, the fungus is then distributed into the haemolymph by formation of blastospores [Bhattacharyya et.al, 2004].

Different works are going on all over the world to distinguish the various enzymes which are required for the mechanism of entomopathogenic *Metarhizium anisopliae, M. flaviviridae, Paecilomyces farinosus, Beauveria bassiana*, and *B. brongniartii*. Host specificity may be associated with the physiological state of the host system (i.e., insect maturation and host plant) [McCoy et.al, 1988], the properties of the insect's integument with the nutritional requirements of the fungus [Kerwin and Washino, 1986], and the cellular defense of the host [Amer et.al, 2008]. In contrast to bacteria and viruses that pass through the gut wall from contaminated food, fungi have a unique mode of infection. They reach the haemocoel through the cuticle.

4. Production of Toxins

A plethora of work with circumstantial evidence is available from deuteromycete pathogens for the involvement of fungal toxins in host death. The action of cytotoxins is suggested by cellular disruption prior to hyphae penetration. Behavioural symptoms such as partial or general paralysis, sluggishness, and decreased irritability in mycosed insects are consistent with the action of neuromuscular toxins [Leger et.al, 1987]. *B. bassiana* and *M. anisopliae* produced significant amounts of toxic compounds within their hosts. For example, the toxins Beauvericin, Bassianolide, Isarolides, and Beauverolides have been isolated from *B. bassiana* infected hosts [Hamil et.al, 1969; Elsworth et.al, 1977], toxins Destruxins (DTXs) and Cytochalasins have been isolated from *M. anisopliae* infected hosts. The toxins have shown to have diverse effects on various insect tissues. DTX depolarizes the *lepidopteran* muscle membrane by activating calcium channels. In addition, function of insect hemocytes can be inhibited by DTX [Bradfisch and Harmer, 1990]. Presumably, there are still many toxins that remain to be isolated from parasitized insects and except DTXs, their relevance in the process of pathogenicity remains to be studied in detail.

c). Viruses

Baculoviruses are specific to individual insect host species and have been shown to be useful in biological pest control. For example, the Lymantria dispar multicapsid nuclear polyhedrosis virus (NPV) has been used to spray large areas of forest in North America where larvae of the gypsy moth are causing serious defoliation. The moth larvae are killed by the virus they have eaten and die, the disintegrating cadavers leaving virus particles on the foliage to infect other larvae.

d). Nematodes

Entomopathogenic nematodes (EPNs) belonging to genera *Steinernema* and *Heterorhabditis* together with their symbiotic bacteria *Xenorhabdus* and *Photorhabdus*, respectively, and slug-parasitic nematodes *Phasmarhabditis* with its symbiotic bacteria *Moraxella* have been considered as promising biocontrol agents (Table. 58) for the management of crop insect pests and slugs.

These nematodes have short life cycle, wide host range, and can resist under unfavourable conditions and environmental extremes. Survival and pathogenicity of these nematodes vary from 5°C to 35°C. They can be mass produced under both in vivo and in vitro conditions. With the realization of these attributes among these bioagents there is a need to search out an ideal formulation and proper application technology to include them in pest management programme.

Table.58: Commercial use of Entomopathogenic nematodes (EPNs) against some pest insects.

Crops (targeted)	Pest common name	Pest scientific name	Effective nematodes[b]
Artichokes	Artichoke plume moth	*Platyptilia carduidactyla*	Sc
Vegetables	Armyworm	Lep: Noctuidae	Sc, Sf, Sr
Ornamentals	Banana moth	*Opogona sacchari*	Hb, Sc
Bananas	Banana root borer	*Cosmopolites sordidus*	Sc, Sf, Sg
Turf	Billbug	*Sphenophorus* spp. (Col: Curculionidae)	Hb, Sc
Turf, vegetables	Black cutworm	*Agrotis ipsilon*	Sc
Berries, ornamentals	Black vine weevil	*Otiorhynchus sulcatus*	Hb, Hd, Hm, Hmeg, Sc, Sg
Fruit trees, ornamentals	Borer	*Synanthedon* spp. and other sesiids	Hb, Sc, Sf
Home yard, turf	Cat flea	*Ctenocephalides felis*	Sc
Citrus, ornamentals	Citrus root weevil	*Pachnaeus* spp. (Col: Curculionidae)	Sr, Hb
Pome fruit	Codling moth	*Cydia pomonella*	Sc, Sf
Vegetables	Corn earworm	*Helicoverpa zea*	Sc, Sf, Sr
Vegetables	Corn rootworm	*Diabrotica* spp.	Hb, Sc
Cranberries	Cranberry girdler	*Chrysoteuchia topiaria*	Sc
Turf	Crane fly	Dip: Tipulidae	Sc
Citrus, ornamentals	Diaprepes root weevil	*Diaprepes abbreviatus*	Hb, Sr
Mushrooms	Fungus gnat	Dip: Sciaridae	Sf, Hb
Grapes	Grape root borer	*Vitacea polistiformis*	Hz, Hb
Iris	Iris borer	*Macronoctua onusta*	Hb, Sc
Forest plantings	Large pine weevil	*Hylobius abietis*	Hd, Sc
Vegetables, ornamentals	Leafminer	*Liriomyza* spp. (Dip: Agromyzidae)	Sc, Sf
Turf	Mole cricket	*Scapteriscus* spp.	Sc, Sr, Sscap
Nut and fruit trees	Navel orangeworm	*Amyelois transitella*	Sc
Fruit trees	Plum curculio	*Conotrachelus nenuphar*	Sr
Turf, ornamentals	Scarab grub[c]	Col: Scarabaeidae	Hb, Sc, Sg, Ss, Hz
Ornamentals	Shore fly	*Scatella* spp.	Sc, Sf
Berries strawberry	Root weevil	*Otiorhynchus ovatus*	Hm
Bee hives	Small hive beetle	*Aethina tumida*	Hi, Sr
Sweet potato	Sweetpotato weevil	*Cylas formicarius*	Hb, Sc, Sf

[b] Abbreviations of nematode species; Hb: *Heterorhabditis bacteriophora*, Hd: *H. downesi*, Hi: *H. indica*, Hm: *H. marelata*, Hmeg: *H. megidis*, Hz: *H. zealandica*, Sc: *Steinernema carpocapsae*, Sf: *S. feltiae*, Sg: S. *glaseri*, Sk: *S. kushidai*, Sr: *S. riobrave*, Sscap: *S. scapterisci*, Ss: *S. scarabaei*.

Source: Ugur Gozel and Cigdem Gozel, 2016.

29.3.4. Combined use of Parasitoids and Pathogens

In cases of massive and severe infection of invasive pests, techniques of pest control are often used in combination. An example is the emerald ash borer, *Agrilus planipennis*, an invasive beetle from China, which has destroyed tens of millions of ash trees in its introduced range in North America. As part of the campaign against it, from 2003 American scientists and the Chinese Academy of Forestry searched for its natural enemies in the wild, leading to the discovery of several parasitoid wasps, namely *Tetrastichus planipennisi*, a gregarious larval endoparasitoid, *Oobius agrili*, a solitary, parthenogenic egg parasitoid, and *Spathius agrili*, a gregarious larval ectoparasitoid. These have been introduced and released into the United States of America as a possible biological control of the emerald ash borer. Initial results for *Tetrastichus planipennisi* have shown promise, and it is now being released along with *Beauveria bassiana*, a fungal pathogen with known insecticidal properties.

29.4. Advantage and Constraints of Myco-Biocontrol Agents

The advantages of using fungi as myco-biocontrol agents are as follows

(1) Their high degree of specificity for pest control. Fungi can be used to control harmful insect pests without affecting beneficial insect predators and non-harmful parasites.

(2) The absence of effects on mammals and thus the reduction of the hazards normally encountered with insecticide applications, such as pollution of the environment.

(3) The lack of problems caused to insect resistance and prolonged pest control.

(4) A high potential for further development by biotechnological research.

(5) High persistence in the environment provides long-term effects of entomopathogenic fungi on pest suppression.

However, there are also a number of constraints on the use of fungi as insecticides:

(1) 2-3 weeks are required to kill the insects whereas chemical insecticides may need only 2-3 hours.

(2) Application needs to coincide with high relative humidity, low pest numbers, and a fungicide free period.

(3) Due to the high specificity additional control agents are needed for other pests.

(4) Their production is relatively expensive and the short shelf life of spores necessitates cold storage.

(5) The persistence and efficacy of entomopathogenic fungi in the host population varies among different insects species, thus insect-specific application techniques need to be optimised to retain long-term impacts.

(6) A potential risk to immunosuppressive people.

Entomopathogenic fungi are important as they are virulent, infect by contact, and persist in environment for a long period of time. These can be mass produced in liquid or solid media. Most of the entomopathogenic fungi are facultative parasites which exist as saprotrophs and therefore can be grown apart from living hosts. Few groups are obligate parasites which must be reared in living hosts. Introduction of fungal pathogens into the host population initiates epizootic and prevents or reduces damage by the pest.

The initiation of artificial epizootics has been accomplished for long-term control especially in areas where high humidity condition prevails. There are several defense mechanisms in insect which prevents the penetration and the growth of the fungus. Entomopathogenic fungi can display either a very broad host spectrum like *M. anisopliae, B. bassiana* or have a very narrow host range like *Aschersonia* spp..

29.4.1. Difficulties Encounter in Biological Control

Many of the most important pests are exotic, invasive species that severely impact agriculture, horticulture, forestry, and urban environments. They tend to arrive without their co-evolved parasites, pathogens and predators, and by escaping from these, populations may soar. Importing the natural enemies of these pests may seem a logical move but this may have unintended consequences; regulations may be ineffective and there may be unanticipated effects on biodiversity, and the adoption of the techniques may prove challenging because of a lack of knowledge among farmers and growers.

29.4.2. Side Effects of Biocontrol Approach

Biological control can affect biodiversity through predation, parasitism, pathogenicity, competition, or other attacks on non-target species. An introduced control does not always target only the intended pest species; it can also target native species. In Hawaii during the 1940s parasitic wasps were introduced to control a lepidopteran pest and the wasps are still found there today. This may have a negative impact on the native ecosystem; however, host range and impacts need to be studied before declaring their impact on the environment.

Vertebrate animals tend to be generalist feeders, and seldom make good biological control agents; many of the classic cases of "biocontrol gone awry" involve vertebrates. For example, the cane toad (*Rhinella marina*) was intentionally introduced to Australia to control the greyback cane beetle (*Dermolepida albohirtum*), and other pests of sugar cane. 102 toads were obtained from Hawaii and bred in captivity to increase their numbers until they were released into the sugar cane fields of the tropic north in 1935. It was later discovered that the toads could not jump very high and so were unable to eat the cane beetles which stayed on the upper stalks of the cane plants. However, the toad thrived by feeding on other insects and soon spread very rapidly; it took over native amphibian habitat and brought foreign disease to native toads and frogs, dramatically reducing their populations. Also, when it is threatened or handled, the cane toad releases poison from parotoid glands on its shoulders; native Australian species such as goannas, Tiger snake, dingos and northern quolls that attempted to eat the toad were harmed or killed. However, there has been some recent evidence that native predators are adapting, both physiologically and through changing their behaviour, so in the long run, their populations may recover.

Rhinocyllus conicus, a seed-feeding weevil, was introduced to North America to control exotic musk thistle (*Carduus nutans*) and Canadian thistle (*Cirsium arvense*). However, the weevil also attacks native thistles, harming such species as the endemic Platte thistle (*Cirsium neomexicanum*) by selecting larger plants (which reduced the gene pool), reducing seed production and ultimately threatening the species' survival. Similarly, the weevil *Larinus planus* was also used to try to control the Canadian thistle, but it damaged other thistles as well.

The sturdy and prolific eastern mosquitofish (*Gambusia holbrooki*) is a native of the southeastern United States and was introduced around the world in the 1930s and '40s to feed on mosquito larvae and thus combat malaria. However, it has thrived at the expense of local species, causing a decline of endemic fish and frogs through competition for food resources, as well as through eating their eggs and larvae. In Australia, control of the mosquitofish is the subject of discussion; and researchers stated that "biological population control is well beyond present capabilities.

29.4.3. Obstacles

A potential obstacle to the adoption of biological pest control measures is that growers may prefer to stay with the familiar use of pesticides. However, pesticides have undesired effects, including the development of resistance among pests, and the destruction of natural enemies; these may in turn enable outbreaks of pests of other species than the ones originally targeted, and on crops at a distance from those treated with pesticides. One method of increasing grower adoption of biocontrol methods involves letting them learn by doing, for example showing them simple field experiments, enabling them to observe the live predation of pests, or demonstrations of parasitised pests. In the Philippines, early-season sprays against leaf folder caterpillars were common practice, but growers were asked to follow a 'rule of thumb' of not spraying against leaf folders for the first 30 days after transplanting; participation in this resulted in a reduction of insecticide use by 1/3 and a change in grower perception of insecticide use.

29.5. Strategies of Biological Pest Control

There are three basic strategies for successful biological pest control:

Importation

Importation involves the introduction of a pest's natural enemies to a new locale where they do not occur naturally. Early instances were often unofficial and not based on research, and some introduced species became serious pests themselves.

To be most effective at controlling a pest, a biological control agent requires a colonizing ability which allows it to keep pace with changes to the habitat in space and time. Control is greatest if the agent has temporal persistence so that it can maintain its population even in the temporary absence of the target species, and if it is an opportunistic forager, enabling it to rapidly exploit a pest population.

One of the earliest successes was in controlling *Icerya purchasi* (cottony cushion scale) in Australia, using a predatory insect *Rodolia cardinalis* (the vedalia beetle). This success was repeated in California using the beetle and a parasitoidal fly, *Cryptochaetum iceryae*. Other successful cases include the control of *Antonina graminis* in Texas by *Neodusmetia sangwani* in the 1960s.

Damage from *Hypera postica*, the alfalfa weevil, a serious introduced pest of forage, was substantially reduced by the introduction of natural enemies. 20 years after their introduction the population of weevils in the alfalfa area treated for alfalfa weevil in the Northeastern United States remained 75 percent down.

Alligator weed was introduced to the United States from South America. It takes root in shallow water, interfering with navigation, irrigation, and flood control. The alligator weed flea beetle and two other biological controls were released in Florida, greatly reducing the amount of land covered by the plant. Another aquatic weed, the giant salvinia (*Salvinia molesta*) is a serious pest, covering waterways, reducing water flow and harming native species. Control with the salvinia weevil (*Cyrtobagous salviniae*) and the salvinia stem-borer moth (*Samea multiplicalis)* is effective in warm climates, and in Zimbabwe, a 99% control of the weed was obtained over a two-year period.

Small commercially reared parasitoidal wasps, *Trichogramma ostriniae*, provide limited and erratic control of the European corn borer (*Ostrinia nubilalis*), a serious pest. Careful formulations of the bacterium *Bacillus thuringiensis* are more effective. The O. nubilalis integrated control releasing Tricogramma brassicae (egg parasitoiod) and later Bacillus thuringiensis subs. kurstaki (larvicidae effect) reduce pest damages as better than insecticide treatments.

The population of *Levuana iridescens*, the Levuana moth, a serious coconut pest in Fiji, was brought under control by a classical biological control program in the 1920s.

Augmentation

Augmentation involves the supplemental release of natural enemies that occur in a particular area, boosting the naturally occurring populations there. In inoculative release, small numbers of the control agents are released at intervals to allow them to reproduce, in the hope of setting up longer-term control and thus keeping the pest down to a low level, constituting prevention rather than cure. In inundative release, in contrast, large numbers are released in the hope of rapidly reducing a damaging pest population, correcting a problem that has already arisen. Augmentation can be effective, but is not guaranteed to work, and depends on the precise details of the interactions between each pest and control agent.

An example of inoculative release occurs in the horticultural production of several crops in greenhouses. Periodic releases of the parasitoidal wasp, *Encarsia formosa*, are used to control greenhouse whitefly, while the predatory mite *Phytoseiulus persimilis* is used for control of the two-spotted spider mite.

The egg parasite *Trichogramma* is frequently released inundatively to control harmful moths. New way for inundative releases are now introduced i.e. use of drones. Egg parasitoids are able to find the eggs of the target host by means of several cues. Kairomones were found on moth scales. Similarly, *Bacillus thuringiensis* and other microbial insecticides are used in large enough quantities for a rapid effect. Recommended release rates for *Trichogramma* in vegetable or field crops range from 5,000 to 200,000 per acre (1 to 50 per square metre) per week according to the level of pest infestation. Similarly, nematodes that kill insects (that are entomopathogenic) are released at rates of millions and even billions per acre for control of certain soil-dwelling insect pests.

Conservation

The conservation of existing natural enemies in an environment is the third important step involved in biological pest control. Natural enemies are already adapted to the habitat and to the target pest, and their conservation can be simple and cost-effective, as when nectar-producing crop plants are grown in the borders of rice fields. These provide nectar to support parasitoids and predators of planthopper pests and have been demonstrated to be so effective (reducing pest densities by 10- or even 100-fold) that farmers sprayed 70% less insecticides and enjoyed yields boosted by 5%. Predators of aphids were similarly found to be present in tussock grasses by field boundary hedges in England, but they spread too slowly to reach the centers of fields. Control was improved by planting a meter-wide strip of tussock grasses in field centers, enabling aphid predators to overwinter there.

Cropping systems can be modified to favor natural enemies, a practice sometimes referred to as habitat manipulation. Providing a suitable habitat, such as a shelterbelt, hedgerow, or beetle bank where beneficial insects such as parasitoidal wasps can live and reproduce, can help ensure the survival of populations of natural enemies. Things as simple as leaving a layer of fallen leaves or mulch in place provides a suitable food source for worms and provides a shelter for insects, in turn being a food source for such beneficial mammals as hedgehogs and shrews. Compost piles and stacks of wood can provide shelter for invertebrates and small mammals. Long grass and ponds support amphibians. Not removing dead annuals and non-hardy plants in the autumn allow insects to make use of their hollow stems during winter. In California, prune trees are sometimes planted in grape vineyards to provide an improved overwintering habitat or refuge for a key grape pest parasitoid. The providing of artificial shelters in the form of wooden caskets, boxes or flowerpots is also sometimes undertaken, particularly in gardens, to make a cropped area more attractive to natural enemies. For example, earwigs are natural predators that can be encouraged in gardens by hanging upside-down flowerpots filled with straw or wood wool. Green lacewings can be encouraged by using plastic bottles with an open bottom and a roll of cardboard inside. Birdhouses enable insectivorous birds to nest; the most useful birds can be attracted by choosing an opening just large enough for the desired species.

In cotton production, the replacement of broad-spectrum insecticides with selective control measures such as Bt cotton can create a more favorable environment for natural enemies of cotton pests due to reduced insecticide exposure risk. Such predators or parasitoids can control pests not affected by the Bt protein. Reduced prey quality and abundance associated increased control from Bt cotton can also indirectly decrease natural enemy populations in some cases, but the percentage of pests eaten or parasitized in Bt and non-Bt cotton are often similar.

Advance techniques in biocontrol

An advance technique to biological pest control is the technique of introducing sterile individuals into the native population of some pest. This technique is widely practised with insects where a large number of males sterilized by radiation are released into the environment, where these proceed to compete with the native males for females. Those females that copulate with the sterile males will lay infertile eggs, resulting in a decrease in the size of the pest population. Over time, with repeated introductions of sterile males, this could result in a significant decrease in the size of the targeted pest population. A similar technique has recently been applied to weeds using irradiated pollen, resulting in deformed seeds that do not sprout.

There are several commercial Biological control agents in the form of bacterial, fungal and nematodes are available in the market for control of various diseases and pests (table 59, a, b, c).

Table 59.a. Commercial Biological Control agents against plant pathogens

Biological control Agents (type of organism)	Trade Name	Target pathogens
Agrobacterium radiobacter strain K 84 (bacterium)	Galltrol A	*Agrobacterium tumefacians* (crown gall)
Ageobacterium radiobacter strain K 1026 (bacterium)	Nagol	*Agrobacterium tumefaciens*
Bacillus subtilis (bacterium)	Companion	*Pythium, Fusarium, Phytophthora,*
Steptomyces griseoviridis (bacterium)	Mycostop	*Botrytis, Alternaria, Phomopsis, Pythium, Fusarium, Phytophthora*
Pseudomonas fluorescence (bacterium)	Frostban	*Sclerotinia and Rhizoctonia, Pythium, Alternaria, Ascochyta, Cercospora, Macrophomina, Myrothecium & Ramularia, Downy mildews* and *Powdery mildews*
Pseudomonas aureofaciens strain TX-1 (bacterium)	Bio–jet, spot less	*Pythium, Rhizoctonia* solani
Ampelomyces quisqualis (fungus)	AG10	*Powdery Mildew*

Biological control Agents (type of organism)	Trade Name	Target pathogens
Trichoderma harzianum (fungus)	PlantShield or RootShield	*Pythium, Fusarium, Phytophthora, Rhizoctonia, Botrytis*, Powdery mildew, downy mildew, *Sclerotinia*
Trichoderma virens (fungus)	SoilGard	*Pythium, Fusarium, Phytophthora, Rhizoctonia, Botrytis*, Powdery mildew, downy mildew, *Sclerotinia*
Aspergillus flavus AF36 (fungus)	Alfa guard	*Aspergillus flavus*
Gliocladium catenulatum strain JI446 (fungus)	Prima stop, soil guard	*Pythium, Fusarium, Phytophthora, Rhizoctonia*

b. Commercial Entomopathogenic fungi as Biological Control agents against insect pest.

Product	Fungal biocontrol agent	Target Insect Pest
Mycotal	*Verticillium lecanii*	Sucking pest
Pfr21	*Paecilomyces fumosoroseus*	Whiteflies
Verelac	*Verticillium lecanii*	Sucking pests
Beevicide	*Beauveria bassiana*	Borer type pests
Grubkill	Selected fungus and bacteria	Borers and sucking pest
Pelicide	*Paecilomyces lilacinus*	Effective against nematode
Biologic Bio 1020	*Metarhizium anisopliae*	Mycelium granules as pesticide against arthopods
Bioter	*Verticillium lecanii*	Effective against termites
Brocaril	*Beauveria bassiana*	Wettable powder used against borer pest
Ostrinil	*Beauveria bassiana*	Microgranules of mycelium used against borer pest
Boverol	*Beauveria bassiana*	Dry pellets used against borer pest
Naturalis	*Beaveria bassiana*	Liquid formulation as pesticide
Mycontrol-WP	*Beauveria bassiana*	Wettable powder as pesticide
Betel	*Beauveria brongniartii*	Microgranules of mycelium used as pesticide
Engerlingspilz	*Beauveria brongniartii*	Barley kernels colonized with fungus used as pesticide
Biopath	*Metarhizium anisopliae*	Conidia on a medium used as pesticides
Biomite	*Verticillium lecanii* and other entomopathogenic organisms	Effective against mites
Biogreen	*Metarhizium anisopliae*	Conidia produced on grain used as pesticide against teak pest and plant hoppers.
Naturalis-O and BotaniGard	*Beauveria bassiana*	Effective against whiteflies
Trypae Mix	*Trichoderma* and *Paecilomyce*	Effective against fungal pathogens and nematodes in soil

C. Commercial entomopathogenic nematode (EPN) as biocontrol agent of Insect Pest.

	Commercial name	EPN	Manufacturer	Formulation
1.	Gnat Not	*Steinernema feltiae*	Integrated Bio System, Inc, Greendale, IN	25 milliom IJs (Spong)
2.	GrubStake-Hi	*Heterorhabdits indica*	Integrated Bio Control System, Inc.	25 million IJs (Spong)
3.	Horticultural Scanmask	*S. feltiae*	Biologic Co., Willow Hill, PA.	25 million IJs (Spong)
4.	Nema Shield	*S. feltiae*	Bioworks, Inc., Fairport, NY	50 million IJs (gel)
5.	Nemasys	*S. feltiae*	Becker Underwood, Inc, Ames, IA.	50 million IJs (gel)

EPN, entomopathogenic nematode; IJs, infective juveniles.

Source: Caamano et.al, 2008

30

Scenario of Banned and Restricted Pesticides

The United States of America (USA), European Union (EU), Brazil and China are four of the largest agricultural producers and users of agricultural pesticides in the world. Comparing the inclination and ability of different regulatory agencies to ban or eliminate pesticides that have the most potential for harm to humans and the environment can provide a glimpse into the effectiveness of each nation's pesticide regulatory laws and oversight.

The approval status of more than 500 agricultural pesticides was identified in the USA, EU, Brazil and China and compared between nations. The Pesticides used in one country are Baned in at Least Two of Three Other Agricultural Nations and vice-versa (fig.5).

There are 72, 17, and 11 pesticides approved for outdoor agricultural applications in the USA that are banned or in the process of complete phase out in the EU, Brazil, and China, respectively. Of the pesticides used in USA agriculture in 2016, 322 million pounds were of pesticides banned in the EU, 26 million pounds were of pesticides banned in Brazil and 40 million pounds were of pesticides banned in China. Pesticides banned in the EU account for more than a quarter of all agricultural pesticide use in the USA. The majority of pesticides banned in at least two of these three nations have not appreciably decreased in the USA over the last 25 years and almost all have stayed constant or increased over the last 10 years. Of the pesticides banned in at least two of these nations, many have been implicated in acute pesticide poisonings in the USA and some are further restricted by individual states. The United States Environmental Protection Agency (US EPA) has all but abandoned its use of non-voluntary cancellations in recent years, making pesticide cancellation in the USA largely an exercise that requires consent by the regulated industry.

Pesticide	List	EU	USA	BRA	CHN
2,4-DB		3	3	1	1
Bensulide		1	3	1	0
Chloropicrin		1	3	0	2
Dichlobenil		1	3	1	4
Dicrotophos	W	1	3	1	0
EPTC		1	3	1	0
Norflurazon		1	3	1	0
Oxytetracycline	A	1	3	1	4
Paraquat	R2	1	3	2	1
Phorate	W, R2	1	3	1	2
Streptomycin	A	1	3	1	3
Terbufos	W	1	3	3	1
Tribufos (DEF)		1	3	1	0

Fig. 5: Country Dependent Variation in Ban status of Pesticide.

The first column gives the common pesticide name. The second column indicates whether the pesticide is on an international list of concern

Where, W=World Health Organization (WHO) "extremely" or "highly" hazardous pesticide [79];

R2=Rotterdam Convention Annex III list, Recommended [73];

A=WHO "critically" or "highly" important antibiotics [53].

Columns 3–6 indicate the pesticide status in the European Union (EU), the United States of America (USA), China (CHN) or Brazil (BRA).

1=Banned; 2=In process of complete phase out; 3=Approved; 4=Not approved/voluntarily withdrawn; 0=Not in database/unknown.

Red=banned/phasing out; Green=approved; Orange=Not approved/unknown

30.1. Pesticide Substances Banned and Authorised in the European Union

The European Commission started a review process for all active ingredients used in plant protection products (pesticides) within the European Union in 1992. In a review process based on scientific assessments, each applicant had to prove that a substance could be used safely regarding a set of criteria comprehending human health, environment, ecotoxicity and residues in the food chain. If a pesticide is approved under 91/414/EEC it is placed on Annex

1 of the Directive, and may be used throughout Member States. If it is not approved it might, nevertheless, be granted "essential use" derrogations for some crops and Member States. In other cases, the substance is banned or severely restricted without any space for "essential use" derrogations.

The number of substances banned and authorised has been updated by PAN Europe on the basis of decisions taken by the Standing Committee on the Food Chain and Animal Health in October 2006 and summarised in table.60.

Tabl 60: Status of the pesticide situation in PAN Europe

Status	Number of pesticide substances
Banned	84
Banned with "essential use" derogations	70
Registration withdrawn	362
Authorised	147
Pending	478
Total	1141

(*Source*: DG Health and Consumer Protection)

30.1.1. Substances Banned in the EU Market

Table 61 presents pesticides banned or severely restricted in EU as a consequence of the application of Directive 79/117/EEC, Council Regulation 805/2004/EC and Directive 91/414/EEC (49 in total). Directive 79/117/EEC can be regarded as the predecessor of Directive 91/414/EEC and concerns the prohibition of active substances which, even if applied in an appropriate manner could give rise to harmful effects. Council Regulation 850/2004/EC regards persistent organic pollutants (POPs) and implements the Stockholm Convention in EU.

Council Regulation 304/2003/EC concerns the export and import of dangerous chemicals and implements the Rotterdam Convention. It provides for the notification of importers of any products banned under Directive 79/117/EEC or included in a Prior Informed Consent (PIC) list of the Rotterdam Convention. This Regulation does not itself ban any chemicals, but with amending Directives (1212/2003 and 775/2004) reflects the regulatory status of chemicals under other EU legislation (79/117/EEC, 91/414/EEC, 850/2004/EEC or the Biocides Directive).

Table 61: Pesticides banned or severely restricted in the European Union.

Substance	Use limitation	Regulation/Directive (Regulatory Decision excluding substance from Annex I of Directive 91/414)
Acephate	Ban	03/219/EEC
1,2-Dibromoethane	Ban	79/117/EEC
1,2-Dichloroethane	Ban	79/117/EEC
8-Hydroxyquinokine	Ban	Voted out 04/04/2006
Alachlor	Ban	Voted out 04/04/2006
Alanycarb	Ban	Incomplete dossier
Aldicarb	Ban	03/199/EC
Aldrin	Ban and export ban	79/117/EEC (1991) + 850/2004 (1)
Alkyl mercury compounds	Ban	79/117/EEC
Alkyloxyl and aryl mercury compounds	Ban	79/117/EEC
Amitraz	Severe restriction	775/2004 (04/247)
Ammonium sulphamate	Ban	Votes out 04/04/2006
Atrazine	Ban	04/247/EC
Azafenidin	Ban	02/949/EC
Azamethiphos	Ban	02/949/EC
Azinphos ethyl	Ban	95/276/EC
Binapacryl	Ban	79/117/EEC (1991)
Camphechlor	Ban	79/117/EEC
Captafol	Ban	79/117/EEC (1991)
Carbaryl	Ban	Voted out 29/09/2006
Chlordane	Ban and export ban	79/117/EEC (1981) + 850/2004
Chlorfenapyr	Severe restriction	01/697/EC
Chlorfenprop	Ban	01/697/EC
Chlozolinate	Ban	00/626/EC
Choline, K and Na salts of makeic acid	Ban	79/117/EEC
Cresylic acid	Ban	2005/303/EC
Cyhalothrin	Ban	94/643/EC
DDT	Ban and export ban	79/117/EEC (1986) + 850/2004
Diazinon	Ban	Voted out 29/09/2006
Dichlorvos	Ban	Voted out 29/09/2006
Dicofol containing more than 78% p,p*-Dicofol or 1 g/kg of DDT and DDT related compounds	Severe restriction	79/117/EEC (1991)

Dieldrin	Ban and export ban	79/117/EEC (1981) + 850/2004
Dimethenamide	Ban	Voted out 23/05/2006
Dinoseb, its acetate and salts	Ban	79/117/EEC (1991)
Dinoterb	Ban	Noted in 304/2003 (98/269)
Endosulfan	Ban	05/864/EC
Endrin	Ban and export ban	79/117/EEC (1991) + 850/2004
Ethylene oxide	Ban	79/117/EEC (86/355)
Fenitrothion	Ban	Voted out 14/07/2006
Fenthion	Severe restriction	775/2004 (04/140)
Fentin acetate	Ban	Noted in 304/2003 (02/478)
Fentin hydroxide	Ban	Noted in 304/2003 (02/479)
Fenvalerate	Ban	Noted in 304/2003 (98/270)
Ferbam	Ban	Noted in 304/2003 (95/276)
Flusulfamide	Ban	Application withdrawn
HCH containing less than 99.0% of the gamma isomer	Ban	79/117/EEC (1981)
Heptachlor	Ban and export ban	79/117/EEC (1984) + 850/2004
Hexachlorobenzene	Ban and export ban	79/117/EEC (1981) + 850/2004
Hexaconazole	Ban	Voted out 04/04/2006
Imazamethabenz	Ban	2005/303/EC
Kasugamycin	Ban	2005/303/EC
Lindane (gamma-HCH)	Ban	Noted in 304/2003 (00/801/EC)
Malathion	Ban	Voted out 29/09/2006
Maleic hydrazide and its salts, other than choline, potassium and sodium salts; choline, potassium and of sodium salts maleic hydrazide containing more than 1 mg/kg of freehydrazine expressed on the basis of the acid equivalent	Ban	79/117/EEC (1991)
Mefluidide	Ban	2004/401/EC
Mephospholan	Ban	2004/401/EC
Mercuric oxide	Ban	79/117/EEC
Mercurous chloride	Ban	79/117/EEC
Metalaxyl	Ban	03/308/EC
Methabenzthiazuron	Ban	2006/302
Monolinuron	Ban	00/234/EC
Naled	Ban	2005/788
Nitrofen	Ban	79/117/EEC (1988)

Other inorganic mercury compouds	Ban	79/117/EEC
Oxydemeton-methyl	Ban	Voted out 29/09/2006
Parathion-ethyl	Severe restriction	Noted in 304/2003 (01/520)
Parathion methyl (methyl parathion)	Severe restriction	Noted in 304/2003 (03/166)
p-Chloronitrobenzene	Ban	03/166/EEC
Permethrin	Ban	Noted in 304/2003 (00/817)
Phosalone	Ban	Voted out 14/07/2006
Polyoxin	Ban	2005/303/EC
Propham	Ban	Noted in 304/2003 (96/586)
Pyrazophos	Ban	Noted in 304/2003 (00/233)
Pyradafol	Ban	Application withdrawn
Quintozene	Ban	79/117/EEC (1991) (00/816)
Sodium trithiocarbonate	Ban	Voted out 04/04/2006
Tecnazene	Ban	Noted in 304/2003 (00/725)
Temephos	Ban	00/725/EC
Thiodicarb	Ban	Voted out 14/07/2006
Zineb	Ban	Voted out 29/09/2006
Zucchini yellow mosaic virus	Ban	Application withdrawn

Source: DG Health and Consumer Protection

*Pentachlorophenol: Ban of concentration equal to or greater than 0.1% by mass, except in substances and preparations intended for use in industrial installations: in the treatment of wood; in the impregnation of heavy-duty textiles; as a synthesising and/or processing agent in industrial processes. EU production is banned under Directive 76/769/EEC.

30.1.2. Substances Banned from the EU Market with "Essential use" Derogation

Table 62 presents substances that have been phased out in the EU under the authorisations Directive 91/414/EEC. These active substances have been granted a limited extension of use on certain crops most until the end of December 2007 when they must come off the market.

Pesticides with essential uses have not been given EU-wide approval. Certain countries have been given an extention (or derogation) for one or more essential uses for the active ingredient on specific crops. Member States are expected to explore alternatives to their use and to report on progress in substituting less harmful products or methods.

Table 62: Pesticides withdrawn from the EU with "essential use" derogations

Substance	Type	Decision date	Decision on essential use
Alkyltrimethylbenzyl ammonium chloride	HB	Out 7/03	essential use 2076/2002
2-Aminobutane (aka sec-butylamine)	FU	Out 7/03	essential use 2076/2002
4-CPA (4-chlorophenoxyacetic acid = PCPA)	PG	Out 7/03	essential use 2076/2002
Acifluorfen	HB	Out 7/03	essential use 2076/2002
Anthracene oil	IN, AC, HB, RO	Out 7/03	essential use 2076/2002
Atrazine	HB	Out 10/04	essential use 835/2004, 04/247
Azaconazole	IN,FU	Out 7/03	essential use 835/2004, 2076/2002
Benfuresate	HB	Out 7/03	essential use 2076/2002
Benomyl	FU	Out 05/03	derogation HU (771/2004) essential use 835/04, 02/928
Bensultap	IN	Out 7/03	essential use 835/04, 2076/2002
Bromacil	HB	Out 7/03	essential use 2076/2002
Bromopropylate	AC	Out 7/03	essential use 835/04, 2076/2002
Butylate	HB	Out 07/03	derogation HU (771/2004) essential use 2076/2002
Calcium hydroxide (aka slake lime)		Out 7/03	essential use 2076/2002
Cartap	IN	Out 7/03	essential use 2076/2002
Chinomethionate (aka quinomethionate)	AC,FU	Out 7/03	essential use 2076/2002
Chlorfenvinphos	IN	Out 7/03	essential use 835/04, 2076/2002
Cyanazine	HB	Out 7/03	essential use 835/04, 2076/2002
Cycloate	HB	Out 07/03	derogation HU (771/2004) essential use 835/04, 2076/2002
Dalapon	HB	Out 7/03	essential use 2076/2002
Dichlorprop	HB	Out 7/03	essential use 835/04, 2076/2002
Dikegulac	PG	Out 7/03	essential use 2076/2002
Dimefuron	HB	Out 7/03	essential use 2076/2002
Dinobuton	AC,FU	Out 7/03	essential use 2076/2002
Disodium octaborate tetrahydrate	HB	Out 7/03	essential use 2076/2002
EPTC (ethyl dipropylthiocarbamate)	HB	Out 07/03	derogation HU (771/2004) 2076/2002
Ethion (aka diethion)	IN,AC	Out 7/03	essential use 2076/2002
Fenpropathrin	IN,AC	Out 7/03	essential use 2076/2002
Fenuron	HB	Out 7/03	essential use 2076/2002

Substance	Type	Decision date	Decision on essential use
Flumethralin	PG	Out 7/03	essential use 2076/2002
Fomesafen	HB	Out 7/03	essential use 2076/2002
Fosamine	E	Out 7/03	essential use 2076/2002
Furalaxyl	FU	Out 7/03	essential use 2076/2002
Furathiocarb	IN	Out 7/03	essential use 835/04, 2076/2002
Haloxyfop	HB	Out 7/03	essential use 2076/2002
Heptenophos	IN	Out 7/03	essential use 2076/2002
Hexazinone	HB	Out 7/03	essential use 835/04, 2076/2002
Imazapyr	HB	Out 7/03	essential use 835/04, 2076/2002
Iminoctadine	FU	Out 7/03	essential use 835/04, 2076/2002
Mepronil	FU	Out 7/03	essential use 2076/2002
Methidathion	IN,AC	Out 12/04	essential use 835/04, 2004/129
Metobromuron	HB	Out 7/03	essential use 2076/2002
Metoxuron	HB	Out 7/03	essential use 2076/2002
Naptalam	HB	Out 7/03	essential use 835/04, 2076/2002
Omethoate	IN,AC	Out 7/03	essential use 2076/2002
Orbencarb	HB	Out 7/03	essential use 2076/2002
Oxadixyl	FU	Out 7/03	essential use 2076/2002
Oxine-copper	FU	Out 7/03	essential use 835/04, 2076/2002
Oxycarboxin	FU	Out 7/03	essential use 835/04, 2076/2002
Pebulate	HB	Out 7/03	essential use 2076/2002
Pentanochlor	HB	Out 7/03	essential use 2076/2002
Prometryne	HB	Out 7/03	essential use 835/04, 2076/2002
Pyridafenthion	IN,AC	Out 7/03	essential 2076/2002
Resmethrin	IN	Out 7/03	essential use 2076/2002
Rock powder		Out 7/03	essential use 2076/2002
Sethoxydim	HB	Out 7/03	essential use 2076/2002
Silver nitrate	PG,FU	Out 7/03	essential use 2076/2002
Simazine	HB	Out 10/04	essential use 835/2004, 04/247
Sodium dimethylarsinate	RO	Out 7/03	essential use 2076/2002
Sodium monochloroacetate	HB	Out 7/03	essential use 2076/2002
Sulfotep	IN,AC	Out 7/03	essential use 2076/2002
Tar acids	IN	Out 7/03	essential use 2076/2002
Temephos	IN	Out 7/03	essential 2076/2002
Terbacil	HB	Out 7/03	essential use 835/04, 2076/2002
Terbufos	IN	Out 7/03	essential use 835/04, 2076/2002
Terbutryn	HB	Out 7/03	essential use 835/04, 2076/2002
Tetradifon	AC,IN	Out 7/03	essential use 2076/2002
Thiocyclam	IN	Out 7/03	essential use 835/04, 2076/2002
Triazophos	IN,AC	Out 7/03	essential use 2076/2002
Triforine	FU,AC	Out 7/03	essential use 835/04, 2076/2002

Source: DG Health and Consumer Protection, Pesticide Safety Directorate, Agrow magazine Abbreviations "for essential use type" : AT= Attractant; ST= Soil treatment; PG= Plant Growth Regulator; HB= Herbicide; FU= Fungicide; IN= Insecticide; NE= Nematicide; AC= Acaricide; BA= Bactericide; RE= Repellent; MO= Molluscicide; RO= Rodenticide; E= External parasite; Bio= Bio-agent; GR=Growth regulator; PA= Plant Activator

30.1.3. Substances Where Registration Has Been Withdrawn Under Directive 91/414/Eec And Its Amendments

A substancial number of substances (362 in total) are no longer used in Member States because no manufacturer applied for registration. Table 63 lists substances that have failed to make it onto Annex 1 of Directive 91/414/EEC and have been withdrawn from the market. In some cases this may be because the manufacturer has not supplied the data required to extend registration.

Table 63: Substances where registration has been withdrawn under Directive 91/414/EEC and its amendments

Substance	Category	91/414 stat	Remark
(4E-7Z)-4,7-Tridecadien-1-yl-acetate	*AT*	*Out 12/04*	*2004/129*
(4Z-9Z)-7,9-Dodecadien-1-ol	*AT*	*Out 12/04*	*2004/129*
(E)-10-Dodecenyl acetate	*AT*	*Out 12/04*	*2004/129*
(Z)-3-Methyl-6-isopropenyl- -3,4-decadien-1-yl	*AT*	*Out 12/04*	*2004/129*
(Z)-3-Methyl-6-isopropenyl--9-decen-1-yl acetate	*AT*	*Out 12/04*	*2004/129*
(Z)-7-Tetradecanole	*AT*	*Out 12/04*	*2004/129*
1,3-Dichloropropene (cis)	*ST*	*Out 7/03*	*2076/2002/EC*
1,3-Diphenyl urea	*PG*	*Out 7/03*	*2076/2002/EC*
2,3,6-TBA	*HB*	*Out 7/03*	*2076/2002/EC*
2-Benzyl-4-chlorophenol	*FU*	*Out 7/03*	*2076/2002/EC*
3,7-Dimethyl-2,6-octadienal	*AT*	*Out 12/04*	*2004/129*
4-Chloro-3-methylphenol	*FU*	*Out 12/04*	*2004/129*
7,8-Epoxi-2-methyl-octadecane	*AT*	*Out 12/04*	*2004/129*
7-Methyl-3-methylene-7--octene-1-yl-propionate	*AT*	*Out 12/04*	*2004/129*
Acephate	*IN*	*out 09/03*	*03/219/EC*
Acridinic bases	*RE*	*Out 12/04*	*2004/129*
Agrotis segetum granulosis virus	*IN*	*Out 12/04*	*2004/129*
Aldicarb	*NE, IN, AC*	*out 09/04*	*03/199/EC*
Aldimorph	*FU*	*Out 7/03*	*2076/2002/EC*
Alkyldimethylbenzyl ammonium chloride	*HB*	*Out 12/04*	*2004/129*
Alkyldimethylethylbenzylammonium chloride	*HB*	*Out 12/04*	*2004/129*
Alkyltrimethyl ammonium chloride	*BA,FU*	*Out 7/03*	*2076/2002/EC*
Allethrin	*IN*	*Out 7/03*	*2076/2002/EC*

Substance	Category	91/414 stat	Remark
Alloxydim	*HB*	*Out 7/03*	*2076/2002/EC*
Allyl alcohol	*HB*	*Out 7/03*	*2076/2002/EC*
Ametryn	*HB*	*Out 7/03*	*2076/2002/EC*
Amitraz	*AC,IN*	*out 08/04*	*04/141/EC*
Ammonium hydroxide	*FU*	*Out 12/04*	*2004/129*
Ammonium sulphate	*HB*	*Out 12/04*	*2004/129*
Ampropylofos	*FU*	*Out 7/03*	*2076/2002/EC*
Ancymidol	*PG*	*Out 7/03*	*2076/2002/EC*
Anilazine	*FU*	*Out 7/03*	*2076/2002/EC*
Aschersonia aleyrodis	*IN*	*Out 12/04*	*2004/129*
Azamethiphos	*IN*	*Out 7/03*	*2076/2002/EC*
Azinphos ethyl	*IN, AC*	*out 1/96*	*95/276/EC*
Aziprotryne	*HB*	*Out 7/03*	*2076/2002/EC*
Barban	*HB*	*Out 7/03*	*2076/2002/EC*
Barium fluosilicate	*IN*	*Out 7/03*	*2076/2002/EC*
Barium nitrate	*RE*	*Out 12/04*	*2004/129*
Barium polysulphide	*IN,FU*	*Out 7/03*	*2076/2002/EC*
Benazolin	*HB*	*Out 7/03*	*2076/2002/EC*
Bendiocarb	*IN*	*Out 7/03*	*2076/2002/EC*
Benodanil	*FU*	*Out 7/03*	*2076/2002/EC*
Bensulide	*HB*	*Out 7/03*	*2076/2002/EC*
Bentaluron	*FU*	*Out 7/03*	*2076/2002/EC*
Benzalkonium chloride	*HB*	*Out 7/03*	*2076/2002/EC*
Benzoximate	*AC*	*Out 7/03*	*2076/2002/EC*
Benzoylprop	*HB*	*Out 7/03*	*2076/2002/EC*
Benzthiazuron	*HB*	*Out 7/03*	*2076/2002/EC*
Bioallethrin	*IN*	*Out 7/03*	*2076/2002/EC*
Bioresmethrin	*IN*	*Out 7/03*	*2076/2002/EC*
Bitumen	*Pruning*	*Out 7/03*	*2076/2002/EC*
Boric acid	*IN*	*Out 12/04*	*2004/129*
Brandol (hydroxynonyl-2,6-dinitrobenzene)	*FU*	*Out 7/03*	*2076/2002/EC*
Bromethalin	*RO*	*Out 12/04*	*2004/129*
Bromocyclen	*IN*	*Out 7/03*	*2076/2002/EC*
Bromofenoxim	*HB*	*Out 7/03*	*2076/2002/EC*
Bromophos	*IN*	*Out 7/03*	*2076/2002/EC*
Bromophos-ethyl	*IN*	*Out 7/03*	*2076/2002/EC*
Bronopol	*FU,BA*	*Out 7/03*	*2076/2002/EC*
Butachlor	*HB*	*Out 7/03*	*2076/2002/EC*
Butocarboxim	*IN*	*Out 7/03*	*2076/2002/EC*
Butoxycarboxim	*IN.AC*	*Out 7/03*	*2076/2002/EC*
Calciferol	*RO*	*Out 12/04*	*2004/129*

Substance	Category	91/414 stat	Remark
Calcium carbonate (aka chalk)		*Out 7/03*	*2076/2002/EC*
Calcium oxide (quick lime)		*Out 7/03*	*2076/2002/EC*
Calcium phosphate	*RO*	*Out 12/04*	*2004/129*
Carbon disulfide	*IN,NE*	*Out 7/03*	*2076/2002/EC*
Carbophenothion	*IN.AC*	*Out 7/03*	*2076/2002/EC*
Cetrimide	*HB*	*Out 7/03*	*2076/2002/EC*
Chlomethoxyfen	*HB*	*Out 7/03*	*2076/2002/EC*
Chloral-bis-acylal	*PG*	*Out 7/03*	*2076/2002/EC*
Chloral-semi-acetal	*HB*	*Out 7/03*	*2076/2002/EC*
Chloramben	*HB*	*Out 7/03*	*2076/2002/EC*
Chlorbromuron	*HB*	*Out 7/03*	*2076/2002/EC*
Chlorbufam	*HB*	*Out 7/03*	*2076/2002/EC*
Chloretazate	*PG*	*Out 7/03*	*2076/2002/EC*
Chlorfenprop	*HB*	*Out 7/03*	*2076/2002/EC*
Chlorfenson (aka chlorfenizon)	*IN,AC*	*Out 7/03*	*2076/2002/EC*
Chlorfluazuron	*IN*	*Out 7/03*	*2076/2002/EC*
Chlorflurenol (chlorflurecol)	*PG*	*Out 12/04*	*2004/129/EC*
Chlorhydrate of poly(iminino imido biguanidine)	*FU,BA*	*Out 12/04*	*2004/129*
Chlormephos	*IN*	*Out 7/03*	*2076/2002/EC*
Chlorobenzilate	*AC*	*Out 7/03*	*2076/2002/EC*
Chlorophylline	*FU,BA*	*Out 12/04*	*2004/129*
Chloropropylate	*AC*	*Out 7/03*	*2076/2002/EC*
Chloroxuron	*HB*	*Out 7/03*	*2076/2002/EC*
Chlorphonium chloride	*PG*	*Out 7/03*	*2076/2002/EC*
Chlorthiamid	*HB*	*Out 7/03*	*2076/2002/EC*
Chlorthiophos	*IN*	*Out 7/03*	*2076/2002/EC*
Chlozolinate	*FU*	*out 4/02*	*00/626/EC*
Cholecalciferol	*RO*	*Out 12/04*	*2004/129*
Choline chloride	*RO*	*Out 12/04*	*2004/129*
Cinosulfuron	*HB*	*Out 12/04*	*2004/129/EC*
Clofencet	*PG*	*Out 12/04*	*2004/129/EC*
Coumachlor	*RO*	*Out 12/04*	*2004/129*
Coumafuryl	*RO*	*Out 12/04*	*2004/129*
Coumatetralyl	*RO*	*Out 12/04*	*2004/129*
Crimidine	*RO*	*Out 12/04*	*2004/129*
Cufraneb	*FU*	*Out 7/03*	*2076/2002/EC*
Cyanides: calcium, hydrogen, sodium	*IN,RO*	*Out 12/04*	*2004/129*
Cycluron	*HB*	*Out 7/03*	*2076/2002/EC*
Cyhalothrin	*IN*	*out 3/95*	*94/643/EC*
Cyprofuram	*FU*	*Out 7/03*	*2076/2002/EC*
DADZ (zinc-dimethylditiocarbamate)	*RE*	*Out 7/03*	*2076/2002/EC*

Substance	Category	91/414 stat	Remark
delta-endotoxin of Bacillus thuringiensis	*IN*	*Out 7/03*	*2076/2002/EC*
Demeton-S-methyl	*IN,AC*	*Out 7/03*	*2076/2002/EC*
Demeton-S-methyl sulphone	*IN*	*Out 7/03*	*2076/2002/EC*
Desmetryne	*HB*	*Out 7/03*	*2076/2002/EC*
Diafenthiuron	*IN,AC*	*Out 7/03*	*2076/2002/EC*
Dialifos	*IN,AC*	*Out 7/03*	*2076/2002/EC*
Di-allate	*HB*	*Out 7/03*	*2076/2002/EC*
Diammonium phosphate	*AC*	*Out 7/03*	*2076/2002/EC*
Dichlofenthion	*IN*	*Out 7/03*	*2076/2002/EC*
Dichlofluanid	*FU*	*Out 7/03*	*2076/2002/EC*
Dichlone	*FU*	*Out 7/03*	*2076/2002/EC*
Diclobutrazol	*FU*	*Out 7/03*	*2076/2002/EC*
Dicrotophos	*IN,AC*	*Out 7/03*	*2076/2002/EC*
Dicyclopentadiene	*PG*	*Out 7/03*	*2076/2002/EC*
Dienochlor	*AC*	*Out 7/03*	*2076/2002/EC*
Diethatyl (-ethyl)	*HB*	*Out 7/03*	*2076/2002/EC*
Difenoxuron	*HB*	*Out 7/03*	*2076/2002/EC*
Difenzoquat	*HB*	*Out 7/03*	*2076/2002/EC*
Difethialone	*RO*	*Out 12/04*	*2004/129*
Dimefox	*IN*	*Out 7/03*	*2076/2002/EC*
Dimepiperate	*HB*	*Out 7/03*	*2076/2002/EC*
Dimethirimol	*FU*	*Out 7/03*	*2076/2002/EC*
Dimexano	*HB*	*Out 7/03*	*2076/2002/EC*
Dinitramine	*HB*	*Out 7/03*	*2076/2002/EC*
Dinoterb	*HB*	*out 10/98*	*98/269/EC*
Dioctyldimethyl ammonium chloride	*FU,BA*	*Out 12/04*	*2004/129*
Dioxacarb	*IN*	*Out 7/03*	*2076/2002/EC*
Dioxathion	*IN*	*Out 7/03*	*2076/2002/EC*
Diphacinone	*RO*	*Out 12/04*	*2004/129*
Diphenamid (aka difenamide)	*HB*	*Out 7/03*	*2076/2002/EC*
Disulfoton	*IN*	*Out 7/03*	*2076/2002/EC*
Ditalimfos	*FU*	*Out 7/03*	*2076/2002/EC*
DNOC	*IN,AC,FU,HB*	*out 6/00*	*99/164/EC*
Drazoxolon	*FU*	*Out 7/03*	*2076/2002/EC*
Endothal	*HB*	*Out 7/03*	*2076/2002/EC*
Etacelasil	*PG*	*Out 7/03*	*2076/2002/EC*
Ethanethiol	*RO*	*Out 12/04*	*2004/129*
Ethidimuron (aka sulfodiazol)	*HB*	*Out 7/03*	*2076/2002/EC*
Ethiofencarb	*IN*	*Out 7/03*	*2076/2002/EC*
Ethirimol	*FU*	*Out 7/03*	*2076/2002/EC*
Ethoate-methyl	*IN*	*Out 7/03*	*2076/2002/EC*

Substance	Category	91/414 stat	Remark
Ethylhexanoate	*FU,BA*	*Out 12/04*	*2004/129*
Etrimfos	*IN,AC*	*Out 7/03*	*2076/2002/EC*
Fenaminosulf	*FU*	*Out 7/03*	*2076/2002/EC*
Fenazaflor	*AC*	*Out 7/03*	*2076/2002/EC*
Fenfuram	*FU*	*Out 7/03*	*2076/2002/EC*
Fenoprop	*PG,HB*	*Out 7/03*	*2076/2002/EC*
Fenothiocarb	*IN,AC*	*Out 7/03*	*2076/2002/EC*
Fenoxaprop	*HB*	*Out 7/03*	*2076/2002/EC*
Fenpiclonil	*FU*	*Out 7/03*	*2076/2002/EC*
Fenridazon	*PG*	*Out 7/03*	*2076/2002/EC*
Fenson (aka fenizon)	*AC*	*Out 7/03*	*2076/2002/EC*
Fenthion	*IN*	*out 08/04*	*04/140/EC*
Fenthiosulf	*IN*	*Out 7/03*	*2076/2002/EC*
Fentin acetate	*FU,HB*	*out 12/02*	*02/478/EC*
Fentin hydroxide	*FU,HB*	*out 12/02*	*02/479/EC*
Fenvalerate	*E*	*out 4/99*	*98/270/EC*
Ferbam	*E*	*out 1/96*	*95/276/EC*
Flamprop	*HB*	*Out 7/03*	*2076/2002/EC*
Flamprop-M	*HB*	*Out 12/04*	*2004/129/EC*
Flocumafen	*RO*	*Out 12/04*	*2004/129*
Fluazifop	*HB*	*Out 7/03*	*2076/2002/EC*
Flubenzimine	*AC*	*Out 7/03*	*2076/2002/EC*
Flucycloxuron	*AC*	*Out 7/03*	*2076/2002/EC*
Flucythrinate	*IN*	*Out 7/03*	*2076/2002/EC*
Flumequine	*BA*	*Out 7/03*	*2076/2002/EC*
Fluoroacetamide	*RO*	*Out 12/04*	*2004/129*
Fluorodifen	*HB*	*Out 7/03*	*2076/2002/EC*
Fluoroglycofene	*HB*	*Out 7/03*	*2076/2002/EC*
Flupoxam	*HB*	*Out 7/03*	*2076/2002/EC*
Flurenol (flurecol)	*HB*	*Out 12/04*	*2004/129/EC*
Fluridone	*HB*	*Out 7/03*	*2076/2002/EC*
Fonofos	*IN*	*Out 7/03*	*2076/2002/EC*
Formothion	*IN,AC*	*Out 7/03*	*2076/2002/EC*
Fosthietan	*NE*	*Out 7/03*	*2076/2002/EC*
Furconazole	*FU*	*Out 7/03*	*2076/2002/EC*
Furfural		*Out 7/03*	*2076/2002/EC*
Furmecyclox	*FU*	*Out 7/03*	*2076/2002/EC*
Gentian violet	*BA*	*Out 7/03*	*2076/2002/EC*
Halfenprox (aka brofenprox)	*AC*	*Out 7/03*	*2076/2002/EC*
Hexachlorophene	*FU*	*Out 7/03*	*2076/2002/EC*
Hexaflumeron	*IN*	*Out 12/04*	*2004/129/EC*
Hydramethylnon	*IN*	*Out 7/03*	*2076/2002/EC*

Substance	Category	91/414 stat	Remark
Hydroxy-MCPA	*PG*	*Out 7/03*	*2076/2002/EC*
Hydroxyphenyl-salicylamide	*FU*	*Out 7/03*	*2076/2002/EC*
Imazethabenz	*HB*	*Out 7/03*	*2076/2002/EC*
Imazethapyr	*HB*	*Out 12/04*	*2004/129/EC*
Iodofenphos	*IN*	*Out 7/03*	*2076/2002/EC*
Isazofos	*IN*	*Out 7/03*	*2076/2002/EC*
Isocarbamide	*HB*	*Out 7/03*	*2076/2002/EC*
Isofenphos	*IN*	*Out 7/03*	*2076/2002/EC*
Isolan	*IN*	*Out 7/03*	*2076/2002/EC*
Isopropalin	*HB*	*Out 7/03*	*2076/2002/EC*
Isoprothiolane	*HB*	*Out 7/03*	*2076/2002/EC*
Isoval	*RO*	*Out 12/04*	*2004/129*
Isoxathion	*IN*	*Out 7/03*	*2076/2002/EC*
Karbutilate	*HB*	*Out 7/03*	*2076/2002/EC*
Kinoprene	*IN*	*Out 7/03*	*2076/2002/EC*
Lactic acid	*PG*	*Out 12/04*	*2004/129*
Lauryldimethylbenzylammonium bromide	*FU,BA*	*Out 12/04*	*2004/129*
Lauryldimethylbenzylammonium chloride	*FU,BA*	*Out 12/04*	*2004/129*
Lime phosphate	*PG*	*Out 12/04*	*2004/129*
Lindane	*IN,RO*	*out 6/02*	*00/801/EC*
Mamestra brassica nuclear polyhedrosis virus	*IN*	*Out 12/04*	*2004/129*
Mancopper	*FU*	*Out 7/03*	*2076/2002/EC*
Mecarbam	*IN,AC*	*Out 7/03*	*2076/2002/EC*
Mefenacet	*HB*	*Out 7/03*	*2076/2002/EC*
Mefluidide	*PG*	*Out 10/04*	*2004/401/EC*
Mephospholan	*IN*	*Out 7/03*	*2076/2002/EC*
Merphos (aka tributylphosphorotrithioite)	*PG*	*Out 7/03*	*2076/2002/EC*
Methacrifos	*IN*	*Out 7/03*	*2076/2002/EC*
Methazole	*HB*	*Out 7/03*	*2076/2002/EC*
Methfuroxam	*FU*	*Out 7/03*	*2076/2002/EC*
Methoprene	*IN*	*Out 7/03*	*2076/2002/EC*
Methoprothryne	*HB*	*Out 7/03*	*2076/2002/EC*
Methoxychlor	*IN*	*Out 7/03*	*2076/2002/EC*
Methylenebisthiocyanate	*FU*	*Out 7/03*	*2076/2002/EC*
Methylisothiocyanate	*FU, NE, HB, IN*	*Out 7/03*	*2076/2002/EC*
Methylnaphthylacetamide	*PG*	*Out 7/03*	*2076/2002/EC*
Methylnaphthylacetic acid	*PG*	*Out 7/03*	*2076/2002/EC*
Methyl-trans-6-nonenoate	*AT*	*Out 12/04*	*2004/129*

Substance	Category	91/414 stat	Remark
Metolachlor	*HB*	*Out 7/03*	*2076/2002/EC*
Metsulfovax	*FU*	*Out 7/03*	*2076/2002/EC*
Mevinphos	*IN,AC*	*Out 7/03*	*2076/2002/EC*
Monalide	*HB*	*Out 7/03*	*2076/2002/EC*
Monocrotophos	*AC,IN*	*Out 7/03*	*2076/2002/EC*
Monolinuron	*HB*	*out 9/01*	*00/234/EC*
Monuron	*HB*	*Out 7/03*	*2076/2002/EC*
MSMA (methyl arsonic acid)	*HB*	*Out 7/03*	*2076/2002/EC*
Nabam	*FU, HB*	*Out 7/03*	*2076/2002/EC*
Naphtalene	*RE*	*Out 12/04*	*2004/129*
Naphtylacetic acid hydrazide	*PG*	*Out 7/03*	*2076/2002/EC*
Neburon	*HB*	*Out 7/03*	*2076/2002/EC*
Nitralin	*HB*	*Out 7/03*	*2076/2002/EC*
Nitrogen	*IN*	*Out 12/04*	*2004/129*
Nitrothal	*FU*	*Out 7/03*	*2076/2002/EC*
Nonylphenol ether polyoxyethyleneglycol	*PG*	*Out 7/03*	*2076/2002/EC*
Nonylphenol ethoxylate	*FU*	*Out 7/03*	*2076/2002/EC*
Norflurazon	*HB*	*Out 7/03*	*2076/2002/EC*
Noruron	*HB*	*Out 7/03*	*2076/2002/EC*
Nuarimol	*FU*	*Out 12/04*	*2004/129/EC*
Octhilinone	*FU*	*Out 7/03*	*2076/2002/EC*
Octyldecyldimethyl ammonium chloride	*FU,BA*	*Out 12/04*	*2004/129*
Ofurace	*FU*	*Out 7/03*	*2076/2002/EC*
Onion extract		*Out 12/04*	*2004/129*
Oxytetracycline	*BA*	*Out 7/03*	*2076/2002/EC*
Papaine	*RO*	*Out 12/04*	*2004/129*
Paraformaldehyde	*IN*	*Out 7/03*	*2076/2002/EC*
Parathion-ethyl	*IN,AC*	*out 1/03*	*01/520/EC*
Parathion-methyl	*IN,RE*	*Out*	*03/166/EC*
p-Chloronitrobenzene	*IN*	*Out 7/03*	*2076/2002/EC*
p-Cresyl acetate	*RE*	*Out 12/04*	*2004/129*
p-Dichlorobenzene	*RO*	*Out 12/04*	*2004/129*
Pentachlorophenol	*HB*	*Out 7/03*	*2076/2002/EC*
Perfluidone	*HB*	*Out 7/03*	*2076/2002/EC*
Permethrin	*IN*	*out 12/03*	*00/817/EC*
Phenols	*HB,ST*	*Out 7/03*	*2076/2002/EC*
Phenothrin	*IN*	*Out 7/03*	*2076/2002/EC*
Phenthoate	*IN*	*Out 7/03*	*2076/2002/EC*
Pherodim	*AT*	*Out 12/04*	*2004/129*
Phorate	*IN*	*Out 7/03*	*2076/2002/EC*
Phosametine	*HB*	*Out 7/03*	*2076/2002/EC*

Substance	Category	91/414 stat	Remark
Phosphamidon	*IN,AC*	*Out 7/03*	*2076/2002/EC*
Phosphoric acid		*Out 12/04*	*2004/129*
Pirimiphos-ethyl	*IN*	*Out 7/03*	*2076/2002/EC*
Plant oils / Soybean oil, epoxylated	*IN*	*Out 12/04*	*2004/129*
Potassium silicate		*Out 7/03*	*2076/2002/EC*
Potassium sorbate	*FU*	*Out 12/04*	*2004/129*
Pretilachlor	*HB*	*Out 12/04*	*2004/129/EC*
Primisulfuron	*HB*	*Out 12/04*	*2004/129/EC*
Profenofos	*IN*	*Out 7/03*	*2076/2002/EC*
Promecarb	*IN*	*Out 7/03*	*2076/2002/EC*
Pronumone	*AT*	*Out 12/04*	*2004/129*
Propazine	*HB*	*Out 7/03*	*2076/2002/EC*
Propetamphos	*IN*	*Out 7/03*	*2076/2002/EC*
Propham	*HB PG*	*out 4/97*	*96/586/EC*
Propionic acid	*FU,BA*	*Out 12/04*	*2004/129*
Propoxur	*IN*	*Out 7/03*	*2076/2002/EC*
Propyl-3-t-butylphenoxyacetate	*PG*	*Out 7/03*	*2076/2002/EC*
Prothiofos	*IN*	*Out 7/03*	*2076/2002/EC*
Prothoate	*IN,AC*	*Out 7/03*	*2076/2002/EC*
Pyraclofos	*AC,IN*	*Out 7/03*	*2076/2002/EC*
Pyranocumarin	*RO*	*Out 12/04*	*2004/129*
Pyrazophos	*FU*	*out 9/01*	*00/233/EC*
Pyrazoxyfen	*HB*	*Out 7/03*	*2076/2002/EC*
Pyrifenox	*FU*	*Out 7/03*	*2076/2002/EC*
Pyroquilone	*FU*	*Out 7/03*	*2076/2002/EC*
Quinalphos	*IN*	*Out 7/03*	*2076/2002/EC*
Quinclorac	*HB*	*Out 12/04*	*2004/129/EC*
Quintozene	*FU*	*out 6/02*	*00/816/EC*
Quizalofop	*HB*	*Out 7/03*	*2076/2002/EC*
Scilliroside	*RO*	*Out 12/04*	*2004/129*
Sebacic acid	*RE*	*Out 12/04*	*2004/129*
Secbumeton	*HB*	*Out 7/03*	*2076/2002/EC*
Seconal (aka 5-allyl-5-(1'-methylbutyl) barbituric acid)	*OT*	*Out 7/03*	*2076/2002/EC*
Serricornin	*AT*	*Out 12/04*	*2004/129*
Siduron	*HB*	*Out 7/03*	*2076/2002/EC*
Sodium arsenite	*FU,IN*	*Out 7/03*	*2076/2002/EC*
Sodium carbonate		*Out 12/04*	*2004/129*
Sodium chloride	*HB*	*Out 12/04*	*2004/129*
Sodium diacetoneketogulonate	*PG*	*Out 7/03*	*2076/2002/EC*
Sodium dichlorophenate	*FU,BA*	*Out 7/03*	*2076/2002/EC*
Sodium dimethyldithiocarbamate	*FU*	*Out 7/03*	*2076/2002/EC*

Substance	Category	91/414 stat	Remark
Sodium dioctyl sulfosuccinate	*AC*	*Out 7/03*	*2076/2002/EC*
Sodium fluosilicate	*IN*	*Out 7/03*	*2076/2002/EC*
Sodium hydroxide	*HB*	*Out 12/04*	*2004/129*
Sodium o-benzyl-p-chlorphenoxide	*FU*	*Out 12/04*	*2004/129*
Sodium pentaborate	*PG*	*Out 7/03*	*2076/2002/EC*
Sodium propionate	*FU*	*Out 12/04*	*2004/129*
Sodium p-t-amylphenate	*FU,BA*	*Out 7/03*	*2076/2002/EC*
Sodium p-t-amylphenoxide	*FU*	*Out 12/04*	*2004/129*
Sodium tetraborate	*IN,HB,MO*	*Out 12/04*	*2004/129*
Sodium tetrathiocarbamate	*NE*	*Out 7/03*	*2076/2002/EC*
Sodium thiocyanate	*HB*	*Out 7/03*	*2076/2002/EC*
Soybean extract		*Out 12/04*	*2004/129*
Streptomycin	*BA*	*Out 12/04*	*2004/129/EC*
Strychnine	*RO*	*Out 12/04*	*2004/129*
Sulprofos	*IN*	*Out 7/03*	*2076/2002/EC*
Tar oils	*IN,HB*	*Out 12/04*	*2004/129*
TCA	*HB*	*Out 7/03*	*2076/2002/EC*
TCMTB	*FU*	*Out 7/03*	*2076/2002/EC*
Tebutam (aka butam)	*HB*	*Out 7/03*	*2076/2002/EC*
Tebuthiuron	*HB*	*Out 7/03*	*2076/2002/EC*
Tecnazene	*FU,PG*	*out 1/03*	*00/725/EC*
Terbumeton	*HB*	*Out 7/03*	*2076/2002/EC*
Tetrachlorvinphos	*IN*	*Out 7/03*	*2076/2002/EC*
Tetramethrin	*IN*	*Out 7/03*	*2076/2002/EC*
Tetrasul	*AC*	*Out 7/03*	*2076/2002/EC*
Thallium sulphate	*RO*	*Out 12/04*	*2004/129*
Thiazafluron	*HB*	*Out 7/03*	*2076/2002/EC*
Thiazopyr	*HB*	*Out 7/03*	*2076/2002/EC*
Thiofanox	*IN*	*Out 7/03*	*2076/2002/EC*
Thiometon	*IN,AC*	*Out 7/03*	*2076/2002/EC*
Thionazin	*NE*	*Out 7/03*	*2076/2002/EC*
Thiophanate (ethyl)	*FU*	*Out 7/03*	*2076/2002/EC*
Thiourea	*RO*	*Out 12/04*	*2004/129*
Tiocarbazil	*HB*	*Out 7/03*	*2076/2002/EC*
Tolylphtalam	*PG*	*Out 7/03*	*2076/2002/EC*
Tomato mosaic virus	*VI*	*Out 12/04*	*2004/129*
Tralomethrin	*IN*	*Out 7/03*	*2076/2002/EC*
trans-6-Nonen-1-ol	*AT*	*Out 12/04*	*2004/129*
Triadimefon	*FU*	*Out 12/04*	*2004/129/EC*
Triapenthenol	*PG*	*Out 7/03*	*2076/2002/EC*
Triazamate	*IN*	*To be withdrawn*	
Triazbutyl	*FU*	*Out 7/03*	*2076/2002/EC*

Substance	Category	91/414 stat	Remark
Tribufos (s,s,s-tributyl-phosphorotrithioate)	*PG*	*Out 7/03*	*2076/2002/EC*
Tributyltinoxyde	*FU*	*Out 7/03*	*2076/2002/EC*
Trichloronate	*IN*	*Out 7/03*	*2076/2002/EC*
Tridemorph	*FU*	*Out 12/04*	*2004/129/EC*
Tridiphane	*HB*	*Out 7/03*	*2076/2002/EC*
Trietazine	*HB*	*Out 7/03*	*2076/2002/EC*
Trifenmorph	*MO*	*Out 7/03*	*2076/2002/EC*
Trimedlure	*AT*	*Out 12/04*	*2004/129*
Trioxymethylen	*FU*	*Out 7/03*	*2076/2002/EC*
Validamycin	*FU*	*Out 7/03*	*2076/2002/EC*
Vernolate	*HB*	*Out 7/03*	*2076/2002/EC*
Zineb	*FU*	*out 3/03*	*01/245/EC*

Abbreviations "for substance category" : AT= Attractant; ST= Soil treatment; PG= Plant Growth Regulator; HB= Herbicide; FU= Fungicide; IN= Insecticide; NE= Nematicide;AC= Acaricide; BA= Bactericide; RE= Repellent; MO= Molluscicide;RO= Rodenticide; E= External parasite; Bio= Bio-agent; GR=Growth regulator; PA= Plant Activator

30.1.4. Substances Authorised in EU

Since the beginning of the revision process, 147 substances were approved and placed in Annex 1 of Directive 91/414/EEC. Table 64 depicts the substances approved and the category.

Table 64: Substances authorised in the EU market

Substance	EU Decision	Use
1-Methyl-cyclopropene	06/19/EC	PG
2,4-D	01/103/EC	HB, PG
2,4-DB	03/31/EC	HB
Acetamiprid	04/99/EC	IN
Acibenzolar-S-Methyl (benzothiadiazole)	01/87/EC	PA
alpha-Cypermethrin (aka alphamethrin)	04/58/EC	IN
Amitrole (aminotriazole)	01/21/EC	HB
Ampelomyces quisqualis	05/2/EC	Bio
Azimsulfuron	99/80/EC	HB
Azoxystrobin	98/47/EC	FU
Benalaxyl	04/58/EC	FU
Bentazone	00/68/EC	HB
Benzoic acid	04/30/EC	BA, FU, OT
beta-Cyfluthrin	03/31/EC	IN
Bifenazate	05/58/EC	AC
Bromoxynil	04/58/EC	HB
Captan		FU

Substance	EU Decision	Use
Carfentrazone-ethyl	03/68/EC	HB
Chlorothalonil	05/53/EC	FU
Chlorpropham	04/20/EC	PG, HB
Chlorpyrifos	05/72/EC	IN
Chlorpyrifos-methyl	05/72/EC	IN
Clothianidin	06/41/EC	IN
Cinidon ethyl	02/64/EC	HB
Clodinafop	06/39/EC	HB
Clopyralid	06/64/EC	HB
Chlortoluron	05/53/EC	HB
Coniothyrium minitans	03/79/EC	FU
Cyazofamid	03/23/EC	FU
Cyclanilide	01/87/EC	PG
Cyfluthrin	03/31/EC	IN, AC
Cyhalofop-butyl	02/64/EC	HB
Cypermethrin	05/53/EC	IN
Cyprodinil	06/64/EC	FU
Daminozide	05/53/EC	GR
Deltamethrin	03/5/EC	IN
Desmedipham	04/58/EC	HB
Dichlorprop-P	06/74/EC	HB
Dimethenamid-P	03/84/EC	HB
Dimoxystrobin		FU
Diquat (dibromide)	01/21/EC	HB
Esfenvalerate	00/67/EC	IN
Ethephon		PG
Ethofumesate	02/37/EC	HB
Ethoxysulfuron	03/23/EC	HB
Etoxazole	05/34/EC	AC, IN
Famaxadone	02/64/EC	FU
Fenamidone	03/68/EC	FU
Fenamiphos		NE
Fenhexamid	01/28/EC	FU
Ferric phosphate	01/87/EC	MO
Flazasulfuron	04/30/EC	HB
Florasulam	02/64/EC	HB
Florchlorfenuron	06/10/EC	PG
Flufenacet	03/84/EC	HB
Flumioxazine	02/81/EC	HB
Flupyrsulfuron methyl	01/49/EC	HB
Fluroxypyr	00/10/EC	HB
Flurtamone	03/84/EC	HB

Substance	EU Decision	Use
Folpet		FU
Foramsulfuron	03/23/EC	HB
Formetanate		IN, AC
Fosetyl	06/64/EC	FU
Fosthiazate	03/94/EC	NE
Gliocladium catenulatum	05/2/EC	Bio
Glyphosate (incl trimesium aka sulfosate)	01/99/EC	HB
Imazalil (aka enilconazole)	97/73/EC	FU
Imazamox	03/23/EC	HB
Imazosulfuron	05/3/EC	HB
Indoxacarb	06/10/EC	IN
Iodosulfuron-methyl-sodium	03/84/EC	HB
Ioxynil	04/58/EC	HB
Iprodione	03/31/EC	FU
Iprovalicarb	02/48/EC	FU
Isoproturon	02/18/EC	HB
Isoxaflutole	03/68/EC	HB
Kresoxim-methyl	99/01/EC	FU
lambda-Cyhalothrin	00/80/EC	IN
Laminarim	05/3/EC	ST
Linuron	03/31/EC	HB
Maleic hydrazide	03/31/EC	PG
Mancozeb	05/72/EC	FU
Maneb	05/72/EC	FU
MCPA	05/57/EC	HB
MCPB	05/57/EC	HB
Mecoprop	03/70/EC	HB
Mecoprop-P	03/70/EC	HB
Mepanipyrim	04/62/EC	FU
Mesosulfuron	03/119/EC	HB
Mesotrione	03/68/EC	HB
Metalaxyl-M	02/64/EC	FU
Metconazole	06/74/EC	FU
Metiram	05/72/EC	FU
Metrafenone		FU
Methiocarb		IN, MO, RE
Methylcyclopropene		PT
Methoxyfenozide	05/3/EC	IN
Metsulfuron	00/49/EC	HB
Milbemectin	05/58/EC	AC, IN
Molinate	03/81/EC	HB
Oxadiargyl	03/23/EC	HB

Substance	EU Decision	Use
Oxamyl	06/16/EC	IN, NE
Oxasulfuron	03/23/EC	HB
Paecilomyces fumosoroseus	01/47/EC	IN
Paraquat	03/112/EC	HB
Pendimethalin	03/31/EC	HB
Pethoxamide	06/41/EC	HB
Phenmedipham	04/58/EC	HB
Picolinafen	02/64/EC	HB
Picoxystrobin	03/84/EC	FU
Pirimicarb	06/39/EC	IN
Prohexadione calcium	00/50/EC	PG
Propiconazole	03/70/EC	FU
Propineb	03/39/EC	FU
Propoxycarbazone	03/119/EC	HB
Propyzamide	03/39/EC	HB
Prosulfuron	02/48/EC	HB
Pseudomonas chlororaphis	04/71/EC	FU
Pymetrozine	01/87/EC	IN
Pyraclostrobin	04/30/EC	FU
Pyrafluten-ethyl	01/87/EC	HB
Pyridate	01/21/EC	HB
Pyrimethanil	06/74/EC	FU
Quinoxyfen	04/60/EC	FU
Rimsulfuron	06/39/EC	HB
Silthiofam	03/84/EC	FU
S-metolachlor	05/3/EC	HB
Spinosad		IN
Spiroxamine	99/73/EC	FU
Sulfosulfuron	02/48/EC	HB
Tepraloxydim	05/34/EC	HB
Thiabendazole	01/21/EC	FU
Thiacloprid	04/99/EC	IN
Thiamethoxam	00/181/EC	IN
Thifensulfuron (aka thiameturon)	01/99/EC	HB
Thiofanate-methyl	05/53/EC	FU
Thiram	03/81/EC	FU
Tolclophos-methyl	06/39/EC	FU
Tolyfluanid	06/06/EC	FU
Triasulfuron	00/66/EC	HB
Tribenuron	05/54/EC	HB
Triclopyr	06/74/EC	HB
Trifloxystrobin	03/68/EC	FU

Substance	EU Decision	Use
Trinexapac	06/64/EC	PG
Triticonazole	06/39/EC	FU
Warfarin	06/05/EC	RO
Ziram	03/81/EC	FU, RE
Zoxamide	03/119/EC	FU

Source: DG Health and Consumer Protection, Pesticide Safety Directorate, Agrow magazine Abbreviations " for substance use category" : AT= Attractant; ST= Soil treatment; PG= Plant Growth Regulator; HB= Herbicide; FU= Fungicide; IN= Insecticide; NE= Nematicide; AC= Acaricide; BA= Bactericide; RE= Repellent; MO= Molluscicide; RO= Rodenticide; E= External parasite; Bio= Bio-agent; GR=Growth regulator; PA= Plant Activator

30.1.5. Substances with Decision Pending

Besides the 40 substances approved, there are 489 substances in the market waiting for a decision and being used throughout the EU. Table 65 depicts the substances with decision pending, category and status under Directive 91/414/ EEC.

Table 65: Pesticides with decision pending

S. No.	Pesticide substance	Category	91/414 status
1.	(1R)-1,3,3-Trimethyl-4,6-dioxatricyclo[3.3.1.02,7] nonane (lineatin)	'	New' existing
2.	(2E, 13Z)-Octadecadien-1-yl acetate		Notified
3.	(3-Benzyloxycarbonyl-methyl)-2-benzothiazolinone (Benzolinone)		'New' existing
4.	(3E, 13Z)-Octadecadien-1-yl acetate		Notified
5.	(3Z, 13Z)-Octadecadien-1-yl acetate (formerly (Z,Z) Octadienyl acetate)	AT	Notified
6.	(7E, 9E)-Dodecadienyl acetate		Notified
7.	(7E, 9Z)-Dodecadienyl acetate (formerly (E)7-(Z)9-Dodecadienyl acetate)	AT	Notified
8.	(7Z, 11E)-Hexadecadien-1-yl acetate (formerly cis-7,trans-11-Hexadecadienyl acetate)	AT	Notified
9.	(7Z, 11Z)-Hexadecdien-1-yl acetate (formerly (7Z-11Z)-7,11-Hexadien-1-yl-acetate)	AT	Notified
10.	(9Z, 12E)-Tetradecadien-1-yl acetate (formerly (Z,E)-11-Tetradecadien-1-yl acetate) (aka Rimilure-8L)AT		Notified
11.	(Calcium) copper compunds	FU	
12.	(E)-11-Tetradecenyl acetate	AT	Notified
13.	(E)-2-Methyl-6-methylene-2,7-octadien-1-ol(myrcenol)		'New' existing
14.	(E)-2-Methyl-6-methylene-3,7-octadien-2-ol(isomyrcenol)		'New' existing
15.	(E)-8-Dodecenyl acetate		Notified

S. No.	Pesticide substance	Category	91/414 status
16.	(E,E)-8,10-Dodecadien-1-ol (formerly 8,10-Dodecadien-1-ol) (aka codlemone)	AT	Notified
17.	(E,Z)-8,10-Tetradecadienyl		'New' existing
18.	(E,Z)-9-Dodecenyl acetate (formerly trans-9-Dodecyl acetate)	AT	Notified
19.	(E/Z)-8-Dodecenyl acetate	AT	Notified
20	(E/Z)-9-Dodecen-1-ol		Notified
21.	(Z)-11-Hexadecen-1-ol		Notified
22.	(Z)-11-Hexadecen-1-yl acetate		Notified
23.	(Z)-11-Hexadecenal (formerly (Z)-11-Hexadecanole)	AT	Notified
24.	(Z)-11-Tetradecen-1-yl-acetate	AT	Notified
25.	(Z)-13-Hexadecen-1-ynyl acetate		Notified
26.	(Z)-13-Octadecanal (formerly (Z)-13-Octadecanole)	AT	Notified
27.	(Z)-7-Tetradecenal	AT	Notified
28.	(Z)-8-Dodecenol	AT	Notified
29.	(Z)-8-Dodecenyl acetate	AT	Notified
30.	(Z)-9-Dodecenyl acetate	AT	Notified
31.	(Z)-9-Hexadecenal (formerly Z-9-Hexadecenal)	AT	Notified
32.	(Z)-9-Tetradecenyl acetate	AT	Notified
33.	(Z)-9-Tricosene (formerly Z-9-Tricosene)	AT	Notified
34.	(Z,E)-3,7,11-trimethyl-2,6,10-dodecatrien-1-ol (aka Farnesol)	AT	Notified
35.	(Z,Z,Z,Z)-7,13,16,19-Docosatetraen-1-yl butyrate		Notified
36.	1, 3, 5-tri-(2-hydroxyethyl)-hexa-hydro-s-triazyne		'New' existing
37.	1,3-Dichloropropene	NE, HB	Dossier/DAR
38.	1,4-Diaminobutane (aka Putrescine)		Notified
39.	1,7-Dioxaspiro-5,5-undecan	AT	Notified
40.	1-Decanol	PG	Notified
41.	1-Methoxy-4-propenylbenzene (Anethole)		'New' existing
42.	1-Methyl-4-isopropylidenecyclohex-1-ene (Terpinolene)		'New' existing
43.	1-Naphthylacetamide	PG	Notified
44.	1-Naphthylacetic acid	PG	Notified
45.	1-Naphthylacetic acid ethylester	PG	Notified
46.	1-Tetradecanol		Notified
47.	2,6,6-Trimethylbicyclo(3.1.1)hept-2-en-4-ol		Notified
48.	2,6,6-Trimethylbicyclo[3.1.1]hept-2-ene (alpha-Pinen)		'New' existing
49.	2-Ethyl-1,6-dioxaspiro (4,4) nonan (chalcogran)		'New' existing
50.	2-Hydroxyethyl butyl sulfide		'New' existing
51.	2-Mercaptobenzothiazole		'New' existing

S. No.	Pesticide substance	Category	91/414 status
52.	2-Methoxy-5-nitrofenol sodium salt		'New' existing
53.	2-Methoxypropan-1-ol		'New' existing
54.	2-Methoxypropan-2-ol		'New' existing
55.	2-Methyl-6-methylene-2,7-octadien-4-ol (ipsdienol)		'New' existing
56.	2-Methyl-6-methylene-7-octen-4-ol (Ipsenol)		'New' existing
57.	2-Naphthyloxyacetamide	PG,HB	Notified
58.	2-Naphthyloxyacetic acid	PG	Notified
59.	2-Phenylphenol (incl. sodium salt orthophenyl phenol)	FU	Notified
60.	3,7,11-Trimethyl-1,6,10-dodecatrien-3-ol (aka Nerolidol)		Notified
61.	63,7,7-Trimethylbicyclo[4.1.0]hept-3-ene (3-Carene)		'New' existing
62.	63,7-Dimethyl-2,6-octadien-1-ol (aka Nerolidol)	AT	Notified
63.	3-Methyl-3-buten-1-ol		'New' existing
64.	3-phenyl-2-propenal (Cinnamaldehyde)		'New' existing
65.	4,6,6-Trimethyl-bicyclo[3.1.1]hept-3-en-ol,((S)-cis-verbenol)		'New' existing
66.	65-Decen-1-ol	AT	Notified
67.	5-Decen-1-yl acetate	AT	Notified
68.	6-Benzyladenine	PG	Notified
69.	7,11-Dimethyl-3-methylene-1,6,10-dodecatriene (aka (E)-beta-Farnesene)		Notified
70.	Abamectin (aka avermectin)	AC,IN	Dossier
71.	Acetic acid	HB	Notified
72.	Acetochlor	HB	Dossier
73	Aclonifen	HB	Data list
74.	Acrinathrin	AC	Data list
75.	*Agrobacterium radiobacter K 84*		'New' existing
76.	Aluminium ammonium sulfate	RE	Notified
77.	Aluminium phosphide	IN, RO	Data list
78.	Aluminium phosphide incl. phosphine		Notified
79.	Aluminium silicate (aka kaolin)		Notified
80.	Aluminium sulphate	MO,PG	Notified
81.	Amidosulfuron	HB	Dossier
82.	Amino acids: Cystein	PG	Notified
83.	Amino acids: gamma aminobutyric acid	PG	Notified
84.	Amino acids: L-glutamic acid		Notified
85.	Amino acids: L-tryptophan		Notified
86.	Ammonium acetate		Notified
87.	Ammonium carbonate	FU	Notified
88.	Anthraquinone	RE	Notified

S. No.	Pesticide substance	Category	91/414 status
89.	Asphalts		'New' existing
90.	Asulam	HB	Data list
91.	Azadirachtin	IN	Notified
92.	Azocyclotin	AC	Data list
93.	*Bacillus sphaericus*	IN	Notified
94.	*Bacillus thuringiensis*	IN	Notified
95.	*Bacillus thuringiensis subsp. aizawai*		Notified
96.	*Bacillus thuringiensis subsp. israelensis*		Notified
97.	*Bacillus thuringiensis subsp. kurstaki*		Notified
98.	*Bacillus thuringiensis subsp. tenebrionis*		Notified
99.	*Baculovirus GV*		'New' existing
100.	*Beauveria bassiana*	IN	Notified
101.	*Beauveria bassiana bassiana*		Notified
102.	*Beauveria brongniartii (*formerly *Beauveria bassiana brongniartii)*		Notified
103.	Benfluralin	HB	Dossier
104.	Bensulfuron	HB	Data list
105.	Bifenox	HB	Dossier
106.	Bifenthrin	IN, AC	Dossier
107.	Biohumus		'New' existing
108.	Bitertanol	FU	Dossier
109.	Bone Oil	RE	Notified
110.	Brodifacoum	RO	Notified
111.	Bromadiolone	RO	Notified
112.	Bromuconazole	FU	Dossier
113.	Bupirimate	FU	Data list
114.	Buprofezin	IN	Dossier
115.	Butralin	HB,PG	Dossier
116.	Calcium carbide	RE	Notified
117.	Calcium carbonate		'New' existing
118.	Calcium chloride	FU,PG	Notified
119.	Calcium phosphide	RO	Data list
120.	Calcium polysulphid		'New' existing
121.	Carbetamide	HB	Dossier
122.	Carbon dioxide	IN,RO	Notified
123.	Carbon monoxide		'New' existing
124.	Carboxin	FU	Data list
125.	Casein		'New' existing
126.	Chinin hydrochloride		'New' existing
127.	Chitosan		Notified
128.	Chloralose	RO	Notified
129.	Chlorates (incl. Mg, Na, K chlorates)	HB	Data list

S. No.	Pesticide substance	Category	91/414 status
130.	Chloridazon (aka pyrazone)	HB	Dossier
131.	Chlormequat (chloride)	PG	Data list
132.	Chlorophacinone	RO	Notified
134.	Chloropicrin	NE	Dossier
135.	Chlorsulfuron	HB	Data list
136.	Chlorthal-dimethyl	HB	Dossier
137.	cis-Zeatin	PG	Notified
138.	Citrus extract		Notified
139.	Citrus extract/grapefruit extract		'New' existing
140.	Clethodim	HB	Dossier
141.	Clofentezine	AC	Dossier
142.	Clomazone	HB	Dossier
143.	Conifer needle powder		'New' existing
144.	Copper complex: 8-hydroxyquinolin with salicylic acid		'New' existing
145.	Copper compounds		Dossier
146.	Cumylphenol		'New' existing
147.	Cyanamide (H & Ca cyanamide)	PG,HB	Dossier
148.	Cycloxydim	HB	Dossier
149.	*Cydia pomonella granulosis virus*	IN	Notified
150.	Cyhexatin	AC	Data list
151.	Cymoxanil	FU	Data list
152.	Cyproconazole	FU	Data list
153.	Cyromazine	IN	Data list
154.	Dazomet	NE, FU, HB,ST	Data list
155.	Denathonium benzoate	RE	Notified
156.	Di-1-p-menthene B470	PG	'New' existing
157.	Dicamba	HB	Data list
158.	Dichlobenil	HB	Data list
159.	Dichlorobenzoic acid methylester	FU	Data list
160.	Diclofop	HB	Dossier
161.	Dicloran	FU	Dossier
162.	Dicofol	AC	Data list
163.	Didecyl-dimethylammonium chloride	FU	Notified
164.	Diethofencarb	FU	Data list
165.	Difenacoum	RO	Notified
166.	Difenoconazole	FU	Data list
167.	Diflubenzuron	IN	Dossier
168.	Diflufenican	HB	Dossier
169.	Dimethachlor	HB	Data list
170.	Dimethenamide	HB	Dossier/DAR

S. No.	Pesticide substance	Category	91/414 status
171.	Dimethipin	PG,HB	Dossier
172.	Diniconazole	FU	Data list
173.	Diphenylamine	PG	Data list
174.	Dithianon	FU	Dossier
175.	Dodecan-1-yl acetate		'New' existing
176.	Dodecyl alcohol	AT	Notified
177.	Dodemorph	FU	Data list
178.	Dodine	FU	Data list
179.	EDTA and its salts	HB	Notified
180.	Epoxiconazole	FU	Dossier
181.	Ethalfluralin	HB	Data list
182.	Ethanedial (glyoxal)		'New' existing
183.	Ethanol	MO	Notified
184.	Ethoxyquin	PG	Notified
185.	Ethyl 2,4-decadienoate		'New' existing
186.	Ethylene		Notified
187.	Etofenprox	IN	Dossier
188.	Etridiazole	FU	Data list
189.	Extract from tea tree		'New' existing
190.	Extract from Equisetum		'New' existing
191.	Extract from Menta piperata		'New' existing
192.	Extract from plant Red oak, Prickly pear cactus, Fragrant sumac, Red mangrove		'New' existing
193.	Fat destilation residues		'New' existing
194.	Fatty acids / Isobutyric acid		'New' existing
195.	Fatty acids / Isovaleric acid		'New' existing
196.	Fatty acids / Lauric acid		'New' existing
197.	Fatty acids / Valeric acid		'New' existing
198.	Fatty acids: fatty acid methyl ester		Notified
199.	Fatty acids: Heptanoic acid		Notified
200.	Fatty acids: Octanoic acid		Notified
201.	Fatty acids: Decanoic acid		Notified
202.	Fatty acids: Oleic acid incl ethyloleate		Notified
203.	Fatty acids: Pelargonic acid		Notified
204.	Fatty acids: potassium salt (aka potassium soap)	HB,IN	Notified
205.	Fatty alcohols	PG	Notified
206.	Fenazaquin	AC	Dossier
207.	Fenbuconazole	FU	Dossier
208.	Fenbutatin oxide	AC	Data list
209.	Fenoxaprop-P	HB	Dossier
210.	Fenoxycarb	IN	Data list
211.	Fenpropidin	FU	Dossier

S. No.	Pesticide substance	Category	91/414 status
212.	Fenpropimorph	FU	Dossier
213.	Fenpyroximate	AC	Dossier
214.	Fluazifop-P	HB	Dossier
215.	Fluazinam	FU	Dossier
216.	Fludioxonyl	FU	Dossier
217.	Flufenoxuron	IN	Data list
218.	Flufenzin		'New' existing
219.	Flumetsulam		'New' existing
220.	Fluometuron	HB	Dossier
221.	Fluquinconazole	FU	Dossier
222.	Flurochloridone	HB	Data list
223.	Flurprimidole	PG	Data list
224.	Flutolanil	FU	Dossier
225.	Flutriafol	FU	Data list
226.	Folic acid	PG	Notified
227.	Formaldehyde	FU,ST	Notified
228.	Formic acid	IN	Notified
229.	Fuberidazole	FU	Dossier
230.	Garlic extract		Notified
231.	Garlic pulp		'New' existing
232.	Gelatine	IN	Notified
233.	Gibberellic acid	PG	Notified
234.	Gibberellin	PG	Notified
235.	Glutaraldehyde (aka glutardialdehyde)	FU,BA	Notified
236.	Grapefruit seed extract		Notified
237.	Guazatine	FU,RE	Data list
238.	HBTA (high boiling tar acid)		Notified
239.	Hexamethylene tetramine (urotropin)		'New' existing
240.	Hexythiazox	AC,IN	Dossier
241.	Hydrogen peroxide		Notified
242.	Hydrolysed proteins	AT	Notified
243.	Hymexazol	FU	Data list
244.	Ichthyol complex		'New' existing
245.	Imazaquin	PG	Data list
246.	Imidacloprid	IN	Dossier
247.	Indolylacetic acid (aka auxins)	PG	Notified
248.	Indolylbutyric acid	PG	Notified
249.	Iron pyrophosphate		'New' existing
250.	Iron sulphate	HB,MO	Notified
251.	Isoxaben	HB	Data list
252.	Jasmonic acid		'New' existing
253.	Kieselguhr (aka diatomaceous earth)	IN	Notified

S. No.	Pesticide substance	Category	91/414 status
254.	Lactofen		'New' existing
255.	Lanolin		'New' existing
256.	Lecithin	FU	Notified
257.	Lenacil	HB	Data list
258.	Lime sulphur	FU, IN, AC	Notified
259.	Lufenuron	IN	Data list
260.	Magnesium phosphide	IN,RO	Data list
261.	Magnesium phosphide incl. phosphine		Notified
262.	Maltodextrin		Notified
263.	Marigold extract		Notified
264.	Mepiquat	PG	Dossier
265.	Metaldehyde	MO	Dossier
266.	Metam (incl. -potassium and -sodium)	FU, IN, HB, NE	Data list
267.	Metamitron	HB	Data list
268.	*Metarhizium anisopliae*	IN	Notified
269.	Metazachlor	HB	Dossier
270.	Methomyl	IN	Dossier/DAR
271.	Methyl bromide	FU, IN, NE, HB	Dossier
272.	Methyl nonyl ketone	RE	Notified
273.	Methyl p-hydroxybenzoate		'New' existing
274.	Metosulam	HB	Data list
275.	Milk albumin		'New' existing
276.	*Mimosa tenuiflora* extract		Notified
277.	Monocarbamide-dihydrogensulphate	PG,HB	Data list
278.	Mustard powder		'New' existing
279.	Myclobutanil	FU	Dossier
280.	Napropamide	HB	Dossier
	Neodiprion sertifer nuclear		Notified
282.	polyhedrosis virus		
283.	Nicosulfuron	HB	Dossier
284.	Nicotine	IN	Notified
285.	N-phenylphthalamic acid		'New' existing
286.	Olein		'New' existing
287.	Oryzalin	HB	Data list
288.	Oxadiazon	HB	Data list
289.	Oxyfluorfen	HB	Data list
290.	Paclobutrazol	PG	Data list
291.	Paraffin oil	IN,AC	Notified
292.	Penconazole	FU	Data list
293.	Pencycuron	FU	Dossier

S. No.	Pesticide substance	Category	91/414 status
294.	Pepper	IN	Notified
295.	Peracetic acid		Notified
296.	Petroleum oils	FU, HB, IN, AC	Notified
297.	*Phlebiopsis gigantea*	FU	Notified
298.	Phoxim	IN	Notified
299.	p-Hydroxybenzoic acid		'New' existing
300.	Picloram	HB	Data list
301.	Plant oils / Blackcurrant bud oil		Notified
302.	Plant oils / Citronellol	IN,RE	Notified
303.	Plant oils / Clove oil		Notified
304.	Plant oils / Daphne oil	RE	Notified
305.	Plant oils / Etheric oil (Eugenol)	Notified	
306.	Plant oils / Etheric oils	RE	Notified
307.	Plant oils / Eucalyptus oil		Notified
308.	Plant oils / Gaiac Wood oil		Notified
309.	Plant oils / Garlic oil		Notified
310.	Plant oils / Lemongrass oil		Notified
311.	Plant oils / Marjoram oil		Notified
312.	Plant oils / Olive oil		Notified
313.	Plant oils / Orange oil		Notified
314.	Plant oils / Pinus oil		Notified
315.	Plant oils / Rapeseed oil		Notified
316.	Plant oils / Soya oil		Notified
317.	Plant oils / Spearmint oil		Notified
318.	Plant oils / Sunflower oil		Notified
319.	Plant oils / Thyme oil		Notified
320.	Plant oils / Ylang-Ylang oil		Notified
321.	Plant oils:	IN, AC, OT, FU, PG	Notified
322.	Polyvinyl acetate		'New' existing
323.	Potassium hydrogen carbonate		Notified
324.	Potassium permanganate	FU, BA, MO	Notified
325.	Prochloraz	FU	Data list
326.	Propachlor	HB	Data list
327.	Propaquizafop	HB	Dossier
328.	Propargite	AC	Data list
329.	Propisochlor		'New' existing
330.	Propolis		'New' existing
331.	Prosulfocarb	HB	Dossier
332.	Pyrethrins	IN	Notified

S. No.	Pesticide substance	Category	91/414 status
333.	Pyridaben	AC,IN	Data list
334.	Pyriproxyfen	IN	Dossier
335.	*Pythium oligandrum*		'New' existing
336.	Quartz sand	RE	Notified
337.	Quassia	IN,RE	Notified
338.	Quinmerac	HB	Data list
339.	Quinoclamine	HB, AL	Dossier
340.	Quizalofop-P	HB	Data list
341.	Repellant (by taste) of vegetal and animal origin/ extract of food grade/phosphoric acid and fish flour		'New' existing
342.	Repellants: Blood meal		Notified
343.	Repellants: Essential oils		Notified
344.	Repellants: Fatty acids, fish oil		Notified
345.	Repellants: Fish oil		Notified
346.	Repellants: Sheep fat		Notified
347.	Repellants: Tall oil		Notified
348.	Repellants: Tall oil crude		Notified
349.	Resins		'New' existing
350.	Rotenone	IN	Notified
351.	Sea-algae extract	PG	Notified
352.	Seaweed		Notified
353.	Sintofen (aka Cintofen)	PG	Data list
354.	Sodium 5-nitroguaiacolate	PG	Data list
355.	Sodium aluminium silicate		Notified
356.	Sodium hydrogen carbonate	FU	Notified
357.	Sodium hypochlorite	BA	Notified
358.	Sodium lauryl sulfate	FU,BA	Notified
359.	Sodium metabisulphite	FU	Notified
360.	Sodium o-nitrophenolate	PG	Data list
361.	Sodium p-nitrophenolate	PG	Data list
362.	Sodium-p-toluene-sulfonchloramid	BA	Notified
363.	*Streptomyces griseoviridis*	FU	Notified
364.	Sulcotrione	HB	Data list
365.	Sulphur	FU, AC, RE	Notified
366.	Sulphur dioxide		Notified
367.	Sulphuric acid	HB	Notified
368.	tau-Fluvalinate	IN	Data list
369.	Tebuconazole	FU	Data list
370.	Tebufenozide	IN	Dossier
371.	Tebufenpyrad	AC	Data list
372.	Teflubenzuron	IN	Dossier

S. No.	Pesticide substance	Category	91/414 status
373.	Tefluthrin	IN	Data list
374.	Terbuthylazine	HB	Data list
375.	Tetraconazole	FU	Dossier
376.	Thidiazuron	PG	Data list
377.	Thiobencarb	HB	Dossier
378.	Tralkoxydim	HB	Dossier
379.	Triadimenol	FU	Dossier
380.	Tri-allate	HB	Data list
381.	Triazoxide	FU	Data list
382.	Tricalcium phosphate	RO	Notified
383.	*Trichoderma harzianum*	FU	Notified
384.	*Trichoderma polysporum*	FU	Notified
385.	*Trichoderma viride*	FU	Notified
386.	Tricyclazole	FU	Data list
387.	Triflumizole	FU	Dossier
388.	Triflumuron	IN	Dossier
389.	Triflusulfuron	HB	Dossier
390.	Trimethylamine hydrochloride		Notified
391.	Urea	FU	Notified
392.	*Verticillium dahliae Kleb.*	FU	Notified
393.	*Verticillium lecanii*	IN	Notified
394.	Wheat gluten		Notified
395.	zeta-Cypermethrin	IN	Dossier
396.	Zinc phosphide incl. phosphine	RO	Notified

Source: DG Health and Consumer Protection,Pesticide Safety Directorate, Agrow magazine Abbreviations for "pesticide substance category" : AT= Attractant; ST= Soil treatment; PG= Plant Growth Regulator; HB= Herbicide; FU= Fungicide; IN= Insecticide; NE= Nematicide; AC= Acaricide; BA= Bactericide; RE= Repellent; MO= Molluscicide; RO= Rodenticide; E= External parasite; Bio= Bio-agent; GR=Growth regulator; PA= Plant Activator

30.2. Scenario of Banned Pesticdes in USA

United States uses 85 Pesticides Outlawed in Other Countries. *A New study published by* Nathan Donley (2019) of Centre for biological diversity in USA stated *that Harmful Poisons Shunned Elsewhere Account for Quarter of All U.S.Pesticide Use.* According to him The United States allows the use of 85 pesticides that have been banned or are being phased out in the European Union, China or Brazil, (according to a peer-reviewed study published by the academic journal *Environmental Health) are being use in the USA.*.In 2016 the United States used 322 million pounds of pesticides that are banned in the E.U., accounting for more than one-quarter of all agricultural pesticide use in this country. U.S. applicators also used 40 million pounds of pesticides that are banned or being phased out in China and 26 million pounds of pesticides that are banned or being phased out in Brazil. "It's appalling the U.S. lags so far behind these major agricultural powers in banning harmful pesticides," said Nathan Donley, a senior scientist with the Center for Biological Diversity and author of the study. "The study compared the approval status of more than 500 pesticides used in outdoor applications in the world's four largest agricultural economies: the United States, European Union, China and Brazil. According to this report

- The U.S. EPA continues to allow use of 85 pesticides for outdoor agricultural applications that are banned or in the process of being completely phased out elsewhere, including 72 in the E.U., 17 in Brazil and 11 in China.
- The United States has banned only four pesticides still approved for use in the E.U., Brazil or China.
- Pesticides approved in the United States but banned or being phased out in at least two of the three other nations include:
    1. 2, 4-DB,
    2. Bensulide,
    3, Chloropicrin,
    4. Dichlobenil,
    5. Dicrotophos,
    6. EPTC,
    7. Norflurazon,
    8. Oxytetracycline,
    9. Paraquat,
    10. Phorate,
    11. Streptomycin,
    12. Terbufos and
    13. Tribufos.
- The majority of pesticides banned in at least 2 of the 3 nations studied have not appreciably decreased in the United States over the past 25 years and almost all have stayed constant or increased over the past 10

years. Many have been implicated in acute pesticide poisonings in the United States, and some have been further restricted by individual states.

The study concludes that deficiencies in the U.S. pesticide regulatory process are the likely cause of the country failing to ban or phase out pesticides that the E.U., China and Brazil have prohibited. The Federal Insecticide, Fungicide, and Rodenticide Act gives the U.S. EPA significant discretion on which pesticides to cancel and makes the EPA-initiated, nonvoluntary cancellation process particularly onerous and politically fraught. This has, in effect, made pesticide cancellation in the United States largely a voluntary endeavor by the pesticide industry itself. As a result, pesticide cancellations in the U.S. are more often economic decisions rather than decisions made to protect human or environmental health.

Further "A combination of weak laws and the EPA's broken pesticide regulatory process has allowed the pesticide industry to dictate which pesticides stay in use.

- e.g.Ignoring its own established protocols to conclude that glyphosate, the active ingredient in Roundup, does not cause cancer, a finding that is at odds with the World Health organization's International Agency for Research on Cancer,
- Its refusal to protect endangered species from pesticides, even when it has been demonstrated by other federal agencies that use of the chemicals could put certain species at risk of extinction;
- The agency's industry-motivated decision to overturn a long-overdue ban on chlorpyrifos despite compelling evidence that it harms the brains of children;
- The recent approval of the largest ever expansion of medically-important antibiotics for use in plant agriculture, ignoring strong concerns about increased antibiotic resistance from the FDA, CDC and public health officials;
- Having to change the instructions on the dicamba pesticide label twice after the drift-prone pesticide damaged a reported 5 million acres of crops, trees and backyard gardens over the last two years,
- It's liberal use of an "emergency" exemption loophole that allows unapproved pesticides to be used for routine, foreseeable situations for many consecutive years.

Now, Chlorpyrifos on food crop is banned in USA, while it will continue to be permitted for non-food uses like growing cotton or on golf courses.

30.3. Scenario of Banned and Restricted Pesticides in China

In order to safeguard the safety of agricultural production, agricultural product quality and ecological environment and to effectively prevent, control and reduce their application risks, the Ministry of Agriculture (MoA) and other relevant competent authorities released several documents to prohibit/restrict the use of some pesticide products. To make it easier for enterprises to search for banned/restricted pesticide products, CIRS has prepared the *List of Banned and Restricted Pesticide Products* which covers all the pesticide products banned (table.66)/ restricted (table.67) /suspended (table.68) /proposed to be banned (table.69) for use in China (up to 23 Jun. 2017).

Table 66: List of banned pesticides in China (40)

S.N.	Name	Reason for Prohibition	Scope of Prohibition	MoA Announcement No. /Date of Issuance	Effective date of prohibition for Sale and Application of Technical Material and formulation with single active ingredient	Effective date of prohibition for Sale and Application of mixtures
1	BHC/HCH	Persistent organic pollutants (POPs)	18 pesticide products officially prohibited for use	MoA Announcement No. 199/5 Jun. 2002 Nongnongfa (2010) Notification No. 2 /15 Apr. 2010	5 Jun. 2002	/
2	DDT					
3	Camphechlor					
4	HHDN/Aldrin					
5	HEOD/Dieldrin					
6	Nitrofen	Teratogenicity, carcinogenicity and reproductive toxicity				
7	EDB					
8	Chlordimeform					
9	MATDA					
10	Dibromochloropropane					
11	Mercurial	Highly toxic and bio-concentration				
12	Arsenic-containing insecticide					
13	Plumbum					
14	Floroacetamide	Extremely toxic and highly toxic				
15	Tetramine					
16	Sodium Fluoroacetate					
17	Gliftor					
18	Silatrane					

19	Methamidophos		5 pesticide products officially prohibited for use	MoA Announcement No. 274 /30 Dec. 2003 Announcement No. 632 jointly issued by four ministries /4 Apr. 2006 Announcement No. 1 in 2008 jointly issued by six ministries /9 Jan. 2008 Nongnongfa (2010) Notification No. 2 /15 Apr. 2010	9 Jan. 2008	/
20	Parathion-methyl					
21	Parathion					
22	Monocrotophos					
23	Ammonium Phosphate					
24	Fenamiphos		10 pesticide products officially prohibited for use	MoA Announcement No. 1586 /14 Jun. 2011	31 Oct. 2013	/
25	Fonofos					
26	Phosfolan-methyl					
27	Calcium Phosphide					
28	Magnesium Phosphide					
29	Zine phosphide					
30	Cadusafos					
31	Coumaphos					
32	Sulfotep					
33	Terbufos					
34	Chlorsulfuron	Long residual effect which can lead to phytotoxicity	5 pesticide products officially prohibited for use	MoA Announcement No. 2032/9 Dec. 2013	31 Dec. 2015	/
35	Metsulfuron-methyl				31 Jun. 2015 (formulation with single active ingredient)	1 Jul. 2017
36	Ethametsulfuron					1 Jul. 2017
37	Asomate	Bring high risks to human and the environment and their impurities are carcinogenic			31 Dec. 2015	/
38	Urbacide					

39	Pesticide products containing octachlorodipropyl ether	Very harmful to human and animals	1 pesticide product prohibited for use	MoA Announcement No. 747 /20 Nov. 2006	1 Jan. 2008	1 Jan. 2008
40	Paraquat ready-mixture aqueous solution	Very poisonous to human and animals	1 pesticide product prohibited for use	Announcement No. 1745 jointly issued by three ministries /24 Apr. 2012	1 Jul. 2016 (AS)	/

Table 67: List of Restricted Pesticide (21)

S.N.	Name	Reason for restriction	Scope of Prohibition	MoA Announcement No. /Date of Issuance	Date of Prohibition
1	Phorate (3911)	Highly toxic	Prohibition of use on vegetables, fruit trees, tea and Chinese herbal medicine	MoA Announcement No. 199 /5 Jun. 2002 Nongnongfa (2010) Notification No. 2/ 15 Apr. 2010	5 Jun. 2002
2	Isofenphos-methyl				
3	Demeton				
4	Carbofuran				
5	Aldicarb				
6	Ethoprophos				
7	Phosfolan				
8	Isazofos				
9	Chlorpyrifos	Residual exceeding limits	Prohibition of use on vegetables	MoA Announcement No. 2032 /9 Dec. 2013	31 Dec. 2016
10	Triazophos				
11	Dicofol	The impurity is organochlorine and residual exceeds limits	Prohibition of use on tea plants	MoA Announcement No. 199/ 5 Jun. 2002 Nongnongfa (2010) Notification No. 2/15 Apr. 2010	5 Jun. 2002
12	Fenvalerate				
13	Daminozide	Carcinogenicity	Prohibition of use on peanuts	MoA Announcement No. 274/30 Apr. 2003 Nongnongfa (2010) Notification No. 2/15 Apr. 2010	30 Apr. 2003

14	Omethoate	Highly toxic	Prohibition of use on cabbage	MoA Announcement No. 194/24 Apr. 2002	1 Jun. 2002
			Prohibition of use on citrus trees	MoA Announcement No. 1586/15 Jun. 2011	15 Jun. 2011
15	Isocarbophos		Prohibition of use on citrus trees		
16	Methomyl		Prohibition of use on citrus trees, apple trees, tea plants and cruciferous vegetables		
17	Endosulfan	Highly toxic/POPs	Prohibition of use on apple trees and tea plants		
18	Fipronil	Bring high risks to bees and it degrades slowly in water and soil.	For public health use or use as seed coating for corns and other dry farmland seeds only	MoA Announcement No. 1157/25 Feb. 2009 Nongnongfa (2010) Notification No. 2/15 Apr. 2010	1 Oct. 2009
19	Methidathion	Highly toxic	Prohibition of use on citrus trees	MoA Announcement No. 2289/22 Aug. 2015	1 Oct. 2015
20	Methyle Bromide	Highly toxic/controlled products under Montreal Protocol (destroy the ozone layer)	For soil fumigation only (under the instruction of professionals)		
21	Chloropicrin	Highly toxic			

Table 68: List of Pesticide Products Suspended for Registration (13)

S.N.	Name	Reason	MoA Announcement No./Date of Issuance
1	Demeton	Highly toxic	MoA Announcement No. 194 /24 Apr. 2002
2	Phorate (3911)		MoA Announcement No. 194/24 Apr. 2002 MoA Announcement No. 1586/15 Jun. 2011
3	Omethoate		
4	Isocarbophos		
5	Isofenphos-methyl		
6	Methomyl		
7	Aldicarb		
8	Carbofuran		
9	Methidathion		
10	Ethoprophos		
11	Aluminium Phosphide		
12	Methyle Bromide		
13	Endosulfan		

Table 69: List of Newly Added Pesticide Products Proposed to be Banned/Restricted (for Public Comments) (8)

S.N.	Name	Reason for Prohibition/ Restriction	Scope of Prohibition	MoA Announcement No. / Date of Issuance	Date of Prohibition
1	Phorate (3911)	Highly toxic	Prohibition of use on vegetables, melons, tea, sugarcane and Chinese herbal medicine	MoA Announcement (draft) Nongbannonghan (2015) No. 14/2 Jul. 2015	1 Oct. 2015
2	Isofenphos-methyl				
3	Carbofuran				
4	Endosulfan *	Highly toxic/POPs	Banned for use in agriculture	MoA Announcement (draft) Nongbannonghan (2017) No. 6/12 May 2017	27 Mar. 2019
5	Methyle bromide *	Highly toxic/ controlled products under Montreal Protocol (destroy the ozone layer)			1 Jan. 2019
6	Acephate *	Extremely and highly toxic	Prohibition of use on vegetables, melons, tea and Chinese herbal medicine		1 Jul. 2019
7	Carbosulfan *	Highly toxic			
8	Dimethoate *				

30.4. Scenario of Pesticide Ban in Brazil

Brazil is the world's largest user of pesticides, including more than a dozen considered highly hazardous, due to its permissive legislation that allows some of Europe's biggest agrochemical companies to continue selling products that have been banned in their home market. Brazil is not only the world's champion in pesticides consumption (more than US$10 billion annually), but also the largest buyer of Highly Hazardous Pesticides (HHPs) as per United Nations designation for agrochemicals. The toxicity of these pesticides has raised concerns: 22 of them are classified as highly hazardous pesticides (HHPs), by the Pesticide Action Network (PAN), a global coalition that advocates for eco-friendly alternatives to chemical pesticides. The classification is based on criteria developed by the World Health Organization (WHO) and the U.N. Food and Agriculture Organization (FAO): in humans, they can be toxic to the reproductive system, damaging to DNA, or carcinogenic, as well as fatal to bees and other pollinators.

Even though these products have been banned in other countries, companies like Bayer, BASF and Syngenta make millions of dollars selling them in Brazil. According to IBAMA, the Brazilian environmental agency, more than 63,000 tonnes of just 10 of these 22 pesticides were sold in 2018. Sales of the other 12 products were not reported because of commercial confidentiality; IBAMA only discloses data on active ingredients manufactured by three or more companies. It also doesn't break down the amounts sold by each company.

Brazil is an open market for toxic pesticides banned at European countries and the European chemical giants explore the Brazil's pesticide market for their sale. In 2018, Brazil used more than 60,000 tonnes of highly hazardous pesticides banned in the European Union.Three Europe-based multibillion-dollar companies control 54% of the world market.They include German agrochemical giants BASF and Bayer, as well as Swiss company Syngenta, one of whose pesticides still being sold in Brazil has been banned in its home country for more than 30 years.

In 2018, 36.7% and 24.9% of the active ingredients sold worldwide by Bayer and BASF respectively were highly hazardous under the PAN definition, according to a report that lists German agrochemical companies' sales to developing countries. The report was prepared by the Permanent Campaign against Pesticides, INKOTA Network, Rosa Luxemburg Foundation, MISEREOR and South African organization Khanyisa. According to the study, more flexible registration procedures make it easier for highly hazardous pesticides to enter markets in the global South. Brazil is a case in point: 44% of the substances registered here have been banned in the European Union, according to a report released in July by the former president of the Brazilian Association for Agrarian Reform (ABRA), Gerson Teixeira. Alan Tygel, a spokesman for the Campaign against Pesticides and For Life, explains why the study began with Germany: "The country is the world's second-largest pesticide exporter because of these two major manufacturers. It exports 233 active ingredients — nine of which are banned in the EU but produced in Germany and then exported. "Of the 233 active ingredients exported by Germany, 62 are considered highly hazardous," he adds. The report shows that half of the 24 ingredients sold by Bayer and BASF in Brazil are highly hazardous.

One of them is Fipronil, an active ingredient used in insecticides marketed by BASF. The product entered PAN's list for its fatal effects on bees. In the 1990s, it was blamed for a massive bee die-off in France. In 2017, millions of chicken eggs were contaminated by Fipronil in Belgium and the Netherlands. That same year, the product was banned from the entire EU for posing "high acute risks for bees [when used as] seed treatment in maize," according to the European Food Safety Authority (EFSA). In Brazil, beekeepers list it as the main cause of the deaths of more than 500 million bees in 2018-2019. According to IBAMA, 1,600 tonnes were sold in the country in 2018 alone, to be used in the cultivation of cotton, potatoes, soybeans and corn.

Another controversial item on the list is the fungicide Carbendazim from Bayer, which has been banned from the European market since 2016. Its potential harms include genetic defects, impaired fertility, and fetus problems, in addition to being very toxic to bodies of water, according to the Campaign against Pesticides report. The product is also on PAN's list because it can damage DNA and be toxic to the reproductive system. According to IBAMA, Carbendazim sales in Brazil amounted to 4,800 tonnes in 2018. In December last year, the country's National Health Regulatory Agency (ANVISA) started reevaluating it to decide whether it should remain on the market. The process is slow and may take more than a decade, as happened recently with glyphosate, whose registration was renewed after 11 years under reevaluation. In the meantime, Carbendazim continues to be sold for the cultivation of black beans, soybeans, wheat and oranges.

Hoinkes (2018) cites the examples of Fipronil, Paraquat, Atrazine and Thiamethoxam, a baned pesticides sold in Brazil. Due to weaker regulations or poor enforcement in certain political contexts, these companies continue selling highly hazardous pesticides in Brazil which are banned in their own territories because they are acutely toxic to humans, kill bees, persist in drinking water or are suspected of causing cancer, birth defects or other chronic diseases."

The study team cross referenced data for $23.3 billion in pesticide sales to 43 countries in 2018 (about 40% of the global market) along with the Pesticide Action Network (PAN) list of HHPs. The in-depth data set was broken down by active ingredient to ascertain the value and percentage of each sale attributable to highly hazardous pesticides. Analysis showed that 42.4% of total sales ($ 9.9 billion) were in HHPs. Of that group of highly hazardous pesticides, Brazil accounted for more than a fifth (22.2%), totaling $2.2 billion.

In addition, the world's top five pesticide manufacturers — Syngenta (Switzerland), Bayer and BASF (both headquartered in Germany), and Corteva and FMC (both in the United States) — accounted for 80% of all pesticide sales to Brazil; 72% of these sales were HHPs ($1.6 billion). These companies, together with Sumitomo Chemical, combine their political clout as CropLife International, an international Brussels-based pesticide industry trade association and lobbying group. HHP use will likely continue rising in Brazil. In 2019, the Jair Bolsonaro administration approved 474 new pesticides for use — the highest number in 14 years. Pesticide imports to Brazil also broke an all-time record, with almost 335,000 tons of pesticides purchased in 2019, an increase of 16% compared to 2018 and this will goes on till the stringent laws are formulated and implemented to deal with pesticide ban.

30.5. Status of Banned and Restricted Pesticides in India

30.5.1. Pesticides Banned, Refused Registration and Restricted In Use In India: (As on 01.01.2021)

Table 70: Pesticides / Formulations Banned in India

a.	Pesticides Banned for manufacture, import and use.
1.	Alachlor (Vide S.O. 3951(E), dated 08.08.2018)
2.	Aldicarb (vide S.O. 682 (E) dated 17th July2001)
3.	Aldrin
4.	Benzene Hexachloride
5.	Benomyl (vide S.O 3951(E) dated 8th August, 2018)
6.	Calcium Cyanide
7.	Carbaryl (vide S.O 3951(E) dated 8th August, 2018)
8.	Chlorbenzilate (vide S.O. 682 (E) dated 17th July 2001)
9.	Chlordane
10.	Chlorofenvinphos
11.	Copper Acetoarsenite
12.	Diazinon (vide S.O 3951(E) dated 8th August, 2018)
13.	Dibromochloropropane (DBCP) (vide S.O. 569 (E) dated 25th July1989)
14.	Dichlorovos (Vide S.O. 3951(E), dated 08.08.2018)
15.	Dieldrin (vide S.O. 682 (E) dated 17th July 2001)

16.	Endosulfron (vide ad-Interim order of the Supreme Court of India in the Writ Petition (Civil) No. 213 of 2011 dated 13th May, 2011 and finally disposed of dated 10th January, 2017)
17.	Endrin
18.	Ethyl Mercury Chloride
19.	Ethyl Parathion
20.	Ethylene Dibromide (EDB) (vide S.O. 682 (E) dated 17th July 2001)
21.	Fenarimol (vide S.O 3951(E) dated 8th August, 2018)
22.	Fenthion (vide S.O 3951(E) dated 8th August, 2018)
23.	Heptachlor
24.	Lindane (Gamma-HCH)
25.	Linuron (vide S.O 3951(E) dated 8th August, 2018)
26.	Maleic Hydrazide (vide S.O. 682 (E) dated 17th July 2001)
27.	Menazon
28.	Methoxy Ethyl Mercury Chloride (vide S.O 3951(E) dated 8th August, 2018)
29.	Methyl Parathion (vide S.O 3951(E) dated 8th August, 2018)
30.	Metoxuron
31.	Nitrofen
32.	Paraquat Dimethyl Sulphate
33.	Pentachloro Nitrobenzene (PCNB) (vide S.O. 569 (E) dated 25th July 1989)
34.	Pentachlorophenol
35.	Phenyl Mercury Acetate
36.	Phorate (Vide S.O. 3951(E), dated 08.08.2018)
37.	Phosphamidon (Vide S.O. 3951(E), dated 08.08.2018)
38.	Sodium Cyanide (banned for Insecticidal purpose only vide S.O 3951(E) dated 8th August, 2018)*
39.	Sodium Methane Arsonate
40.	Tetradifon
41.	Thiometon (vide S.O 3951(E) dated 8th August, 2018)
42.	Toxaphene (Camphechlor) (vide S.O. 569 (E) dated 25th July 1989)
43.	Triazophos (Vide S.O. 3951(E), dated 08.08.2018)
44.	Tridemorph (vide S.O 3951(E) dated 8th August, 2018)
45.	Trichloro acetic acid (TCA) (vide S.O. 682 (E) dated 17th July 2001)
46.	Trichlorfon (Vide S.O. 3951(E), dated 08.08.2018)

b. Pesticide formulations banned for import, manufacture and use

1.	Carbofuron 50% SP (vide S.O. 678 (E) dated 17th July 2001)
2.	Methomyl 12.5% L
3.	Methomyl 24% formulation
4.	Phosphamidon 85% SL

c. Pesticide / Pesticide formulations banned for use but continued to manufacture for export

1.	Captafol 80% Powder (vide S.O. 679 (E) dated 17th July 2001)
2.	Dichlorvos (vide S.O. 1196 (E) dated 20th March 2020)
3.	Nicotin Sulfate (vide S.O. 325 (E) dated 11th May 1992)
4.	Phorate (vide S.O. 1196 (E) dated 20th March 2020)
5.	Triazophos (vide S.O. 1196 (E) dated 20th March 2020)

d. Pesticides Withdrawn (Withdrawal may become inoperative as soon as required complete data as per the guidelines is generated and submitted by the Pesticides Industry to the Government and accepted by the Registration Committee. (S.O 915(E) dated 15thJun, 2006)

1.	Dalapon
2.	Ferbam
3.	Formothion
4.	Nickel Chloride
5.	Paradichlorobenzene (PDCB)
6.	Simazine
7.	Sirmate (S.O. 2485 (E) dated 24th September 2014)
8.	Warfarin (vide S.O. 915 (E) dated 15th June 2006)

*Regulation to be continued in the extant manner for non-insecticidal uses.

Table.71: Pesticides Refused Registration

S.No.	Name of Pesticides
1.	2, 4, 5-T
2.	Ammonium Sulphamate
3.	Azinphos Ethyl
4.	Azinphos Methyl
5.	Binapacryl
6.	Calcium Arsenate
7.	Carbophenothion
8.	Chinomethionate (Morestan)
9.	Dicrotophos
10.	EPN
11.	Fentin Acetate
12.	Fentin Hydroxide
13.	Lead Arsenate
14.	Leptophos (Phosvel)
15.	Mephosfolan

16.	Mevinphos (Phosdrin)
17.	Thiodemeton / Disulfoton
18.	Vamidothion

Table 72: Pesticides Restricted for use in the Country

S.No.	Name of Pesticides Details of Restrictions
1.	Aluminium Phosphide: The Pest Control Operations with Aluminium Phosphide may be undertaken only by Govt./Govt. undertakings / Govt. Organizations / pest control operators under the strict supervision of Govt. Experts or experts whose expertise is approved by the Plant Protection Advisor to Govt of India except 1. Aluminium Phosphide 15 % 12 g Tablet and 2. Aluminum Phosphide 6 % tablet. [RC dec Decision circular F No. 14-11(2)-CIR-II (Vol. II) dated 21-09-1984 and G.S.R. 371(E) dated 20th may 1999]. 1. Decision of 282nd RC held on 02-11-2007 and, 2. Decision of 326th RC held on 15.2. 2012. The production, marketing and use of Aluminium Phosphide tube packs with a capacity of 10 and 20 tablets of 3 g each of Aluminium Phosphide are banned completely. (S.O.677 (E) dated 17thJuly, 2001)
2.	Captafol: The use of Captafol as foliar spray is banned. Captafol shall be used only as seeddresser. (S.O.569 (E) dated 25thJuly, 1989). The manufacture of Captafol 80 % powder for dry seed treatment (DS) is banned for use in the country except manufacture for export. (S.O.679 (E) dated 17thJuly, 2001)
3.	Cypermethrin: Cypermethrin 3 % Smoke Generator is to be used only through Pest Control Operators and not allowed to be used by the General Public. [Order of Hon,ble High Court of Delhi in WP(C) 10052 of 2009 dated 1407- 2009 and LPA- 429/2009 dated 08-09-2009].
4.	Dazomet: The use of Dazomet is not permitted on Tea. (S.O.3006 (E) dated 31st Dec, 2008)
5.	Dichloro DiphenylTrichloroethane: The use of DDT for the domestic Public Health (DDT) Programme is restricted up to 10,000 Metric Tonnes per annum, except in case of any major outbreak of epidemic. M/s Hindustan Insecticides Ltd., the sole manufacturer of DDT in the country may manufacture DDT for export to other countries for use in vector control for public health purpose. The export of DDT to Parties and State non- Parties shall be strictly in accordance with the par paragraph 2(b) article 3 of the Stockholm Convention on Persistent Organic Pollutants (POPs). (S.O.295 (E) dated 8th March, 2006). Use of DDT in Agriculture is withdrawn. In very special circumstances warranting the use of DDT for plant protection work, the state or central Govt. may purchase it directly from M/s Hindustan Insecticides Ltd. to be used under expert Governmental supervision. (S.O.378 (E) d dated 26thMay, 1989).
6.	Fenitrothion: The use of Fenitrothion is banned in Agriculture except for locust control in scheduled desert area and public health. (S.O.706 (E) dated 3 rd May 2007).
7.	Methyl Bromide: Methyl Bromide may be used only by Govt./Govt. undertakings/ Govt. Organizations / Pest control operators under the strict supervision of Govt. Experts or Experts whose expertise is approved by the Plant Protection Advisor. [G.S.R.371 (E) dated 20thMay, 1999 and earlier RC decision]
8.	Monocrotophos: Monocrotophos is banned for use on vegetables. (S.O.1482 (E) dated 10thOct, 2005)

9.	Trifluralin: (i) The Registration, import, manufacture, formulation, transport, sell and its all uses except use in wheat shall be prohibited and completely banned from 8th August, 2018. (ii) (ii) A cautionary statement has to be incorporated in the label and leaflet that it is toxic to aquatic organism, hence should not be used near water bodies, aquaculture or pisciculture area. (vide S.O 3951(E) dated 8th August, 2018)

30.5.2. Fungicides banned by Govt of India for manufacturing, import and use in India (as on Oct 2020)

Table 73:

i.	**Fungicide banned in India.**
1.	Benomyl
2.	Ethyl Mercury Chloride
3.	Fenarimol
4.	Pentachloro Nitrobenzene(PCNB)
5.	Penta Chlorophenol
6.	Phenyl Mercury Acetate
7.	Tridemorph
ii.	**Fungicide formulation banned in India for Use and application but continued to manufacture for Export from India.**
1.	Captafol 80% powder
iii.	**Fungicide Withdrawn for Use in India**
1.	Ferbam
iv.	**Fungicide Refused registration in India**
1.	Chinomethionate (Morestan)
2.	Fentin Acetate
3.	Fentin Hydroxide
4.	Leptophos(Phosval)

30.5.3. Herbicides banned by Govt of India for Manufacture, import and Use in India

Table 74:

i.	**Herbicides banned in India**
1.	Linuron
2.	Paraquat Dimethyl sulphate
3.	Sodium cyanide
4.	Sodium Methane Arsenate
5.	Trichloro acetic acid (TCA)
6.	Malic hydrazide
7.	Metoxuron
8.	Natrofen

ii.	**Herbicide withdrawn from use in India**
1.	Dalapon
2.	Simazine
3.	Sirmate
4.	Warfarin rodenticide
iii.	**Herbicide refused registration in India**
1.	2,4,5-T (Trichlorophenoxy acetic acid)
2.	Ammonium sulphamate
3.	Dicrotophos
iv.	**Herbicide Restricted for use in India**
1.	Dazomet
2.	Methyl bromide
3.	Trifluralin

30.6. Label Claim Insecticides

Table 75: Label claim insecticides

Product Main Brands	Main Countries of Use	Description
1. ACTARA®	Brazil, USA, Japan, India, South Korea, Mexico, Spain, Cote d'Ivoire, Philippines, Vietnam, Italy	ACTARA® is a second generation neonicotinoid for controlling foliar and soil pests in multiple crops.
2. CHESS®/ PLENUM® / FULFILL®	Philippines, Mexico, USA,Iran, Japan, Germany, Australia, United Kingdom, France	CHESS®/PLENUM®/FULFILL® provides powerful control of aphids and whiteflies (vegetables, potatoes, stone fruits and ornamentals) and of hoppers (rice and mangos). It delivers immediate crop protection through permanent inhibition of feeding.
3. CURACRON®	Pakistan, India, Brazil,Indonesia, Egypt, Colombia, Burkina Faso, USA, Sudan, Japan	The foundation product for the control of lepidoptera in cotton with strong effects against mining and sucking insects as well as mites. The ideal partner for tank-mix or ready-mix with established and new products.
4. DURIVO®	Argentina, Brazil, China, France, India, Indonesia, Japan, Mexico, South Korea, Spain, USA, Vietnam	DURIVO® is innovation in performance and convenience
5. FORCE®	USA, Italy	The premium-performing corn granular insecticide that provides broadspectrum soil insect control and residue activity, with an excellent user and environmental profile.

Product Main Brands	Main Countries of Use	Description
6. INSEGAR®	France, Netherlands,Belgium, Austria, Switzerland, Greece, Spain, Russia, United Kingdom, Australia	Lepidoptera control in fruits and grapes, based on ovicidal activity. Ideally suited for resistance management.
7. KARATE ZEON®	USA, Germany, Brazil, France, India, Mexico, Indonesia, United Kingdom, Canada, Italy	KARATE ZEON® sets a new standard in insecticide technology.
8. MATCH®	Brazil, Japan, South Korea, Italy, Venezuela, Spain, Colombia, Peru, Argentina, France	MATCH® offers reliable and economic insect control.
9. NEMATHORIN (Trademark of Ishihara Sangyo Kaisha Ltd.)	United Kingdom, France, Italy, South Africa, Hungary	Modern contact nematicide, combining powerful effectiveness with a good environmental profile to minimize impact on non-target organisms.
10. PIRIMOR®	Germany, United Kingdom, France, Netherlands, Sweden, Spain, Poland, Italy, Mexico	Selective aphicide for use on a wide range of crops including cereals, fruits, vegetables and potatoes.
11. POLO®	Brazil, Pakistan, Turkey, Indonesia, Sudan, Malaysia, Cuba, Taiwan, Philippines	Excellent whitefly control on cotton and against mites, aphids and jassids through the control of nymphs and adults.
12. PROCLAIM®	Japan, Australia, South Korea, USA, Taiwan, Mexico, Indonesia, Israel, Thailand	Powerful insect control in vegetables and cotton
13. SUPRACIDE®	Japan, China, Spain, Brazil, South Korea, Italy, Greece, Chile, Turkey, Morocco	A fast-acting contact and stomach insecticide, which controls a large number of economically important pests in fruit trees. SUPRACIDE® is used worldwide as the standard scale insect solution in perennial crops.
14. TRIGARD®	USA, Mexico, Brazil, Italy, Spain, Indonesia, Japan, Philippines, Peru, Taiwan	The systemic dipterous leafminer specialist product, which is selective towards beneficials. When used in alternation with VERTIMEC®, it provides the ideal resistance management rotation for vegetable leafminer control.
15. VERTIMEC®	USA, Brazil, Mexico, Italy, Egypt, France, Spain, Indonesia, Argentina, South Korea	The low use rate acaricide/ insecticide used for the control of mites and insects on a number of crops including cotton, citrus, pome fruit, nuts and vegetables.

Table 76: Products with label claims for control of whitefly in New Zealand (September 2002). Not all products listed for each pesticide may have a label claim for all crops indicated

	Type of label claim for each crop group[1]							
Pesticide category and IRAC chemical group Pesticide common and (product) names	Citrus	Beans	Capsicums	Cucurbits	Ornamentals/ flowers	Tamarillo	Tomatoes	Vegetables
Parasites								
Encarsia formosa		GHWF in GH	GHWF in GH	GHWF in GH			GHWF in GH	
Carbamates 1A								
Methomyl			WF in GH	WF in GH			WF in GH	
Organo-phosphates 1B								
Diazinon	XX					GHWF		
Dichlorvos		WF in GH	WF in GH	WF in GH	WF in GH	GHWF	WF in GH	WF in GH
pirimiphos methyl				WF in GH	WF in GH		WF in GH	
Organo-phosphates + Pyrethroids 1B/3								
pirimiphos methyl+ permethrin	XX			WF in GH	WF in GH		WF in GH	
Cyclodiene 2A								
Endosulfan							WF	
Pyrethroids 3								
Deltamethrin						GHWF		
Taufluvalinate					WF	GHWF		
taufluvalinate + fungicide					WF			
Pyrethrins 3								
Pyrethrum					GHWF			GHWF
Chloronicotinyl 4A								
Imidacloprid					WF in TSP			WF in TSP
Thiadiazine 16								
Buprofezin	XX		GHWF in GH	GHWF in GH	GHWF in GH	GHWF	GHWF in GH	GHWF in GH

[1]GHWF = greenhouse whitefly where this species is specified or is the only species on the crop, WF = whitefly, GH = greenhouses, TSP = transplants, XX = no label claim for whitefly, but claim for control of other insects in citrus.

Table 77: Label Claim of Pesticides for Different Vegetable Crops in India

Common Name		Strength	Target Pest	Waitinng period (days)
Tomato				
Insecticides				
1.	Azadirachtin	1%	Fruit borer	3
2.	Azadirachtin	5%	Aphids, Whitefly	5
3.	Carbofuran	3 % G	Whitefly	
4.	Chlorantranilprole	18.5% SC	Fruit borer	3
5.	Dimethoate	30% SC	SC Whitefly	
6.	Imidacloprid	17.8 % SL	Whitefly	3
7.	Indoxacarb	14.5% SC	Fruit borer	5
8.	Lambda Cyhalothrin	5 % EC	Fruit borer	4
9.	Malathion	50% EC	Whitefly	
10.	Methomyl	40 % SP	Pod borer	5-6
11.	Novaluron	10%	EC Fruit borer	1-3
12.	NPV of H armigera	0.43 % - 2.0%	Helicoverpa armigera	
13.	Oxydemeton methyl	25% EC	Whitefly	
14.	Phorate	10% EC	Whitefly	
15.	Phosalone	35 % EC	Fruit borer	
16.	Quinalphos	20 %	AF Fruit borer	7
		25 % EC		
17.	Thiamethoxam	25% WG	Whitefly	5
18.	Trichloforon	5%	GR Fruit borer	
		5% Dust		
		50% EC		
Fungicides				
19.	Azoxystrobin	23% SC	Early/ Late blight	3
20.	Copper Sulphate	2.62% SC	Early/Late blight	3
		75% WP	Damping off (Soil drenching in the nursery)	
		50% WP	Early & Late blight 6	
21.	Iprodione	50% WP	Early blight	15
22.	Kresoxim-methyl	44.3% SC	Early blight	3
23.	Kitazin	48% EC	Early blight	
24.	Mancozeb	75% WG	Early Blight	
		35% SC	Early & Late blight	
		75% WP	Late blight,	
			Buck eye Rot, Leaf spot	
25.	Metarim	70% WG	Alternaria blight	6
26.	Propineb	70% WP	Buck eye rot	
27.	Pyraclostrobin	20% WG	Early blight	3

Common Name		Strength	Target Pest	Waitinng period (days)
28.	Streptomycin	9%+1% SP	Bacterial leaf spot	
29.	Thiophenate Methyl	70% WP	Ring rot	7
30.	Ziram	80% WP	Early blight	3
31.	Zineb	75% WP	Early & Late blight	
32.	Cymoxanil	8% WP	Late blight	
33.	Famoxadone	16.6% SC	Early and Late Blight	3
Brinjal				
Insecticides				
1.	Azadirachtin	1%	Fruit and Shoot borer (FSB)	3
2.	Azadirachtin	0.03% FSB	Beetles	7
3.	Carbofuran	3 % G	Nematodes	
4.	Chlorantranilprole	18.5% SC	FSB	
5.	Chloropyrifos	20% EC	FSB	
6.	Cypermethrin	0.25 % DP	FSB	3
		10 % EC	FSB	3
		25 % EC	FSB, Epilachna beetle	1
7.	Dicofol	18.5 % EC	Mite	15-20
8.	Difenthiuron	50 % WP	Whitefly	3
9.	Dimethoate	30% EC	Jassids, FSB	
10.	Emamectin benzoate	5 % SG	FSB	3
11.	Fenzaquin	10 % EC	Mites	7
12.	Fenpropathrin	30 % EC	Whitefly, FSB, Mites	10
13.	Fenvalerate	20 % EC	FSB	5
14.	Flumite/ Flufenzine	20 % SC	Mites	5
15.	Lambda-Cyhalothrin	5 % EC	FSB	5
16.	Malathion	50% EC	Mites	
17.	Phorate	10% G	Jassids Aphids, Mites, Thrips	
18.	Phosalone	35 % EC	FSB	
19.	Phosphomidon	40% L	Jassids, Aphid, Whitefly	10
20.	Quinalphos	20 % AF	FSB, Jassids, Epilachnna beetle	
		25 % EC	FSB, - Leaf hopper	7
21.	Spiromesifen	22.9% SC	Red spider mite	5
22.	Thiodicarb	75% WP	FSB	6
23.	Thiamethoxam	25% WG	Whiteflies	3
24.	Thiameton	25 % EC	Aphids, Jassids, FSB	
25.	Trichloforon	5% G, 5% Dust, 50% EC	FSB	

Common Name		Strength	Target Pest	Waitinng period (days)
26.	Triazophos	40% EC	FSB, Epilachna beetle	5
27.	Deltamethrin	35 % EC	FSB, Epilachana	21
Fungicides				
29.	Benomyl	50% WP	Powdery mildew	
30.	Carbendazim	50% WP	Leaf spot, Fruit rot	
31.	Captan	75% WP	Damping off	
			(Drench in nursery)	
32.	Zineb	75% WP	Blight	
Chilli				
Insecticides				
1.	Acetamiprid	20% SC	Thrips	3
2.	Buprofezin	25 % SC	Yellow mite	5
3.	Carbofuran	3 % G	Thrips	
4.	Carbosulfan	25 % EC	Whitefly, Aphids	8
5.	Chlorfenpyre	10 % SC	Yellow mite	5
6.	Deltamethrin	2.8 % EC	Fruit borer	5
7.	Difenthiuron	50 % WP	Mites	3
8.	Dimethoate	30% EC	Mites, Thrips	
9.	Emamectin benzoate	5 % SG	Fruit borer, Thrips, mite	3
10.	Endosulfan	35 % EC	Aphids	21
11.	Ethion	50 % EC	Mite, Thrips	05
12.	Fenazaquin	10 % EC	Yellow mite	10
13.	Fenpropathrin	30% EC	Thrips, Whitefly, Mites	7
14.	Fenpyroximate	5 % EC	Yellow mite	7
15.	Fipronil	5 % SC	Fruit borer, Thrips, Aphids	7
16.	Flubendamide	39.35 % SC	Fruit borer	7
17.	Hexythiazox	5.45 % EC	Yellow mite	3
18.	Imidacloprid	70 % WS	Jassids, Aphids,	
		17.8 % SL	Jassid, Aphid, Thrips	40
19.	Indoxacarb	14.5% SC	Fruit borer	5
20.	Lambda Cyhalothrin	5 % EC	Thrips, Mite, Pod borer	5
21.	Methomyl	40 % SP	Pod borer, Thrips	5-6
22.	Milebemectin	1 % EC	Mites	7
23.	Novaluron	10% EC	Fruit borer, Tobacco caterpillar	3
24.	Oxydemeton methyl	25% EC	Aphids, Mites, Thrips	
25.	Phorate	10% GR	Aphids, Mites, Thrips	
26.	Phosalone	35 % EC	Aphid, Mite, Thrips	
27.	Propargite	57% EC	Mite	7
28.	Quinalphos	25%	Gel Aphids	
		25%	EC Aphid- Mite	

Common Name		Strength	Target Pest	Waitinng period (days)
29.	Spinosad	45%	SC Fruit borer	3
30.	Spiromesifen	22.9%	SC Yellow mite	7
31.	Thiacloprid	21.7%	SC Thrips	5
32.	Thiodicarb	75%	WP Fruitborer	6
33.	Indoxacarb	+ 14.5 % + 7.7% SC	Thrips, Fruit borer	5
Fungicides				
34.	Azoxystrobin	23% SC	Fruit rot, Powdery mildew	5
35.	Benomyl	50% WP	Powdery mildew, Fruit rot, Leaf spot	
36.	Copper Sulphate	2.62% SC	Fruit rot, anthracnose	3
37.	Captan	50% WG	Fruit rot, anthracnose	5
		75% WP	Damping off in nursery (soil drench)	
		75% WS	Early blight	8
			Damping off (soil drench)	
38.	Copper Hydroxide	77% WP	Anthraconse, Cercospora leaf spot	
39.	Chlorothalonil	75% WP	Fruit rot	10
40.	Difenoconazole	25% EC	Die-back, Fruit rot	15
41.	Dinocap	48% EC	Powdery mildew	
42.	Fenarimol	12% EC	EC Powdery Mildew	
43.	Flusilazole	40% EC	Powdery Mildew	5
44.	Hexaconazole	2% SC	Powdery mildew & Fruit rot	
45.	Kitazin	48% EC	Fruit rot,	
46.	Mancozeb	75% WP	Damping off (soil drench), Fruit rot, Leaf spot.	
47.	Myclobutanil	10%	Leaf spot & die back	03
48.	Propineb	70% WP	Die back	10
49.	Sulphur	80% WP	Powdery mildew	
		52% SC		
50.	Streptomycin	9%+1% SP	Bacterial leaf spot	
51.	Triadimefon	25% WP	Powdery mildew	15
52.	Tebuconazole	25.9% EC	Fruit rot, Powdery mildew	5
53.	Zineb	75% WP	Fruit rot & Leaf spot	
54.	Captan + Hexaconazole	70%+5% WP	Fruit rot, Anthracnose.	5

Common Name		Strength	Target Pest	Waitinng period (days)
Okra				
Insecticides				
1.	Azadirachtin	0.03 %	FSB, Whiteflies, Jassids	7
2.	Azadirachtin	5%	FSB, Whiteflies 5 Jassids, Aphids	
3.	Carbaryl	5% DP	Jassid, FSB	8
		10 % DP	FSB, Jassids	
4.	Carbofuran	3 % G	Aphids	
5.	Chlorantranilprole	18.5% SC	FSB	5
6.	Cypermethrin	25 % EC	FSB, Jassids	3
7.	Deltamethrin	2.8 % EC	FSB, Jassids	1
8.	Dicofol	18..5 % EC	Red spider mite	15-20
9.	Dimethoate	30% EC	Aphid, - Jassids	
10.	Emamectin benzoate	5 % SG	FSB	5
11.	Endosulfan	35 % EC	Aphids	21
12.	Fenpropathrin	30 % EC	Whitefly, FSB, Mites	7
13.	Fenvalerate	20 % EC	FSB	7
14.	Imidacloprid	70% WG	Jassids, Aphids, Thrips	
		48% FS	Jassids, Aphids	
		70% WS	Jassids, Aphids, Thrips	3
15.	Lambda-Cyhalothrin	5% EC	Jassids, FSB	4
16.	Malathion	50% EC	FSB, Aphid, Jassids	
17.	Oxydemeton-methyl	25% EC	Whitefly, Jassids.	
18.	Permethrin	25 % EC	FSB Aphids, Jassids	
19.	Phosalone	35 % EC	FSB	
20.	Pyridalyl	10% EC	FSB	3
21.	Quinalphos	20 % AF	FSB	7
		25 % EC	FSB Jassids, Mite	
22.	Spiromesifen	22.9% SC	Red spider mite	3
23.	Thiamethoxam	25% WG 70% WDG	Jassid, Aphid, Whitefly Aphids	5
Fungicides				
24.	Dinocap	48% EC	Powdery mildew	
25.	Sulphur	80% WP	Powdery mildew	
Cucurbits				
Insecticides				
1.	Chlorantranilprole	18.5% SC	Fruit borers, Caterpillars	7
2.	Dichlorvos	76% EC	Red pumpkin beetle	
3.	Dicofol	18.5 % EC	Red spider mite	15-20
4.	Imidacloprid	70 % WG	Jassids, Aphids	5

Common Name		Strength	Target Pest	Waitinng period (days)
5.	Trichloforon	5% Dust 5% Gr 50% EC	Red pumpkin beetle	
Fungicides				
6.	Benomyl	50% WP	Powdery Mildew, Anthracnose,	
7.	Carbendazim	50% WP	Powdery mildew	
8.	Thiophanate Methyl	70% WP	Powdery mildew, Anthracnose	
9.	Zineb	75% WP	Downy mildew, Anthracnose, Leaf spot	
10.	Cymoxanil	8% + Mancozeb 64% WP	Downy mildew	10
Cruciferous vegetables (Cabbage & Cauliflower)				
Insecticides				
1.	Acetamiprid	20 % SC	Aphids	7
2.	Azadirachtin	0.03 %	Aphids, DBM**	7
3.	Azadirachtin	5%	DBM, Spodoptera, Aphids	5
4.	Bacillus thuringiensis var. kurstaki	5% WP	DBM	
5.	Carbaryl	10 % DP	DBM, Armyworm, borer	8
6.	Carbofuran	3 % G	Nematodes	
7.	Chlorantranilprole	18.5% SC	DBM	3
8.	Chlorfenpyre	10 % SC	DBM	7
9.	Chlorpyrifos	20% EC	DBM	
10.	Cypermethrin	10 % EC	DBM	7
11.	Difenthiuron	50 % WP	DBM	7
12.	Dimethoate	30% EC	Aphids, Bugs	
13.	Emamectin benzoate	5 % SG	DBM	3
14.	Fenvalerate	20 % EC	DBM, borer	7
15.	Fipronil	5 % SC	DBM	7
16.	Flufenoxuron	10 % DC	DBM	7
17.	Indoxacarb	14.5% SC	DBM	7
		15.8 % SC	DBM	
18.	Lufenuron	5.4% EC	DBM,	5
19.	Malathion	50% EC	Aphids, Head borer	
20.	Metaflumizone	22 % SC	DBM	3
21.	Novaluron	10% EC	DBM	5
22.	Permethrin	25 % EC	DBM	
23.	Phorate	10% G	Aphids	
24.	Phosalone	35 % EC	Aphids	

Common Name		Strength	Target Pest	Waitinng period (days)
25.	Pyridalyl	10% EC	DBM	3
26.	Quinalphos	25 % EC	Aphid Head borer	
27.	Spinosad	2.5% SC	DBM	3
28.	Trichloforon	5% G 5% EC 50% Dust	DBM	
Fungicides				
29.	Captan	75% WP 75% WS	Damping off (Soil drench in nursery)	
30.	Mancozeb	75% WP	Collar rot , Leaf spot	
31.	Zineb	75% WP	Leaf spot	
Pea				
1.	Benomyl	50% WP	Powdery mildew	2
2.	Carbendazim	50% WP	Powdery mildew	
3.	Fenarimol	12% EC	Powdery Mildew	
4.	Sulphur	80% WP	Rust	
		80% WG	Powdery mildew	
		40% WP	Powdery mildew	
		52% SC	Powdery mildew	
		85% DP	Rust, Powdery mildew	
5.	Triadimefon	25% WP	Rust, Powdery mildew	
Legume vegetables				
1.	Benomyl	50% WP	Powdery mildew	
2.	Carbendazim	50% WP	Powdery mildew -	
3.	Captan	75% WP	Damping off (Soil drench in nursery)	
4.	Dinocap	48% EC	Powdery mildew	
5.	Lime Sulphur	22% SC	Rust	
6.	Sulphur	80% WP	Powdery mildew	
		80% WG	Powdery mildew	
		40% WP	Powdery mildew	
		85% DP	Powdery mildew Rust	
7.	Tetracycline Sulphate + Tetracycline Hydrocloride	9%+1% SP	Halo blight	

GR/G= Granules; **WDG/WG**= water dispersible granule; **SC**= suspension concentrate; **EC**= Emulsifiable concentrate; **SG**= soluble granule; **FS**= flowable concentrate for seed treatment; **WS**= water soluble; **DP**=dispersible powder; **AF**= aqueous flowable; **SL**= soluble liquid; **WP**= wettable powder; **SP**= soluble powder.

Source: Compendium on Pesticide Use in Vegetables, Indian Institute of Vegetable Research. (*Source*: CIB& RC), * PHI- Pre Harvest Interval

Literature Cited

Agarwal, M., Walia, S. and S. Dhingra. (1999). Pest control properties of turmeric leaf oil against Spilosoma obliqua, Dysdercus koenigii and Tribolium castaneum. Proceed. 2nd All India People's Congress, Calcutta, pp l–7.

Amer, M.M., T. I. El-Sayed, H. K. Bakheit, S. A. Moustafa, and Y. A. El-Sayed. 2008. Pathogenicity and genetic variability of five entomopathogenic fungi against Spodoptera littoralis. Research Journal of Agriculture and Biological Sciences, vol. 4 (5): 354–367.

Beavers, J.B., C. W. McCoy, and D. T. Kaplan. 1983. Natural enemies of sub-terranean Diaprepes abbbreviatus (Coleoptera: Curculionidae) larvae in Florida. Environmental Entomology, vol. 12: 840–843.

Bergeron V., Bonn D., Martin J-Y., Voyelle L.2000. Controlling droplet deposition with polymer additives. Nature. 405: 772–775.

Bezerra, J and L. Figueiredo. 1982. Ocorrencia de Phytomonas staheli em coqueiro (Coco nucifera) no estado da Bahia, Brazil. Fitopatol.Bras. 7: 139-143.

Bhattacharyya, A; A. C. Samal, and S. Kar.2004. Entomophagous fungus in pest management. News Letter, 5: 12.

Bohannan, D.R. and T.N. Jordan: 1995. Effects of ultra-low volume application on herbicide efficacy using oil diluents as carriers. Weed Tech. 9: 682-688.

Boucias, D.G. et al., 1988. Nonspecific factors involved in the attachment of entomopathogenic deuteromycetes to host insect cuticle. Applied and Environment Microbiology, vol. 54 (7): 1795–1805.

Bradfisch, G.A and S. L. Harmer. 1990. Omega-conotoxin GVIA and nifedipine inhibit the depolarizing action of the fungal metabolite, destruxin B on muscle from the tobacco budworm (Heliothis virescens). Toxicon. 28 (11): 1249–1254.

Brent, K. J., and Hollomon, D. W. 2007. Fungicide Resistance in Crop Pathogens: How Can It Be Managed? 2nd Rev. Edn. Online. Fungicide Resistance Action Committee (FRAC). CropLife Int'l., Brussels, Belgium.

Bridges, D.C. 1994. Impact of weeds on human endeavours. Weed Technol. 8:392–395

Brook, F., F. Rindi., Y. Suto., S. Ohtani and M. Green. 2015. The Trentepohliales (Ulvophyceae, Chlorophyta): An unusual Algal order and its Noval plant pathogen, Cephaleuros. Plant Disease. 99(6): 740-753.

Brown, R. H., & B.R, Kerry. 1987. Principles and practice of nematode control in crops.

Burondkar, M.M and R.T, Gunjate .1993. Control of vegetative growth and induction of regular and early cropping in Alphonso mango with paclobutrazol. Acta Horticulture. 341:206-215.

Caamano, E.X., R.A, Cloyd; L.F. Solter.,and D.J. Fallon. 2008. Quality assessment of two commercially available species of Entomopathogenic nematode: Steinernema feltiae and Heterorhabditis indica. Hort. Technology. 18(1): 84-89.

Cancini, N., J. Borrazzo., J. Menezes., D. Cortez., R. Valdez., F.P.Garcia., S. Ogatta., B. Filho., T. Nakamura., C. Nakamura. 2020. Activity of piperaceae extracts and fractions in the control of Phytomonas serpens . Crop protection. Cenc.Rural. 50(10): http/doi.org/10.1590/0103-8478cr20200343.

Chacko EK, Kohli RR, Doreswamy R, and G.S, Randhawa.1976. Growth regulators and flowering in juvenile mango (*Mangifera indica* L.) Seedlings. Acta Hortic. 56: 173–181.

Chahal, K.K., Arora, M., Joia, B.S. and B.R, Chhabra. 2005. Bioefficacy of turmeric oil against Tribolium castaneum (Herbst) under laboratory conditions. In V.K. Dilawari, G.S. Deol, B.S. Joia and P.K. Chuneja (eds.), Proc. 1st Congress on Insect Science: Contributed Papers, PAU Ludhiana, pp. 147–148.

Chakrabarti, A. K., and B, Choudhury. 1975. Breeding brinjal resistant to little leaf disease. Proc. Ind. Nat. Sci. Acad. 41: 379–385.

Coats, J.R., Karr, L.L. and C.D, Drewes. 1991. Toxicity and neurotoxic effects of monoterpenoids in insects and earthworms. In P. A. Hedin (ed.), Naturally Occurring Pest Bioregulators, ACS Symposium Series 449, American Chemical Society, Washington DC, pp. 306–316.

Collin, J.C. 2007. Challenges and opportunities in crop production over the next decade. Pesticide chemistry, Crop protection, Public health, Environmental safety, pp 3-12.In: H. Ohkawa, H.Miyagawa and P.W. Lee eds. Wiley VCH Verlag, Weinheim, Germany.

Coret J., Gambonnet B., Brabet F., A, Charnel. 1993. Diffusion of three ethoxylated octylphenols across isolated plant cuticles. Pestic Sci. 38: 201–209.

Davenport, T.L and R. Nunez-Elisea. 1990. The role of ethylene in mango flower induction. In: Proceeding of the plant growth Regulator Society of America. 17th Annual Meeting, st.Paul, Minnesota, USA, 5-9 Aug 1990. Pp.22-24, Plant Growth Society of America.

Davenport, T.L. and R, Nunez-Elisea.1997. Reproductive physiology In: The Mango (Ed., R. Litz.). CAB International, pp. 66-146

Dietrich, G., Dolan, M.C., Peralta-Cruz, J., Schmidt, J., Piesman, J., Eisen, R.J. and J.J, Karchesy.2006. Repellent activity of fractioned compounds from Chamaecyparis nootkatensis essential oil against nymphal Ixodes scapularis (Acari: Ixodidae). J. Med. Entomol., 43, 957–961.

Dollet, M. 1984. Plant diseases caused by flagellate protozoa (Phytomonas). Annual Review of Phytopathology. 22(1): 115-132.

Donley, N., 2019. The USA lags behind other agricultural nations in banning harmful pesticides. Environmental Health, 18(1):1-12.

Downer R.A., Mack R.E., Hall F.R. and A.K, Underwood .1998. Roundup Ultra with drift management adjuvants. Proceedings of the Fifth International Symposium on Adjuvants for Agrochemicals; 1998 August 17–21; Memphis, TN. Memphis: ISAA,

Elsworth, J.F and J. F. Grove.1977. Cyclodepsipeptides from Beauveria bassiana bals. Part 1. Beauverol-ides H and I. Journal of the Chemical Society, Perkin Transactions 1, (3): 270–273.

Faraday, M., 1825. XX. On new compounds of carbon and hydrogen, and on certain other products obtained during the decomposition of oil by heat. Philosophical Transactions of the Royal Society of London, 115, pp.440-466.

Faria, M.de and S. P. Wraight.2001. Biological control of Bemisia tabaci with fungi," Crop Protection. 20 (9): 767–778.

Ferron, P. 1981. Pest control by the fungi Beauveria and Metarhizium," in Microbial Control of Insects and Mites, H. D. Burgess, Ed., pp. 465–482, Academic Press, New York, NY, USA.

Firrao, G., Conci, L., & R, Locci. 2007. Molecular identification and diversity of Phytoplasmas. Biotechnology and Plant Disease Management, p.250.

Flor, H. H. 1955. Host parasite interaction in flax rust – its genetics and other implications. Phytopathology 45:680-685.

Frazer, A. C. 1963. Balance of pesticides: Benefits and risks. Pages 3-11 in: Proc. 2nd British Crop Protection Conference.

Foy C.L., Smith L.W. 1969. The Role of Surfactants in Modifying the Activity of Herbicide Sprays. In: Pesticidal Formulations Research, Washington, DC: American Chemical Society.

Gautham, K. C, and J.S,Mishra .1995. Problems, prospects and new approaches in weed management. Pestic Info 21:7–19

Gauvrit, C. and F, Cabanne. 1993. Oils for weed control: uses and mode of action. Pesticide science, 37(2):147-153.

Gella, R., and P. Errea, 1998. Application of in vitro therapy for ilarvirus elimination in three Prunus species. Journal of Phytopathology, 146(8-9), 445-449.

Gianessi, L., and N, Reigner. 2006. The importance of fungicides in U.S. crop production. Outlook on Pest Management 10:209-213.

Giovanni, A., T. Di Lucca., E. Fernando., T. Chipana., M. John., T. Albujar., W. Peraita., Y. Catalina., M. Piedra and J. Zelada. 2013. Slow wilt: another form of Marchitez in oil palm associated with Trypanosomatids in Peru. Tropical Plant Pathology. 38(6): 522-533.

Górski, R., 2004. Effectiveness of natural essential oils in the monitoring of greenhouse whitefly (Trialeurodes vaporariorum Westwood). Folia Hort, 16:183-187.

Green J.M. 2001a. Factors that influence adjuvant performance. Proceedings of the Sixth International Symposium on Adjuvants for Agrochemicals; 2001 August 13–17; Amsterdam. Amsterdam, ISAA.

Green J.M. 2001b. Weed specificity of alcohol ethoxylate surfactants applied with rimsulfuron. Weed Technol. 15: 79–83.

Hackman, R.H.1984. Cuticle: biochemistry," In: Biology of the Integument, J. Bereiter-Hahn, A. G. Matolts, and K. S. Richards, Eds., pp. 626–637, Springer, Berlin, Germany.

Hamill, R.L., H. R. Sullivan, and M. Gorman.1969. Determination of pyrrolnitrin and derivatives by gas-liquid chromatography. Applied microbiology, vol. 18 (3): 310–312.

Hamlen, R.A. 1979. Biological control of insects and mites on European greenhouse crops: research and commercial implementation," Proceedings of the Florida State Horticultural Society. 92: 367–.368.

Hammack, L., 1996. Corn volatiles as attractants for northern and western corn rootworm beetles (Coleoptera: Chrysomelidae: Diabrotica spp.). Journal of chemical ecology. 22(7):1237-1253.

Hartzler, B. 2001. Role of AMS with glyphosate products. Iowa State University Extension Agronomy. http://www.weeds.iastate.edu/mgmt/2001/ams.htm

Hasan, S., A. K. Bhamra, K. Sil, R. C. Rajak, and S. S. Sandhu.2002. Spore production of Metarhizium anisopliae (ENT-12) By Solid State fermentation. Journal of Indian Botanical Society. 8: 85–88.

Hazen, J.L., 2000. Adjuvants—terminology, classification, and chemistry. Weed technology, 14(4):773-784.

Hess, F.D. 1999. Surfactants and additives. In: Proceedings of the California Weed Science Society .51: 156-172.

Hewitt, A.J. 1998. The effect of tank mix and adjuvants on spray drift. In: McMullan, P.M. (ed.) Adjuvants for Agrochemicals: Challenges and Opportunities. Proceedings of the Fifth International Symposium on Adjuvants for Agrochemicals, Chemical Producers Distributors Association, Memphis, TN. Pp. 451-462.

Hoinkes, U. 2018. Regionaler Sprachgebrauch in Fachkontexten: Lexikologische und textuelle Untersuchungen zu Gestaltung und Dynamik fachsprachlicher Kompetenz in zweisprachigen Regionen der europäischen Romania.

Hooker, H. D. 1923. Colloidial copper hydroxide as a fungicide. Indust. Engin. Chem. 15:1177-1178.

Horsfall, J. G. 1975. Fungi and fungicides: The story of a nonconformist. Ann. Rev. Phytopathol. 13:1-14.

Howell, W.E., J. Burgess., J.I, Mink; L.J. Skrzeczkowski and Y.P. Zhang. 1998. Elimination of apple fruit and bark deforming agents by heat therapy. Acta Hortic. 472: 641-664.

Ignoffo, C.M. 1981. The fungus Nomuraea rileyi as a microbial insecticide: fungi," In: Microbial Control of Pests and Plant Diseases, H. D. Burges, Ed., pp. 513–538, Academic Press, London, UK.

Isman, M.B. and Machial, C.M., 2006. Pesticides based on plant essential oils: from traditional practice to commercialization. Advances in phytomedicine, 3: 29-44.

Jaenson, T.G.T., Garboul, S. and K, Pålsson. 2006. Repellency of oils of lemon, eucalyptus, geranium, and lavender and the mosquito repellent MyggA natural to Ixodes ricinus (Acari: Ixodidae) in the laboratory and field. J. Med. Entomol., 43:731– 736.

Jain, N; I. S. Rana, A. Kanojiya, and S. S. Sandhu. 2008. Characterization of Beaveria bassiana strains based on protease and lipase activity and their role in pathogenicity. Journal of Basic & Applied Mycology, vol. I-II: 18–22.

Jaskowska, E., C. Butler., G. Preston and S.Kelly. 2015. Trypanosomatids adopted to plant environment. PLoS Pathogens. 11(1): e 1004484.

Jones, L. R. 1914. Problems and Progress in Plant Pathology. Am. J. Bot. 1:97-111.

Katerinopoulos, H.E., Pagona, G., Afratis, A., Stratigakis, N. and N, Roditakis. 2005. Composition and insect attracting activity of the essential oil of Rosmarinus officinalis. Journal of Chemical Ecology. 31(1):111-122.

Kelman, A., and P.D, Peterson. 2002. Contributions of plant scientists to the development of the germ theory of disease. Microbes Infect. 4:257-260.

Kerwin, J.L and R. K. Washino. 1986. Oosporogenesis by Lagenidium giganteum: induction and maturation are regulated by calcium and calmodulin. Canadian Journal of Microbiology, vol. 32. (8): 663–672.

Kerry, B. 1997. Biological control of nematodes: prospects and opportunities. Plant nematode problems and their control in the near east region, pp.79-92.

Kerry, B.R., D. Crumph, and A. Mullen.1982. Studies of the cereal-cyst nematode, Heterodera avenue under continuous cereals, 1975–1978. II. Fungal parasitism of nematode females and eggs. Annals of Applied Biology. 100: 489–499.

Kim, J.J., M. H. Lee, C. S. Yoon, H. S. Kim, J. K. Yoo, and K. C. Kim. 2002. Control of cotton aphid and greenhouse whitefly with a fungal pathogen. Journal of National Institute of Agricultural Science and Technolog, pp. 7–14.

Kirkwood, R.C.1999. Recent developments in our understanding of the plant cuticle as a barrier to the foliar uptake of pesticides. Pestic Sci. 55: 69–77.

Krämer, W., and U, Schirmer. eds. 2007. Modern Crop Protection Compounds, Vol. 2. Wiley-VCH Verlag, Weinheim, Germany.

Kunkel, L. O. 1936. Heat treatments for the cure of yellows and other virus diseases of Peach. Phytopathology .26 (9):809-830.

Kögl F, Haagen-Smit AJ, Erxleben H. 1933. Über ein phytohormon der zellstreckung. Reindarstellung des auxins aus menschlichem harn. 4. Mitteilung über pflanzliche Wachstumsstoffe. Hoppe-Seyler's Zeitschrift fur Physiologische Chemie. 214: 241–261.

Latge, J.P., M. Monsigny, and M. C. Prevost.1988. Visualization of exocellular lectins in the entomopathogenic fungus Conidiobolus obscurus," Journal of Histochemistry and Cytochemistry. 36 (11): 1419–1424.

Leadbeater, A., and T, Staub. 2007. Exploitation of induced resistance: A commercial perspective. pp 229-242 In: Induced Resistance for Plant Defence. D. Walter, A. Newton, and G. Lyon, eds. Blackwell, Oxford, UK.

Leger, R., J.S., A. K. Charnley, and R. M. Cooper. 1987. Characterization of cuticle-degrading proteases produced by the entomopathogen Metarhizium anisopliae," Archives of Biochemistry and Biophysics. 253 (1): 221–232.

Locke, T., Bobbin, P., Atwood, J. and J, Owen, J. 2002. Effect of strobilurin fungicides on disease control and yield in blackcurrants. Acta Hortic. 585: 375-380. DOI:10.17660/ ActaHortic.2002.585.63

Louise, C., M. Dollet and D. Mariau. 1986. Recherches sur le Hartrot du cocotier maladie a Phytomonas et sur son vecteur lincus sp (Pentatomidae) en Guyane. Oleagineux. 41: 437-446.

Leonhardt, W., Wawrosch, C.H., Auer A., and B,Kopp. 1998. Monitoring of virus diseases in Austrian grapevine varieties and virus elimination using in vitro thermotherapy. Plant Cell Tissue and Organ Culture, 52: 71–74.

Lingua, G., D'Agostino, G., Massa, N., Antosiano, M., & G, Berta. 2002. Mycorrhiza-induced differential response to a yellows disease in tomato. Mycorrhiza, 12(4), 191-198.

Magan. R., C. Marin., J.M.Salas., M. Barrera-Perez., M.J. Rosales., and M. Sanchez-Moreno. 2004. Cytotoxicity of three new triazolo-pyrimidine derivatives against plant Trypanosomatids: Phytomonas sp isolated from Euphorbia characias. Memorias do Instituto Oswaldo Cruz. 99: 651-656.

Mallory-Smith, C.A. and E.J, Retzinger. 2003. Revised classification of herbicides by site of action for weed resistance management strategies. Weed Technology, 17(3) :605-619.

Manganaris, G.A., Economou, A.S., I.N, Boubourakas. 2003. Elimination of PPV and PNRSV through thermotherapy and meristem-tip culture in nectarine. Plant Cell Rep 22: 195–200. https://doi.org/10.1007/s00299-003-0681-y

Matysiak R., J.D, Nalewaja. 1999. Temperature, adjuvants, and UV light affect sethoxydim phytotoxicity. Weed Technol. 13: 94–99.

Mhaske, B.M., Bhoite, K.D., Borkar, S.G and P.N, Rasal. 2005. Eco-friendly control of slug by tobacco stem dust. SAIC Newsletter. 15(3): 12.

McCallan, S. E. A. 1930. Studies on Fungicides II. Testing protective fungicides in the laboratory. Cornell Agric. Exp. Stn. Memoirs 128:8-24.

McCallan, S. E. A. 1967. History of fungicides. Pages 1-37 In: Fungicides, An Advanced Treatise, Vol. I. Academic Press, New York, NY.

McCoy, G.W., R. A. Samson, and D. G. Boucias. 1988. Entomogenous fungi. In: CRC Handbook of Natural Pesticides, Part A. Entomogenous Protozoa and Fungi, C. M. Ignoffo, Ed., CRC Press.

McLaughlin, G.A., 1973. History of pyrethrum. Pyrethrum; The Natural Insecticide, p.3.

McNeil Donald Jr. G. 2005. Fungus fatal to mosquito may aid global war on malaria. The New York Times, 2005.

Mcleod, D.M. 1954. Investigations on the genera Beauveria Vuill. and Tritirachium Limber. Canadian Journal of Botany. 32: 818–890.

McMullan, P.M. 2000. Utility adjuvants. Weed Technology 14: 792-797.

McWhorter, C.G. 1982. The use of adjuvants. pp. 10-25. In: Hodgson, R.H. (Ed.). Adjuvants for herbicides. Published by the Weed Science Society of America. Champaign, Illinois, USA.

Mathew, S.O., S. S. Sandhu, and R. C. Rajak.1998. "Bioactivity of Nomuraea rileyi against Spilosoma obliqua: effect of dosage, temperature and relative humidity," Journal of Indian Botanical Society. 77: 23–25.

Miller, P. and P. Westra. 1996. Herbicide surfactants and adjuvants, no. 0.559. Colorado State University Cooperative Extension, Production Crop Series

Mink, G.I., 1992. Ilarvirus vectors. In Advances in disease vector research. pp. 261-281. Springer, New York, NY.

Mink, G.I., R, Wample and W.E, Howell. 1998. Heat treatment of perennial plants to eliminate phytoplasmas, viruses and viroids while maintaining plant survival. Pp. 332-345. In: Plant Virus disease control. A. Hadidi, R.K.khetarpal and H. Koganezawa eds. APS Press, st.Paul, MN, USA.

Miller, P. and P. Westra. 1998. How surfactants work (Doctoral dissertation, Colorado State University. Libraries).

Millardate. P.A. 1885. Traitement du mildiou par le melange de sulphate de cuivre et de chaux. J.agr. Prat. 2: 707-710. (English translation by F.J. Schneiderhan in Phytopathol. Classics No.3, 1933).

Muhammad, A., I, Muhammad, A. Attique and H, Sher. 2005. Elimination of citrus tristeza closterovirus from citrus budwood through thermotherapy. Pakistan J. Bot. 37(2):423.

Nalewaja, J.D. and R. Matysiak. 1993a. Optimizing adjuvants to overcome glyphosate antagonistic salts. Weed Tech. 7:337-342.

Nalewaja, J.D. and R. Matysiak. 2000. Spray deposits from nicosulfuron with salts that affect efficacy. Weed Technology 14: 740-749.

Nelson, S.C. 2008. Cephaleuros species, the plant parasitic green algae. Plant Disease. 43: 1-6.

Neville, A.C. 1984. Cuticle: organisation, In: Biology of the Integument, J. Bereiter-Hahn, A. G. Matolts, and K. S. Richards, Eds., pp. 611–625, Springer, Berlin, Germany.

Nunez, E., J. Iannacone, and H. Gómez.2008. "Effect of two entomopathogenic fungi in controlling aleurodicus cocois (Curtis, 1846) (Hemiptera: Aleyrodidae)," Chilean Journal of Agricultural Research. 68. (1): 21–30.

Nyland G, and A C Goheen. 1969. Heat Therapy of Virus Diseases of Perennial Plants. Annual Review of Phytopathology. 7: 331-354.

Leger, R.J.S., T. M. Butt, M. S. Goettel, R. C. Staples, and D. W. Roberts.1989. Production, in vitro, of appressoria by the entomopathogenic fungus *Metarhizium anisopliae*. Experimental Mycology. 13(3):274-288.

Leger, R.J.S., D. W. Roberts, and R. C. Staples.1991. A model to explain differentiation of appressoria by germlings of Metarhizium anisopliae," Journal of Invertebrate Pathology. 57(3): 299–310.

Leger, R.J.S; T. M. Butt, R. C. Staples, and D. W. Roberts. 1989. Synthesis of proteins including a cuticle-degrading protease during differentiation of the entomopathogenic fungus Metarhizium anisopliae. Experimental Mycology. 13 (3): 253–262.

Letham, D.S. 1963. Zeatin, a factor inducing cell division isolated from Zea mays. Life Sciences.8: 569-573, Pergamon press, USA.

Leroux, P., R.Fritz, D. Debieu, C. Albertini, C.Lanen, J. Bach, M.Gredt and F. Chapeland. 2002. Mechanism of resistance to fungicides in field strains of Botrytis cinerea. Pest Managemnet Science. 58(9): 876-888, Special Issue: Resistance 2001.

Li, Z.Z, C. R. Li, B. Huang, and M. Z. Meizhen. 2001. "Discovery and demonstration of the teleomorph of Beauveria bassiana (Bals.) Vuill, an important entomogenous fungus," Chinese Science Bulletin. 46 (9): 751–753.

Nene and Thiapyial .2010. Fungicides in plant disease control.Meditech publication,ISBN: 978-93-86479-84-6.

Ohkawa, H. Miyagawa, and P. W. Lee, eds. Wiley VCH Verlag, Weinheim, Germany.

Oyedela, A.O., Gbolade, A.A., Sosan, M.B., Adewoyin, F.B., Soyely, O.L. and O.O, Orafidiya, 2002. Formulation of an effective mosquito repellent topical product from lemon grass oil. Phytomedicine. 9: 259–262.

Osborne, D.J.1962. Effects of kinetin on protein and nucleic acid metabolism in xanthium leaves during senescence. Plant Physiol. 37: 595–602.

Pacanoski, Z. 2007. Herbicide use: Benefits for Society as a Whole: A Review. Pak J Weed Sci Res 13(1-2):135–147

Pair, S. D. and R. J. Horvai. 1997. Volatiles of Japanese honeysuckle flowers as attractant for adult Lepidopteran insects. U.S. Patent 5.665.344. Issued 9 September 1997. United States Patent Ofiice. W'ashinglon, D.C.

Panella, N.A., Dolan, M.C., Karchesy, J.J., Xiong, Y., Peralta-cruz, J., Mohammad Khasawneh, M., Montenieri, J.A. and G.O, Maupin. 2005. Use of novel compounds for pest control: Insecticidal and acaricidal activity of essential oil components from heartwood of alaska yellow cedar. J. Med. Entomol. 42: 352–358.

Panattoni, A., Luvisi, A., and E, Triolo. 2013. Elimination of viruses in plants: twenty years of progress. Spanish Journal of Agricultural Research .1: 173-188.

Penner, D. 2000. Activator adjuvants. Weed technology. 14(4):785-791.

Petroski, R.J. and L, Hammack. 1998. Structure activity relationships of phenyl alkyl alcohols, phenyl alkyl amines, and cinnamyl alcohol derivatives as attractants for adult corn rootworm (Coleoptera: Chrysomelidae: Diabrotica spp.). Environmental entomology, 27(3) :688-694.

Pongsomboon, W; S.Subhadrabandhu , R.A.Stephenson. 1997. Some aspects of the ecophysiology of flowering intensity of mango (Mangifera Indica L.) cv. Nam Dok Mai in a semi-tropical monsoon Asian climate. Scientia Horticulturae 70 (1): 45-56

Phillips McDougall. 2006. Phillips McDougall Agriservice Report. Pathhead, Midlothian, Scotland, UK

Pringnitz, B. 1998. Clearing up confusion on adjuvants and additives. Iowa State University Extension Agronomy. http://www.weeds/iastate.edu/mgmt/qtr98-2/cropoils.htm

Rajak, R.C., S. S. Sandhu, S. Mukherjee, S. Kekre, and A. Gupta. 1991. Natural outbreak of Nomuraea rileyi on Junonia orithyia. Journal of Biological Control. 5 (2): 123–124.

Ramirez, F., and T.L, Devenport. 2010. Mango flowering physiology. Scientia Horticulturiae. 126(2): 65-72.

Randolph, E.M and M.L. Gerardo. 1982. Phytomonas staheli Associated with coconut and oil palm Diseses in Colombia. Plant Disease. 66: 675-677.

Ray, D.P., Walia, S., Dureja, P and R.P, Singh. 2000. Composition and repellent activity of the essential oil of marigold (Tagetes erecta) flower. Ind. Perf. 44: 267–270.

Raychaudhuri, S. P., Varma, A., Chenulu, V. V., Prakash, N., and S, Singh. 1970. Association of mycoplasma- like bodies with little leaf of *Solanum melongena* L. In: Proceedings of the X International Congress of Microbiolpgy Mexico HIV-6. Mexico.

Rehner, S.S and E. Buckley. 2005. A Beauveria phylogeny inferred from nuclear ITS and EF1-α sequences: evidence for cryptic diversification and links to Cordyceps teleomorphs. Mycologia. 97 (1): 84–98.

Retzinger EJ, and C. Mallory-Smith. 1997. Classification of herbicides by site of action for weed resistance management strategies. Weed Technol 11:384–393

Richmond, A.E, and A, Lang. 1957. Effect of kinetin on protein content and survival of detached Xanthium leaves. Science.125:650–651

Rojas E, Leal F, and R.J. Campbell. 1993. Control of flowering and shooting in mango (*Mangifera indica* L.) with various chemical products. Proceedings of the International Society for Tropical Horticulture. 37:142-147.

Rojas, E., and F. Leal. 1997. Effect of pruning and potassium nitrate spray on floral and vegetative bud break of Mango cultivar, Haden. Acta Horticulturae. 455:522-529.

Roggenbuck F.C., Penner D., R.F. Burow, Thomas B. 1993. Study of the enhancement of herbicide activity and rainfastness by an organosilicone adjuvant utilizing radiolabelled herbicide and adjuvant. Pestic Sci. 37: 121–125.

Russell, P. E. 2005. A century of fungicide evolution. Journal of Agricultral Science. 143:11- 25.

Salazar, L.F. 1996. Potato viruses and their control. International Potato Center, Lima, Peru. P. 214. ISBN: 92-9060-184-1.

Salazar-Garcia, S., M.H. Perez-Barraza and V, Vazquez-Vaidivia. 1999. Effect of ammonium nitrate sprays on flowering and harvesting time of Manila Ataulfo and Tommy Atkins Mango in Nayarit, Maxico. In: International Symposium in Mango.509, pp.573-580.

Sandhu, S.S; R. C. Rajak, and G. P. Agarwal.1993. Studies on prolonged storage of Beauveria bassiana conidia: effects of temperature and relative humidity on conidial viability and virulence against chickpea borer Helicoverpa armigera," Biocontrol Science & Technology. 3: 47–53.

Sandhu, S. S; S. E. Unkles, R. C. Rajak, and J. R. Kinghorn. 2001. Generation of benomyl resistant Beauveria bassiana strains and their infectivity against Helicoverpa armigera," Biocontrol Science and Technology, vol. 11(2) : 245–250.

Sandhu, S.S and P. Vikrant. 2004. Myco-insecticides: control of insect pests," In: Microbial diversity, Opportunities & Challenges. S.P. Gautum, S. S. Sandhu, A. Sharma, and A. K. Pandey, Eds., Indica Publishers, New Delhi, India.

Sandhu, S.S., and M. Mishra. 1994. Larvicidal activity of fungal isolates Beaveria bassiana, Metarhizium anisopliae and Aspergillus flavus against mosquito sp. Culex pipiens," in Proceedings of the National Symposium on Advances in Biological Control of Insect Pests, pp. 145–150, Muzaffarnagar, India.

Sandhu, S.S. 1995. Effect of physical factors on germination of entomopathogenic fungus Beauveria bassiana conidia," National Academy of Science Letters. 18 (1-2): 1–5.

Sandhu, S.S., R. C. Rajak, and S. K. Hasija. 2000. Potential of entomopathogens for the Biological management of medically important pest: progress and prospect," In: Glimpses in Plant Sciences. 2000: 110–117.

Sandhu Sardul Singh, Anil K. Sharma, Vikas Beniwal, Gunjan Goel, Priya Batra, Anil Kumar, Sundeep Jaglan, A. K. Sharma, Sonal Malhotra. 2012. Myco-Biocontrol of Insect Pests: Factors Involved, Mechanism, and Regulation. Journal of Pathogens, vol. 2012, Article ID 126819, 10 pages, 2012. https://doi.org/10.1155/2012/126819

Santier, S and A. Chamel. 1996. Penetration of triolein and Methyl oleate through isolated plant cuticles and their effect on penetration of [C-14] quizalofop-ethyl and [C-14] fenoxaprop-ethyl. Weed Research.36: 167-174.

Schmidt, R.R. 1998. Classification of herbicides according to mode of action. Bayer Ag, Leverkusen, p 8

Schlundt, H. 2002. Risks and benefits of biological and chemical plant protection strategies – food safety aspects. Proc. British Crop Protection Conference 2002, pp. 3-21.

Seryczyńska, H and C. Bajan. 1975. Defensive reactions of L3, L4 larvae of the Colorado beetle to the insecticidal fungi Paecilomyces farinosus (Dicks) Brown et Smith, Paecilomyces fumoso-roseus (Wize), Beauveria bassiana (Bols/Vuill.) (Fungi Imperfecti: Moniliales)," Bulletin de l'Academie Polonaise des Sciences. Serie des Sciences Biologiques. 23(4): 267–271.

Sharda, S. and P.J, Rao. 2000. Effect of Ageratum conyzoides on development and reproduction of Spodoptera litura. Indian J. Entomol. 62: 231– 238.

Sharma, V.P. 2001. Health hazards of mosquito repellents and safe alternatives. Current Science. 80(3): 341-343.

Sikora, R. A. 1992. Management of the antagonistic potential in agricultural ecosystems for the biological control of plant parasitic nematodes. Annual review of phytopathology, 30(1): 245-270.

Skoog, F and D.J.Armstrong. 1970. Cytokinins. Annual Review of Plant Physiology. 21: 359-384.

Smith, C. M. 1988. History of benzimidazole use and resistance. Pages 23-24 In: Fungicide Resistance in North America. C. J. Delp, ed. American Phytopathological Society, St.Paul, MN.

Sohi, A. S., Bindra, O. S., and G.S, Deal. 1974. Studies on the control of the brinjal little leaf disease and insect pests of brinjal. Int. J. Entomol. 36: 362–364.

Souza, W.de and M. Attias. 1991. Cell biology of Phytomonas, Trypanosomatids parasites of plants. Memorias do Instituto Oswaldo Cruz. 86(3): 275-284.

Stirling, G. R., and L.M, West. 1991. Fungal parasites of root-knot nematode eggs from tropical and subtropical regions of Australia. Australasian Plant Pathology, 20(4): 149-154.

Tan, R., Wang, L., and N, Hong. 2010. Enhanced efficiency of virus eradication following thermotherapy of shoot-tip cultures of pear. Plant Cell Tiss Organ Cult 101: 229–235. https://doi.org/10.1007/s11240-010-9681-0

Thakur, R; R. C. Rajak, and S. S. Sandhu. 2005. Biochemical and molecular characteristics of indigenous strains of the entomopathogenic fungus Beauveria bassiana of Central India. Biocontrol Science and Technology. 15 (7): 733–744.

Thakur, R; and S. S. Sandhu. 2010. Distribution, occurrence and natural invertebrate hosts of indigenous entomopathogenic fungi of Central India," Indian Journal of Microbiology. 50 (1): 89–96.

Thelen K.D., Jackson E. P., and D, Penner .1995. 2, 4-D Interactions with glyphosate and sodium bicarbonate. Weed Technol. 9: 301–305.

Thimann, K. V. 1963. Plant growth substances; past, present and future. Annual Review of Plant Physiology.14:1–19.

Tillet, M. 1755. Dissertation on the cause of the corruption and smutting of the kernels of wheat in the head. Translated from the French by H.B. Humphrey in Phytopathol. Classics No.5, 1937. American Phytopathological Society, Ithaca, New York, 189 pp.

Tripathi, A.K., Prajapati, V. and S, Kumar. 2003. Bioactivity of l-carvone, d-carvone and dihydrocarvone towards three stored product beetles. J. Econ. Entomol. 96:1594–1601.

Ugur Gozel and Cigdem Gozel. 2016. Entomopathogenic Nematodes in Pest Management. Integrated Pest Management: IntechOpen book Series. DOI.10.5772/63894.

Upadhyay, R. 2016. Varietal susceptibility and effect of antibiotics on little leaf phytoplasma of brinjal (Solanum melongena L). Int. J. Emer. Trends Sci. Technol. 3: 3911–3914.

Van Valkenburg, J.W. 1982. Terminology, classification and chemistry. pp. 1-9. In: Hodgson, R.H. (Ed.). Adjuvant for herbicides. Published by the Weed Science Society of America. Champaign, Illinois, USA.

Vargas, R.I., Stark, J.D., Kido, M.H., Ketter, H.M. and L.C, Whitehand. 2000. Methyl eugenol and cue-lure traps for suppression of male oriental fruit flies and melon flies (Diptera: Tephritidae) in Hawaii: effects of lure mixtures and weathering. Journal of Economic Entomology. 93(1): 81-87.

Varma, A., Raychaudhuri, S. P., Chenulu, V. V., Singh, S., Ghosh, S. K., and N. Prakash.1975. Yellows type of diseases in India: eggplant little leaf. Proc. Nat. Sci. Acad. 41: 355–361.

Vermeulen, H. 1963. A Wilt of coffee liberta in Surinam and its association with a flagellated Phytomonas leptovasorum stahel. The Journal of Protozoology. 10(2): 216-222.

Vermeulen, H. 2005. Investigation into the cause of the phloem necrosis disease of coffee liberta in Surinam, South America. Netherland J. of Plant Pathology. 74: 202-208.

Vrushali T, Tare V and K, Shushil. 2001. Bioactivity of some medicinal plants against chosen insect pests/vectors. In K Sushil, SA Hasan, D Samresh, AK Kukreja, S Ashok, AK Sharma, S Srikant and T Rakesh (eds.), Proceedings of the National Seminar on the Frontiers of Research and Development in Medicinal Plants, CIMAP, Lucknow.

Yadav, A, and R.K, Malik. 2005. Herbicide resistant Phalaris minor in wheat–a sustainability issue. Resource book. Department of Agronomy and Directorate of Extension Education, CCSHAU,Hisar,p.152.

Yashitela, T., P.J. Robbertse and P.J.C, Stassen. 2004. Effect of various inductive periods and chemicals on flowering and vegetative growth of Tommy Atkins and Keitt Mango cultivars. New Zealand J. Crop and Hort.Sci. 32: 209-215.

Walkey, D.G.A., V, Alerie and C. Cooper. 1975. Effect of temperature on virus eradication and growth of infected tissue cultures. Annals of Applied Biology. 80(2): 185-190 https://doi.org/10.1111/j.1744-7348.1975.tb01620.x

Wanamarta G.; Kells J.J.and D, Penner .1993. Overcoming antagonistic effects of sodium bentazon on sethoxydim absorption. Weed Technol.7: 322–325.

Went, F.W. 1928. Wuchsstoff und Wachstum. Extrait du Recueil des Travaux Botaniques Neerlandais. 25:1–116.

Witt, W.W. 2001. Adjuvants. University of Kentucky College of Agriculture, Agripedia. http://www.ca.uky.edu/agripedia/pls404/adjuvant.htm.

Wraight, S.P., R. I. Carruthers, S. T. Jaronski, C. A. Bradley, C. J. Garza, and S. Galaini-Wraight.2000. Evaluation of the entomopathogenic fungi Beauveria bassiana and Paecilomyces fumosoroseus for microbial control of the silverleaf whitefly, Bemisia argentifolii," Biological Control. 17 (3): 203–217.

Wu, W., Ding, Y. A. N. G., Wei, W., Davis, R. E., Lee, I. M., Hammond, R. W., and Y, Zhao. 2012. Salicylic acid-mediated elicitation of tomato defence against infection by potato purple top phytoplasma. Annals of Applied Biology, 161(1): 36-45.

Zaridah, M.Z., M.A.Nor Azah, A.Abu Said, and Z.P. Mohd Faridz. 2003. Larvicidal properties of citronellal and Cymbopogon nardus essential oils from two different localities. Trop. Biomed., 20, 169–174.